Developing an Enterprise Continuity Program

T0383166

RIVER PUBLISHERS SERIES IN SECURITY AND DIGITAL FORENSICS

Developing an Enterprise Continuity Program

Sergei Petrenko

Innopolis University
Russia

LONDON AND NEW YORK

Published 2021 by River Publishers
River Publishers
Alsbjergvej 10, 9260 Gistrup, Denmark
www.riverpublishers.com

Distributed exclusively by Routledge
4 Park Square, Milton Park, Abingdon, Oxon OX14 4RN
605 Third Avenue, New York, NY 10158

First published in paperback 2024

Developing an Enterprise Continuity Program / by Sergei Petrenko.

Routledge is an imprint of the Taylor & Francis Group, an informa business

Publisher's Note
The publisher has gone to great lengths to ensure the quality of this reprint but points out that some imperfections in the original copies may be apparent.

While every effort is made to provide dependable information, the publisher, authors, and editors cannot be held responsible for any errors or omissions.

ISBN: 978-87-7022-397-3 (hbk)
ISBN: 978-87-7004-302-1 (pbk)
ISBN: 978-1-003-33788-1 (ebk)

DOI: 10.1201/9781003337881

Contents

Foreword   ix

Preface   xvii

Acknowledgements   xix

List of Figures   xxi

List of Tables   xxxi

List of Abbreviations   xxxv

Glossary   xliii

Introduction   lxiii

1   ECP Relevance   1
   1.1   Motivation and Achievable Benefits.   1
      1.1.1   Examples of Incidents .   4
      1.1.2   The Main Reasons   8
      1.1.3   Economic Feasibility   15
      1.1.4   Additional Advantages   18
   1.2   ECP Content and Structure   19
      1.2.1   Background   20
      1.2.2   Cloud Perspectives   26
      1.2.3   ECP Practice.   37
   1.3   Example of Task Statement   56
      1.3.1   The Purpose and Objectives of Work   56
      1.3.2   Work Duration.   63
   1.4   Analysis of BCM Technologies   64
      1.4.1   General Approaches and Directions.   65
      1.4.2   Infrastructure Decisions   69
      1.4.3   Software   92

1.5   Business Continuity and Cyber Resilience   98
    1.5.1   Basic Concepts and Definitions of
             Cyber Resilience.   98
    1.5.2   Cyber Transformation Trends   111
    1.5.3   Mathematical Problem Definition   122

2   BCM Best Practice   **137**
2.1   The International ISO 22301:2019 Standard   140
    2.1.1   First Version of the Standard   141
    2.1.2   Second Version of Standard   143
2.2   BCI Practice   153
    2.2.1   Activity Directions   153
    2.2.2   Main Results   154
2.3   DRI Practice   156
    2.3.1   Direction of Activity.   156
    2.3.2   Features of the Approach   161
2.4   SANS Institute Practice   163
    2.4.1   BCP Development.   163
    2.4.2   BCP Testing   172
2.5   AS/NZS 5050:2010 Standard   177
    2.5.1   Basic Recommendations   177
    2.5.2   The Application Specifics   182
2.6   Risk Management Practices   186
    2.6.1   ISO 31000 Family of Standards   186
    2.6.2   Managing Cyber Risks   188
    2.6.3   The NIST SP 800-30 Standard   193
    2.6.4   OCTAVE Methodology   196
    2.6.5   MG-2 Lifecycle   198
    2.6.6   COBIT 2019 Standard.   200
    2.6.7   SA-CMM Maturity Model   200
2.7   Business Process Description Practices   204
    2.7.1   Process Modeling   205
    2.7.2   NGOSS Methodology   208
2.8   COBIT Standard® 2019   218
    2.8.1   Description of the DSS04 process.   218
    2.8.2   DSS04 Maturity Levels   220
2.9   ITIL V4 Library   236
    2.9.1   The ITSCM Process   237
    2.9.2   ITSCM Implementation   243

2.10 ISO/IEC 27001:2013 and ISO/IEC 27031:
 2011 Standards 250
 2.10.1 BCM Aspects 251
 2.10.2 BCP Development and Implementation. 253
2.11 Possible Measures and Metrics 256
 2.11.1 Introducing a Passport System for Programs 256
 2.11.2 Intellectual Cyber Resilience Orchestration. 272

3 BC Project Management   **295**
3.1 Accenture Practice 295
 3.1.1 RA and BIA 304
 3.1.2 Definition of BC Strategy 309
 3.1.3 Improving the BC Strategy 319
3.2 Ernst & Young (E&Y) Experience 322
 3.2.1 ECP Program Maturity Assessment 332
 3.2.2 Developing a BCM Strategy. 340
 3.2.3 Implementing a BCM Strategy 343
3.3 IBM Practice. 345
 3.3.1 Methods of Work Performance 345
 3.3.2 IBM BCRS Approach 348
 3.3.3 Services IBM BCRS. 352
 3.3.4 Example of a Solution Selection 362
 3.3.5 Example of Task Statement 371
3.4 Hewlett-Packard Practice 378
 3.4.1 Evaluating the Current ECP State 378
 3.4.2 Developing a BCM Strategy. 383
 3.4.3 Implementing a BCM Strategy 386
3.5 EMC Practice 394
 3.5.1 Type of Work 395
 3.5.2 EMC Methodology 397
3.6 Microsoft Practice 410
 3.6.1 Characteristics of the Approach 410
 3.6.2 ITCM Function 411

4 ECP Development Samples   **419**
4.1 Characteristics of the Research Object 419
 4.1.1 Current Active Directory Architecture 419
 4.1.2 Target Active Directory Architecture 421

4.2   BIA Example   422
    4.2.1   Classification of Active Directory Processes and
        Services   422
    4.2.2   Calculating RTO and RPO   423
    4.2.3   Active Directory Interrupt Scenarios   423
4.3   Defining BC Strategies   430
    4.3.1   General Requirements   430
    4.3.2   Detailed Reading of RTO and RPO   434
    4.3.3   Selection of Technical Solutions   439
    4.3.4   Possible Recovery Strategies   441
    4.3.5   Restoring the IT Service   444
    4.3.6   The Business Recovery   446
4.4   BCP Example   450
    4.4.1   Requirements Analysis   450
    4.4.2   BCP Content and Structure   452
    4.4.3   Management Procedure   461
    4.4.4   BCP Testing   463

Conclusion   **481**

References   **485**

Index   **509**

About the Author   **513**

Foreword

Dear Readers!

As a Rector of Innopolis University, I pay a lot of attention to business continuity issues in University management as well as in the implementation of IT projects in the interests of our customers. The fact is that, today, business continuity management affects almost every one of us. We are just beginning to fight the global coronavirus pandemic (lat. *Coronaviridae*) *COVID-19* infection, which has already claimed tens of thousands of lives, we are experiencing another global economic crisis equal to which has never been before, and we are only starting to understand new global threats such as climate change, energy security, cyberterror, and cybercrime. Major technogeneous accidents and other emergencies in recent years have become the starting point for revising existing *enterprise continuity programs* and the emergence of a new practice of *cyber resilience management* for digital economics. However, in the professional literature, the issues of *business continuity* and *cyber resilience* have not been fully considered. Therefore, the publication of this monograph is a significant event in this professional field.

"***Developing an Enterprise Continuity Program***" monograph was written by Sergei Petrenko, Prof. Dr.-Ing., Head of the Information Security Center at Innopolis University, on the basis of many years of practical experience of the author in the field of business continuity. In his project portfolio, which was successfully executed under his personal supervision, together with the leading auditors and consultants of the "*Big Four*": *Deloitte Touche Tohmatsu, PricewaterhouseCoopers, Ernst & Young and KPMG*, are the following major projects: "*Development and implementation of the procedures for ensuring the continuity of information system operation of Federal Customs Services of Russia,*" "*Development of the Strategy for Continuity Management of Critical Technological Processes of the Leading Russian Telecommunication Industry "MTS" PJSC (Mobile TeleSystems),*" "*Analysis of the Business Impact on the activities of a large commercial bank of Russia PJSC Gazprombank,*" "*Development of Business Contingency Plan (BCP) for the world's largest aluminum* company *"Russian Aluminum" (United Company Rusal),"* *"Development of Business Continuity Concept of LLC "LUKOIL-Inform"*, a subsidiary of the largest oil company in the world – PJSC "LUKOIL," etc.

The monograph favorably differs from previously published works, both by the method of presentation and by the set of issues considered. Thanks to the successful methodical approach, the author managed to choose the form of presentation of the material, which, without prejudice to the understanding of the merits of the case, found a reasonable balance between the statements of tasks in the terms of business goals and objectives on the one hand, and description of possible engineering tasks and technical solutions for disaster recovery and recovery in emergency situations on the other. The author leads the reader to the intended goal in the shortest way, bypassing the known complexity of describing the numerous components of the process of *business continuity management*, which so often becomes an insurmountable barrier for beginners (*Affiliate Student, CBCI, AMBCI*) and even experienced professionals (*MBCI, FBCI*) in *BCM and DRM* who are not sufficiently trained for a business focused consideration of IT services. This narration provides a clear picture of the proper organization of the *enterprise continuity program (ECP)* and a clear understanding of the *BCP and DRP* plans.

In the monograph, the systemativeness of statement of principles and basic methodical methods of creation and implementation of enterprise programs to ensure business continuity, ECP is achieved by practical application *"Plan-Do-Check-Act"* (*PDCA*) from the international standard ISO 22301:2019 *"Security and resilience – Business continuity management systems – Requirements."* This allowed the author to deal with the proper degree of methodical rigor, and, at the same time, in an accessible way, with the proper organization of business continuity management and disaster recovery in emergency situations. Essentially, the book presents and summarizes the main tasks of developing, implementing, and supporting the *enterprise continuity program (ECP)*, as well as possible ways of solving these problems. In this regard, readers are offered a large number of terms of references and examples of project implementation in the field of BCM.

Significantly, the monograph, for the first time, discusses the issues of building not only qualitative but also quantitative metrics and measures to ensure business continuity, BCM. The author proposes a number of original models and methods to ensure business continuity and sustainability in the face of growing cybersecurity threats. At the same time, the author's methodological and scientific results correspond to the strategic goals and objectives of the technical committee ISO/TC 292 *"Security and resilience"* of the international organization ISO[1]. This committee is responsible for the methodological support and timely updating of ISO 22301:2019 – *Security and resilience – Business continuity management systems – Requirements*, ISO 22300:2018 – *Security and resilience – Vocabulary*, ISO 22313:2020 – *Security and resilience – Business continuity management systems – Guidance on the use of ISO 22301*, etc.

It is especially important to emphasize the importance and depth of study as well as a novelty for the reader of such fundamental issues as business impact analysis (BIA) and business interruption risk management (RM). For example, Chapters 2 and 3, which provide a comparative analysis of best practices for business continuity management and describe the main components of the ECP enterprise program, are an example of methodical and practical study of BCM and provide an idea of a possible form of business continuity methodology in medium and large IT-dependent companies. Undoubtedly, a positive quality of the book is its good methodological study, consistency and simplicity of presentation, a large number of examples and illustrations, the presence of an extensive bibliography, etc. All this helps the reader to study the material independently, and the availability of practical recommendations; for example, the option of developing a program ECP in Chapter 4 contributes to the rapid consolidation of knowledge.

The monograph will certainly interest specialists involved in the development and implementation of modern business continuity programs, ECP, project managers, auditors, and managers, responsible for business continuity. It can serve as an excellent basic tool for undergraduate and postgraduate students studying in general management programs, including *MBA* and *CIO* programs and *CEO*, *CSO*, and *CFO* programs focused on practical application of business continuity issues, and will also be useful for teachers and business trainers to prepare lectures, seminars, and workshops. Significantly, the book contains the results of not only qualitative but also *quantitative study of cyber resilience*, which allows us to discover, for the first time, the *limiting law of the effectiveness of protecting* critical information infrastructure.

The book *"Developing an Enterprise Continuity Program"* is written by Sergei Petrenko, Ph.D. (Eng., Grand Doctor), Full Professor Prof., Head of the Information Security Center at Innopolis University. The work

---

[1] https://www.iso.org/committee/5259148.html

of this author has significantly contributed to the creation of a national training system for highly qualified employees in the field of computer and data security technologies. This book sets out a concept of responsibility in training highly qualified specialists at the international level and in establishing a solid scientific foundation, which is a prerequisite for any effective application of cyber resilience technologies.

Rector of Innopolis University,
Dr. Sci. in Physics and Mathematics,
Professor Alexander Tormasov

Dear Readers!

I am very pleased to contribute to this very relevant and useful monograph.

The beginning of 2020 was marked by an outbreak of extremely dangerous and previously unknown coronavirus (lat. *Coronaviridae*) COVID-19 infection, which was first reported on December 31, 2019 in Wuhan, Hubei Province, China. During 2019, France and Spain experienced a traffic collapse due to strikes at gas stations and on public transport. In winter 2020, Bulgaria had seriously aggravated transport problems due to heavy snowfall in the north of the Balkan Peninsula. These and other events have once again demonstrated to us how vulnerable we are to such threats and how interconnected today's world is.

It should be noted that company management often mistakenly believes that business continuity management (BCM) processes are too complex for the scale of their businesses. This is a serious misconception – threats to which any organization is exposed are similar, regardless of the scale and type of its activity, differences are manifested only in the available powers, means, and resources that can be allocated to ensure business continuity and, accordingly, to respond quickly to security incidents. It is clear that at small enterprises, they are much lower. It should be borne in mind that many assumptions on which traditional risk management (*assessment, reduction, transfer, acceptance,* etc.) are based have certain disadvantages. The fact is that the identification of risks and the assessment of the likelihood of their occurrence are not so important. What matters is the business impact of a security incident, not the likelihood of a security incident. In practice, it is recommended to highlight the following areas of possible impact of security incidents on business: people, facilities and indoor space, technology, supply chains, customers, liquidity, and reputation. Focusing on the possible consequences of losses in these areas, as opposed to a detailed study of each specific risk, allows increasing the sustainability of the organization, which, in its, turn leads to improved business efficiency as a whole.

Business continuity management (BCM) is the only management trend that ensures a high level of protection and sustainability of the enterprise, which is inextricably linked to the issues of security and management and communications in emergency and crisis situations. Many aspects of BCM have always been present in organizations under different names. And, now, it is important to bring them together in a single structure of the continuity management process in order to clarify and form a common course on this issue. For example, we follow the recommendations of the well-known international standard ISO 22301:2019 *"Security and resilience – Business* continuity management systems – Requirements," as well as recommendations of other known standards ISO 9001 *"Quality management systems,"* ISO 14001 *"Environmental management systems,"* ISO 31000 *"Risk management,"* ISO/IEC 20000-1: *Information technology – Service management"*, ISO/IEC 27001:2013 *"Information security management systems,"* ISO 28000 *"Specification for security management systems for the supply chain,"* recommendations of a number of national standards *ASIS ORM.1-2017, NIST SP800-34, NFPA 1600:2019*, and best practices *COBIT®2019, RESILIA 2015, ITIL V4, MOF 4.0* in part BCM, etc.

The book ***"Developing an Enterprise Continuity Program"*** is written by Sergei Petrenko, Ph.D. (Eng., Grand Doctor), Full Professor Prof., Head of the Information Security Center at Innopolis University. The author personally led a number of major ECP projects successfully completed with Deloitte Touche Tohmatsu's leading auditors and consultants, PricewaterhouseCoopers, Ernst & Young, and KPMG. The author supervised such projects as: *"Development of the Strategy for Continuity Management of Critical Technological Processes of the Leading Russian Telecommunication Company MTS PJSC" (Mobile TeleSystems), "Analysis of the impact of downtime of business processes (Business Impact Analysis) on the activities of a large commercial bank of Russia PJSC Gazprombank", "Development of ECP for Home Credit and Finance Bank LLC – a Russian commercial bank, one of the leaders of the Russian consumer lending market," "Development of a Business Continuation Plan (Business) Contingency Plan (BCP) for the world's largest aluminum company "Russian Aluminium" (United Company Rusal),"* etc.

Essentially, the book corresponds to the main strategic trends of the work of the well-known technical committee ISO/TC 292 *"Security and resilience"*[2], responsible for the development and improvement of ISO 22300 *"Security and resilience – Business continuity management systems"* family standards.

The foregoing raises the urgency of the presented monograph ***"Developing an Enterprise Continuity Program."*** I consider that this book will be a very valuable tool for the development and formation of highly qualified specialists of a new class in the field of information technology, cybersecurity, and cyber resilience.

Deputy Director General SAP CIS
Dmitry Shepelyavyi

---

[2]https://www.iso.org/committee/5259148.html

Preface

"He who has not first laid his foundations may be able with great ability to lay them afterward, ... but they will be laid with trouble to the architect and danger to the building."

Niccolo Machiavelli, XV century

The book discusses a number of possible solutions in business continuity management (BCM) and disaster recovery management (DRM). A number of problematic issues for the creation of quantitative metrics and measures of business continuity management, BCM, are discussed. The book discusses a number of copyright models and methods that correspond to the goals and objectives of the well-known technical committee ISO/ TC 292 *"Security and resilience"* of the international organization ISO[3]. Significantly, the book contains the results of not only qualitative but also *quantitative studies of cyber resilience*, which allows us to discover, for the first time, the limiting law of the effectiveness of protecting critical information infrastructure.

The book discusses the recommendations of the ISO 22301: 2019 standard *"Security and resilience – Business continuity management systems – Requirements" for improving business continuity management systems (BCMS)* of organizations based on the well-known *"Plan-Do-Check-Act" (PDCA)* model. The book also discusses the recommendations of the ISO 9001 standards *"Quality Management Systems,"* ISO 14001 *"Environmental Management Systems,"* ISO 31000 *"Risk Management,"* ISO/IEC 20000-1 *"Information Technology – Service Management,"* ISO/IEC 27001 *"Information management security systems,"* ISO 28000 *"Specification for security management systems for the supply chain,"* ASIS ORM.1-2017, NIST SP800-34, NFPA 1600: 2019, COBIT® 2019, RESILIA 2015, ITIL V4, MOF 4.0, etc. in the part of *Business continuity management*.

---

[3]https://www.iso.org/committee/5259148.html

The book presents the best practices of the *British Business Continuity Institute (BCI)*[4] *(Good Practice Guidelines 2018 Edition), Disaster Recovery Institute International (DRI)*[5] *(The Professional Practices for Business Continuity Management 2017 Edition)*, and *SANS*[6]. Possible methods of conducting projects in the field of *business continuity management (BCM)* as well as methods of development and improvement of *enterprise continuity programs (ECP)* are considered in detail. This considers the specifics of *business continuity management (BCM)* and *disaster recovery management (DRM)*. On the basis of the practical experience of the author, there are examples of *risk assessment (RA)* and *business impact analysis (BIA)*, examples of *business continuity plan (BCP)* and *Disaster Recovery Plan (DRP)*, and relevant *BCP and DRP test plans*.

The book will be useful to the *Chief Information Officers (CIO)* and *Chief Information Security Officers (CISO), internal and external Certified Information Systems Auditors (CISA)*, top managers of companies responsible for ensuring business continuity and cyber stability, as well as teachers and students of *MBA, CIO and CSO* programs, and students and postgraduates of relevant specialties.

---

[4] www.thebci.org
[5] www.drii.org
[6] www.sans.org

Acknowledgements

The author would like to thank Professor *Alexander Tormasov* (Innopolis University) and Deputy Director General SAP CIS *Dmitry Shepelyavyi* for the foreword and support.

The author sincerely thanks Prof. *Alexander Lomako* and Prof. *Igor Sheremet* (Russian Foundation for Basic Research, RFBR) for their valuable advice and comments on the manuscript, the editing of which contributed to the improvement of its quality.

The author would like to thank Prof. *Alexander Lomako* and Dr. *Alexey Markov* (Bauman Moscow State Technical University) for the positive review and semantic editing of the monograph.

The author thanks his friends and colleagues: *Kirill Semenikhin, Iskander Bariev*, and *Zurab Otarashvili* (Innopolis University) for their support and attention to the work.

The author expresses special gratitude to *Nikolai Nikiforov* – Minister of Informatization and Communication of Russian Federation, *Roman Shayhutdinov* – Deputy Prime Minister of the Republic of Tatarstan, Minister of Informatization and Communication of the Republic of Tatarstan, and *Igor Kaliayev* – Academician of the Russian Academy of Sciences (RAS).

The author would also like to thank *Khismatullina Elvira* for translating the original text into English as well as *Rajeev Prasad* – Publisher at River Publishers for providing us the opportunity to publish the book and *Junko Nagajima* – Production coordinator who tirelessly worked through several iterations of corrections for assembling the diverse contributions into a homogeneous final version.

The reported study was funded by RFBR, project number 20-04-60080 *"Models and methods for ensuring the sustainability of society's social and technical systems in the face of viral epidemics such as the COVID-19 pandemic based on acquired immunity."*

Professor Sergei Petrenko
s.petrenko@rambler.ru

List of Figures

Figure 1.1	Basic concepts and definitions of business resilience.	2
Figure 1.2	Maturity level of BCMS, Gartner.	3
Figure 1.3	Modern challenges of the digital organizations. . .	4
Figure 1.4	Emergency scenarios of a catastrophic nature. . . .	5
Figure 1.5	Improving the organization's BCMS maturity. . . .	9
Figure 1.6	Main threats to business interruption..	10
Figure 1.7	Certification according to ISO 27001:2013.	12
Figure 1.8	Certification according to ISO 22301:2013.	13
Figure 1.9	Evaluation of performance/price solutions in the field of BCM..	15
Figure 1.10	Cost effectiveness of BCM technical measures. . .	16
Figure 1.11	Value of the ECP program for the organization. . .	17
Figure 1.12	The main stages of development of the ECP program..	19
Figure 1.13	An example of a financial services client solution..	22
Figure 1.14	An example of a travel services client solution. . .	22
Figure 1.15	An example of a transportation services solution. .	23
Figure 1.16	BC best practices reference architecture, 2000.. . .	23
Figure 1.17	Attacker types.	24
Figure 1.18	Hype cycle for business continuity management and IT resilience, Gartner.	26
Figure 1.19	Red Hat OpenShift Container Platform.	28
Figure 1.20	Structure of the open source platform.	28
Figure 1.21	Classic three-level architecture of data center networks..	29
Figure 1.22	Leaf–Spine network architecture..	29
Figure 1.23	Leaf–Spine network architecture of the open source platform..	30
Figure 1.24	Open source platform server infrastructure.	31
Figure 1.25	Ceph storage system operation..	33

Figure 1.26 *Docker and Kubernetes* container management systems. 34

Figure 1.27 Cassandra-based large data storage architecture. . . 35

Figure 1.28 Support and development of the BCM program. . . 40

Figure 1.29 Example of classification of business processes and IT services.. 49

Figure 1.30 Example of defining a BC strategy.. 53

Figure 1.31 Example of data representation for selecting a BC strategy 54

Figure 1.32 Example of calculating the required bandwidth of communication channels. 54

Figure 1.33 Process approach to BCM. 56

Figure 1.34 Example of a solution for resource virtualization . 67

Figure 1.35 Data processing center design. 69

Figure 1.36 Typical scheme of a fault-tolerant data center. . . . 71

Figure 1.37 Comprehensive approach to data storage. 73

Figure 1.38 Possible solutions for network storage infrastructure. 74

Figure 1.39 Integrated data storage solutions. 77

Figure 1.40 Hyper-converged storage solutions data. 77

Figure 1.41 Backup solutions. 80

Figure 1.42 HyperFlex as a platform for Veeam. 80

Figure 1.43 Option for a comprehensive data protection solution. 82

Figure 1.44 Version of the platform for storing "secondary" data. 83

Figure 1.45 Classic three-level model of a multiservice data center network.. 84

Figure 1.46 Option for storing large amounts of data.. 87

Figure 1.47 Possible solution for Hadoop platforms. 89

Figure 1.48 Possible platform for software-defined data warehouses. 91

Figure 1.49 Cyber resilience is multidisciplinary. 100

Figure 1.50 The main cyber resilience discipline goals and objectives. 101

Figure 1.51 The focus on the previously unknown cyber-attacks detection and neutralization. 101

Figure 1.52 The enterprise cyber resilience management program components. 102

Figure 1.53 The main risks of the Industry 4.0 enterprise
business interruption. 102
Figure 1.54 Prospects for state and business digital t
ransformation.. 103
Figure 1.55 The research object characteristics.. 103
Figure 1.56 The complexity of the research object structure. . . 104
Figure 1.57 The research object behavior complexity. 104
Figure 1.58 The diversity of the research object
representation levels. 105
Figure 1.59 Recommended MITRE 2015.. 109
Figure 1.60 Cyber resilience lifecycle, NIST SP 800-160. . . . 109
Figure 1.61 New cybersecurity challenges and threats. 111
Figure 1.62 Who is most interested in a cyber resilience?. . . . 112
Figure 1.63 Approach to ensure cyber resilience, IBM.. 114
Figure 1.64 From backup and disaster recovery to
intelligent cyber resilience services, IBM. 117
Figure 1.65 Cyber system phase behavior.. 129
Figure 1.66 Cycle generation bifurcation. 130
Figure 1.67 Critical points representation when $n = 1$. 131
Figure 1.68 Critical points representation when $n = 2$. 132
Figure 1.69 Function behavior under disturbance.. 133
Figure 1.70 The resilience control unit. 134
Figure 1.71 Cyber system self-recovery algorithm fragment. . . 135
Figure 2.1 An example of the conformity assessment
of a BCMS to the requirements of
ISO 22301:2019. 141
Figure 2.2 BCMS readiness for security incidents. 142
Figure 2.3 The role and place of ISO 22301.. 143
Figure 2.4 Harmonization of ISO 22301:2012 with ISO/IEC
27001:2005 (A. 14).. 146
Figure 2.5 The structure of ISO technical committee
ISO/TC 292.. 146
Figure 2.6 Deliverables of ISO/TC 292 in the field of BCM. . 147
Figure 2.7 ISO 22301 changes.. 147
Figure 2.8 The PDCA model of ISO 22301:2012. 148
Figure 2.9 The content of PDCA model.. 149
Figure 2.10 Possible algorithm for improving BCMS. 151
Figure 2.11 The development of *Good Practice Guidelines
2010–2018*. 153

Figure 2.12   Business continuity management (BCM) lifecycle. .   154
Figure 2.13   BCI areas of activity.   154
Figure 2.14   Stages of BCP development.   162
Figure 2.15   ECP components.   164
Figure 2.16   The BCM lifecycle.   164
Figure 2.17   The scheme of the RA and BIA.   167
Figure 2.18   BCP project details.   170
Figure 2.19   Defining BC strategies.   171
Figure 2.20   BCP/DRP testing.   173
Figure 2.21   BCMS maturity assessment.   173
Figure 2.22   ECP development project management.   175
Figure 2.23   ANAO recommendations on risk management and
business continuity.   178
Figure 2.24   ISO 31000:2018 recommendations for risk
management.   187
Figure 2.25   Evolution of cyber risk management standards. . .   189
Figure 2.26   Risk management   191
Figure 2.27   Selecting tools and services for ITRM.   191
Figure 2.28   IT risk assessment.   195
Figure 2.29   IT risk estimator.   197
Figure 2.30   OCTAVE-method structure.   198
Figure 2.31   Lifecycle model of cyber risks management,
MG-2   199
Figure 2.32   Possible scheme of the business support model. . .   205
Figure 2.33   Existing problems with the description of business
processes.   207
Figure 2.34   An example of the justification of the method for
describing business process.   209
Figure 2.35   Content of the NGOSS methodology.   210
Figure 2.36   ETOM structure.   211
Figure 2.37   Structure of the SID model.   212
Figure 2.38   Structure of the TNA architecture.   213
Figure 2.39   Example of matching a business process map
and a business process map of an enterprise.   214
Figure 2.40   Separation of strategic and operational processes. .   214
Figure 2.41   Focus on the delivery of services.   215
Figure 2.42   Example of product-service-resource classification. .   216
Figure 2.43   Comparison of ITIL V3 and ITIL V4.   216
Figure 2.44   Role and place of the COBIT 2019 methodology. .   218
Figure 2.45   The business-oriented view of IT.   219

Figure 2.46   COBIT 2019 standard. 219
Figure 2.47   COBIT® process 2019.. 220
Figure 2.48   Development of the ITIL library. 237
Figure 2.49   ITSCM process lifecycle on OGC model. 240
Figure 2.50   The nature of the balance between preventive and
                  recovery measures. 241
Figure 2.51   ITIL risk management lifecycle. 243
Figure 2.52   The lifecycle of development of the ICT readiness
                  for business continuity program (IRBC). 250
Figure 2.53   The mapping diagram of the computation
                  correctness recovery. 260
Figure 2.54   Arithmetic expression generation tree. 265
Figure 2.55   Computations representation by similarity
                  equations. 268
Figure 2.56   Distortion control and computation process
                  recovery scheme. 269
Figure 2.57   The correct computation scheme.. 269
Figure 2.58   The passport program formation scheme in the
                  invariant similarity. 270
Figure 2.59   The similarity invariants scheme under exposure. . 271
Figure 2.60   The incorrect computations scheme. 271
Figure 2.61   The similarity invariants database formation
                  scheme. 271
Figure 2.62   The computational processes validation scheme.. . 272
Figure 2.63   Multilevel representation of computer
                  computations. 275
Figure 2.64   Control flow graph of the information and
                  computation task. 278
Figure 2.65   Decomposition of the control flow program
                  graph. 279
Figure 2.66   Graph representation of the computation
                  structure. 280
Figure 2.67   Checkpoint implementation mechanism
                  using IRIDA. 281
Figure 2.68   Block diagram of software and application
                  systems. 285
Figure 2.69   Block diagram of software levels, application
                  systems, and complexes. 285
Figure 2.70   Multilayer similarity invariant. 287
Figure 2.71   Cyber resilience management system. 288

Figure 2.72   Device diagram to control the computations'
semantics. 289

Figure 2.73   Circuit device interaction with the central
processor. 290

Figure 2.74   Architecture of the cyber resilience
management PAC.. 291

Figure 2.75   Example of a decision levels' hierarchy for cyber
resilience management. 291

Figure 2.76   Supervisor decision criteria to ensure the
required cyber resilience. 292

Figure 3.1   Accenture's approach to creating a successful
ECP program. 301

Figure 3.2   An illustration of Accenture's approach to ECP
development. 304

Figure 3.3   *e*TOM model for the telecom operator. 306

Figure 3.4   Definitions of the cyber risk structure. 307

Figure 3.5   Data processing diagram. 313

Figure 3.6   Three-level model of the organization and
its business continuity management system,
BCMS. 315

Figure 3.7   An illustration of a dynamic model
system element and a vulnerability diagram for
this object.. 316

Figure 3.8   Diagram of the information infrastructure
object's risk zones. 317

Figure 3.9   A step change in the level of service, with
an increase of destructive factors.. 318

Figure 3.10   Building a typical BC solution for an organization.. . 320

Figure 3.11   Relationship between the emergency situation
and the organization's financial indicators. 320

Figure 3.12   Calculating the positive effect of implementing
standard BC solutions. 321

Figure 3.13   Magic quadrant for business continuity
management planning software, Worldwide 2016
and Worldwide 2019, Gartner. 322

Figure 3.14   E&Y methodology for ECP development. 330

Figure 3.15   Sample of an ECP maturity level assessment
representation.. 333

Figure 3.16   The prioritization of risk management business
interruption. 338

Figure 3.17 Example of an RTO definition for IT services. . . . 339

Figure 3.18 Choosing a business continuity strategy. 340

Figure 3.19 Presentation of an incident response strategy. . . . 342

Figure 3.20 Possible composition of organizational and
administrative documentation on BCM issues. . . . 345

Figure 3.21 Typical IBM BCM services. 346

Figure 3.22 Regulatory requirements for the ECP program. . . 348

Figure 3.23 Characteristics of the IBM BCRS approach. 349

Figure 3.24 Example of business continuity management
process decomposition. 350

Figure 3.25 Map of business continuity management
maturity levels. 350

Figure 3.26 Main stages of development of the continuity
concept. 351

Figure 3.27 GAP analysis and setting goals for the
development of the BCM process. 351

Figure 3.28 Ways to manage the residual risk of business
interruption. 351

Figure 3.29 Typical IBM BCM services. 352

Figure 3.30 IBM's core services under the ECP program. . . . 353

Figure 3.31 Characteristics of the IBM BCRS approach. 354

Figure 3.32 Example of business interruption threat
assessment. 354

Figure 3.33 Example of possible damage assessment. 355

Figure 3.34 Example of an IBM BCM training workshop
scenario. 355

Figure 3.35 Example of evaluating the current state of
the BCM enterprise program. 357

Figure 3.36 RTO evaluation sample.. 358

Figure 3.37 Example of BCM project portfolio justification. . . 359

Figure 3.38 Criteria for selecting BCM architectural
solutions. 359

Figure 3.39 Example of a multicriteria evaluation of BCM
countermeasures. 360

Figure 3.40 Impact of new threats and technologies on the
BCM program and architecture. 362

Figure 3.41 Example of BCM process decomposition. 363

Figure 3.42 Possible ways to provide BC.. 363

Figure 3.43 Example of ranking business processes by
recovery level.. 364

Figure 3.44   Example of detailing requirements for the BCM
                  process. 364
Figure 3.45   Typical BC software algorithm.. 365
Figure 3.46   Role and place of BC decision selection
                  methodology. 366
Figure 3.47   The "hourglass" concept. 366
Figure 3.48   Algorithm for setting clarifying questions. 367
Figure 3.49   Example of a matrix of possible solutions BC. . . . 368
Figure 3.50   Excluding inappropriate BC solutions. 368
Figure 3.51   The structure of the project team.. 373
Figure 3.52   BCM lifecycle model.. 381
Figure 3.53   Role and place of ECP in the organization.. 381
Figure 3.54   The assessment of the ECP program current state. . . 382
Figure 3.55   The assessment of the improvement of the ECP
                  program.. 382
Figure 3.56   Examples of ECP maturity survey results. 389
Figure 3.57   Example of a fragment of a disaster recovery
                  scenario.. 389
Figure 3.58   An example of the distribution of responsibility . . 390
Figure 3.59   Example of a possible backup scheme for a
                  class RC1 system.. 393
Figure 3.60   ECP program lifecycle. 395
Figure 3.61   Development of EMC BC/DR technologies. 396
Figure 3.62   Features of the EMC approach.. 397
Figure 3.63   Making decisions based on GAP analysis. 397
Figure 3.64   Main components of the BCM EMC
                  methodology. 398
Figure 3.65   Example of a BIA rating. 402
Figure 3.66   Defining BC strategies. 405
Figure 3.67   Example of defining a BC strategy.. 406
Figure 3.68   IT service lifecycle model, Microsoft. 410
Figure 3.69   Interaction between MSF and MOF. 411
Figure 3.70   MOF process model. 412
Figure 3.71   Ratio of MOF V4 methodology to COBIT 2019
                  and ITIL V4. 412
Figure 3.72   Algorithm for implementing the "IT service
                  continuity management" function. 415

Figure 4.1   Current architecture of the *Active Directory*
IT service. 421

Figure 4.2   General architecture model for the *Active
Directory* IT service. 426

Figure 4.3   Typical event schema for Active Directory. 435

Figure 4.4   Active Directory backup automation scheme. . . . 453

Figure 4.5   Approximate functionality of automation tools –
Active Directory backup. 453

Figure 4.6   Interaction diagram of the organization's
emergency recovery team. 455

Figure 4.7   Phase 0 algorithm. 463

Figure 4.8   Phase 1 algorithm. 464

Figure 4.9   The algorithm of phases 2 and 3. 466

List of Tables

Table 1.1   Example of calculating RTO and RPO for BC 2000
             solutions. 21
Table 1.2   Business continuity management system
             components. 38
Table 1.3   Comparison of micro and macro continuity
             management. 39
Table 1.4   Main stages of business impact assessment and
             their results. 41
Table 1.5   Classification of business processes and
             IT services. 44
Table 1.6   Example of comparison and selection of ways |
             to reserve key IS. 55
Table 1.7   Characteristics of data processing center solutions. . . 70
Table 1.8   Complexity factors in ensuring cyber resilience. . . 106
Table 2.1   Harmonization of ISO 22301:2012 with other
             ISO standards. 144
Table 2.2   Comparison of ISO 22301:2012 with other
             well-known standards. 145
Table 2.3   Audit of an BCMS organization. 152
Table 2.4   Professional training levels in BCI. 157
Table 2.5   Main areas of BCM certification 160
Table 2.6   Structure of the DRII specialist certification
             system. 161
Table 2.7   Example of asset classification. 169
Table 2.8   Possible criteria for criticality and impact. 169
Table 2.9   Risk management principles. 179
Table 2.10   Risk assessment methodology. 181
Table 2.11   Features of the BCM enterprise program. 183
Table 2.12   Cyber risks computation example. 196
Table 2.13   COBIT 2019 guidelines for IT risks management . 201
Table 2.14   ITIL practices. 217
Table 2.15   ITIL4® – COBIT® 2019 mapping. 222

Table 2.16   Description of the *DSS04 – Managed*
                  Continuity process. 232
Table 2.17   Detailed description of the *DSS04.01* stage. 233
Table 2.18   Assignment of responsible people and defining
                  their roles.. 234
Table 2.19   Example risk profile. 244
Table 2.20   Comparative analysis of possible recovery
                  methods.. 245
Table 2.21   Examples of roles and responsibilities in the
                  ITSCM process.. 249
Table 2.22   Examples of BCM controls according to
                  ISO/IEC 27001 (A. 14 or A. 17). 257
Table 2.23   The operations on the program variables
                  dimensions. 266
Table 2.24   Sets of non-terminal symbols. 267
Table 2.25   Ways to modify computations. 274
Table 2.26   Scale of relative importance. 292
Table 2.27   Getting priorities vector. 293
Table 3.1   The practice of Accenture in the field of BCM.. . . 296
Table 3.2   Sample plan for creating and implementing ECP. . 301
Table 3.3   Analysis of the dependence of services
                  provided on infrastructure elements. 305
Table 3.4   Assessing the stability of the information
                  infrastructure. 306
Table 3.5   Analysis of possible damage.. 308
Table 3.6   Recommendations on the ECP program
                  improvement. 310
Table 3.7   Project documentation for the stage. 319
Table 3.8   Comparative analysis of BCM software 323
Table 3.9   Project stage documentation. 326
Table 3.10   Characteristics of some E&Y projects in
                  the BCM area.. 327
Table 3.11   The main approaches adopted by E&Y. 328
Table 3.12   Characteristics of ECP development stages. 331
Table 3.13   Example of presenting the results of risk analysis. . 336
Table 3.14   Accounting for industry requirements. 348
Table 3.15   Decision matrix for business continuity. 370
Table 3.16   Example of choosing appropriate BC solutions. . . 371
Table 3.17   Project execution schedule. 377
Table 3.18   A list of some of the projects HP BCM. 379

Table 3.19   Clarification of the results of the previous
stage of work. 384

Table 3.20   Example of a survey questionnaire for the
subject area (processes). 386

Table 3.21   Information on application and data. 387

Table 3.22   Example of a survey questionnaire for the
subject area (map of processes). 388

Table 3.23   Sample set of reporting materials for the stage. . . 391

Table 3.24   Example of a reporting table and distribution of
responsibility areas for groups. 391

Table 3.25   Example of reporting materials for a project stage. . 392

Table 3.26   Example of classification of IT services. 392

Table 3.27   System characteristics based on the example
of the RC1 class (very high speed). 393

Table 3.28   ECP practice in the field of business continuity. . . 394

Table 3.29   Methodological basis of EMC projects in the
field of BCM. 398

Table 3.30   Example of identification of an organization's
business processes. 399

Table 3.31   Example of fixing the IT component of
each process. 400

Table 3.32   Identifying and ranking threats to business
continuity. 401

Table 3.33   Example of a BIA rating. 403

Table 3.34   Application recovery time objective matrix. 404

Table 3.35   Sample plan of the EMC project for creating an
enterprise BCM program. 407

Table 3.36   Examples of indicators of functional stability
of a certain business process. 409

Table 3.37   Comparison of MOF v4 and MOF v3 versions. . . 413

Table 3.38   Example of an IT service risk assessment. 416

Table 3.39   An example of emergency classification. 417

Table 4.1   List of domains of the current Active Directory
architecture. 420

Table 4.2   Role of domains of the current Active Directory
architecture. 422

Table 4.3   Classification of criticality of Active Directory IT
services. 423

Table 4.4   List of business services that depend on Active
Directory. 424

Table 4.5   *Active Directory* matrix of resources and services. .   426

Table 4.6   *Active Directory* criticality classificator of
resources and services.   427

Table 4.7   *Active Directory* uncorrelated scenario of
failure of the resource.   428

Table 4.8   *Active Directory* uncorrelated scenario of
failure of the service.   430

Table 4.9   Active Directory correlated interrupt scenarios. . .   431

Table 4.10   Typical Active Directory information flows.   434

Table 4.11   Typical Active Directory business day.   434

Table 4.12   Possible strategies for restoring the Active
Directory.   442

Table 4.13   Current continuity characteristics of related
services.   444

Table 4.14   Calculating the cost of hardware.   445

Table 4.15   Summary table of the decision characteristics. . . .   446

Table 4.16   Analysis of strategy options.   447

Table 4.17   Analysis of options for Active Directory recovery
strategies.   449

Table 4.18   Frequency of updating continuity documentation. .   454

Table 4.19   List of typical actions when detecting a
security incident.   462

Table 4.20   Order of actions in case of a security incident. . . .   465

Table 4.21   List of typical actions when restoring the service. .   467

Table 4.22   List of typical actions when entering normal
operation mode.   468

Table 4.23   Documenting the security incident and recovery
process.   468

List of Abbreviations

AC	Access Control
ADH	Architectural Diversity/Heterogeneity
AES	Advanced Encryption Standard
AM	Asset Mobility
AMgt	Adaptive Management
AO	Authorizing Official
APT	Advanced Persistent Threat
AS&W	Attack Sensing & Warning
ASLR	Address Space Layout Randomization
AT	Awareness and Training
ATM	Asynchronous Transfer Mode
AU	Audit and Accountability
BCP	Business Continuity Plan
BIA	Business Impact Analysis
BYOD	Bring Your Own Device
BV	Behavior Validation
C3	Command, Control, and Communications
CA	Security Assessment and Authorization
CAL	Cyber-Attack Lifecycle
CAP	Cross Agency Priority
CAPEC	Common Attack Pattern Enumeration and Classification[7]
C&CA	Coordination and Consistency Analysis
CC	Common Criteria
CCoA	Cyber Course of Action
CE	Customer Edge
CEF	Common Event Format
CES	Circuit Emulation Service
CEO	Chief Executive Officer
CIKR	ritical Infrastructure and Key Resources
CIO	Chief Information Officer

---

[7] https://capec.mitre.org/

CIP	Critical Infrastructure Protection
CIS	Center for Internet Security
CISO	Chief Information Security Officer
CKMS	Cryptographic Key Management System
CM	Configuration Management
CMVP	Cryptographic Module Validation Program
CND	Computer Network Defense
CNSS	Committee on National Security Systems
CNSSI	CNSS Instruction
COBIT	Control Objectives for Information and Related Technology
COOP	Continuity of Operations Plan
COP	Common Operational Picture
COTS	Commercial Off The Shelf
CP	Contingency Plan/Contingency Planning
CPS	Cyber–Physical System(s)
CREF	Cyber Resiliency Engineering Framework
CRITs	Collaborative Research Into Threats[8]
CS	Core Segment
CSC	Critical Security Control
CSP	Cloud Service Provider
CSRC	Computer Security Resource Center
CUI	Controlled Unclassified Information
CVE	Common Vulnerabilities and Exposures[9]
CWE	Common Weakness Enumeration[10]
CybOX	Cyber Observable eXpression[11]
CyCS	Cyber Command System[12]
DASD	Direct Access Storage Device
DDH	Design Diversity/Heterogeneity
DF	Distributed Functionality
DiD	Defense-in-Depth
Dis	Dissimulation/Disinformation
DISN	Defense Information Systems Network

---

[8] https://crits.github.io/
[9] https://cve.mitre.org/
[10] https://cwe.mitre.org/
[11] https://cybox.mitre.org/
[12] http://www.mitre.org/research/technology-transfer/technology-licensing/cyber-command-system-cycs

DivA	Synthetic Diversity system[13]
DM&P	Dynamic Mapping and Profiling
DMZ	Demilitarized Zone
DoD	Department of Defense
DHS	Department of Homeland Security
DNS	Domain Name System
DRA	Dynamic Resource Allocation
DReconf	Dynamic Reconfiguration
DRP	Disaster Recovery Plan
DS	Digital Signal
DSI	Dynamic Segmentation/Isolation
DTM	Dynamic Threat Modeling
DVD	Digital Video Disc
DVD-ROM	Digital Video Disc-Read-Only Memory
DVD-RW	Digital Video Disc-Rewritable
EA	Enterprise Architecture
EAP	Employee Assistance Program
FCD	Federal Continuity Directive
FDCC	Federal Desktop Core Configuration
FIPS	Federal Information Processing Standards
FIRMR	Federal Resource Management Regulation
FIRST	Forum for Incident Response Teams
FISMA	Federal Information Security Management Act
FOIA	Freedom of Information Act
FOSS	Free and Open Source Software
FTE	Full-Time Equivalent
FW	FireWall
GOTS	Government Off-The-Shelf
FRA	Functional Relocation of Cyber Assets
HA	High Availability
HSPD	Homeland Security Presidential Directive
HTML	Hypertext Markup Language
HTTP	Hypertext Transfer Protocol
HVAC	Heating, Ventilation, And Air Conditioning
I&W	Indications & Warning
ICS	Industrial Control Systems
ICT	Information and Communications Technology

[13] https://www.atcorp.com/technologies/verifiable-computing/synthetic-diversity

IdAM	Identity and Access Management
IDS	Intrusion Detection System
IEC	International Electrotechnical Commission
IoT	Internet of Things
I/O	Input/Output
IR	Interagency Report
IS	Information System
ISA	Interconnection Security Agreement
ISAC	Information Sharing and Analysis Center
ISAO	Information Sharing and Analysis Organization
ISO	International Organization for Standardization
ISP	Internet Service Provider
ISCP	Information System Contingency Plan
ISSM	Information System Security Manager
ISSO	Information System Security Officer
IA	Identification and Authentication
ICS	Industrial Control System
ICT	Information and Communications Technology
IDS	Intrusion Detection System
IEC	International Electrotechnical Commission
IMS IP	Multimedia Subsystem
InfoD	Information Diversity
IQC	Integrity/Quality Checks
IP	Internet Protocol
IPSec	IP Security
IR	Incident Response
IRM	Information Resource Management
ISAC	Information Sharing and Analysis Center
ISCM	Information Security Continuous Monitoring
ISO	International Standards Organization
IT	Information Technology
ITL	Information Technology Laboratory
JTF	Joint Task Force
L2TP	Layer 2 Tunneling Protocol
LAN	Local Area Network
LDAP	Lightweight Directory Access Protocol
LTE	Long Term Evolution
M&DA	Monitoring and Damage Assessment
M&FA	Malware and Forensic Analysis
MA	Maintenance

MAC	Message Authentication Code
MAEC	Malware Attribute Enumeration and Characterization[14]
MAO	Maximum Allowable Outage
MB	Megabyte
Mbps	Megabits Per Second
MD&SV	Mission Dependency and Status Visualization
MEF	Mission Essential Functions
MOA	Memorandum Of Agreement
MOU	Memorandum Of Understanding
MP	Media Protection
MPLS	MultiProtocol Label Switching
MTD	Maximum Tolerable Downtime/Moving Target Defense
NARA	National Archives and Records Administration
NAS	Network-Attached Storage
NCF	NIST Cybersecurity Framework
NE	Network Edge
NEF	National Essential Functions
NGN	Next Generation Network
NIPP	National Infrastructure Protection Plan
NIST	National Institute of Standards and Technology
NOFORN	Not Releasable to Foreign Nationals
NPC	Non-Persistent Connectivity
NPI	Non-Persistent Information
NPS	Non-Persistent Services
NSP	Network Service Provider
NSPD	National Security Presidential Directive
NVD	National Vulnerability Database
O/O	Offloading/Outsourcing
OAI-ORE	Open Archives Initiative Object Reuse and Exchange
OEP	Occupant Emergency Plan
OMB	Office of Management and Budget
OPM	Open Provenance Model[15]
OPSEC	Operations Security
OS	Operating System
OSS	Operations Support System
OT	Operational Technology

---

[14] https://maec.mitre.org/

[15] http://openprovenance.org/

OTN	Optical Transport Network
P2P	Peer-to-Peer
PA	Personal Authorization
PB&R	Protected Backup and Restore
PBX	Private Branch Exchange
PDH	Plesiochronous Digital Hierarchy
PE	Physical and Environmental Protection
PGP	Pretty Good Privacy
PI	Pandemic Influenza
PII	Personally Identifiable Information
PIN	Personal Identification Number
PKI	Public Key Infrastructure
PL	Planning
PM	Project Management/Privilege Management
PMEF	Primary Mission Essential Functions
P.L.	Public Law
POC	Point Of Contact
POET	Political, Operational, Economic, and Technical
PON	Passive Optical Network
PPTP	Point-to-Point Tunneling Protocol
PROV	W3C Provenance family of specifications[16]
PS	Predefined Segmentation/Personnel Security
PT	Provenance Tracking
PUR	Privilege-Based Usage Restrictions
QoS	Quality of Service
RA	Risk Assessment
RAdAC	Risk-Adaptable (or Adaptive) Access Control
RAID	Redundant Array of Independent Disks
RAR	Risk Assessment Report
RBAC	Role-Based Access Control
RFI	Request for Information
RMF	Risk Management Framework
RMP	Risk Management Process
RPO	Recovery Point Objective
RTO	Recovery Time Objective
S/MIME	Secure/Multipurpose Internal Mail Extension
SA	Situational Awareness/Systems and Services Acquisition

---

[16]http://www.w3.org/TR/prov-dm/

SAISO	Senior Agency Information Security Officer
SAN	Storage Area Network
SAOP	Senior Agency Official for Privacy
SARA	Situational Awareness Reference Architecture
SC	System and Communications Protection/Surplus Capacity
SCAP	Security Content Automation Protocol
SCD	Supply Chain Diversity
SCP	System Contingency Plan
SCRM	Supply Chain Risk Management
SD	Synthetic Diversity
SDLC	System Development Lifecycle
SDH	Synchronous Digital Hierarchy
SDN	Software-Defined Networking
SF&A	Sensor Fusion and Analysis
SI	System and Information Protection
SIEM	Security Information and Event Management
Sim	Misdirection/Simulation
SLA	Service-Level Agreement
SOA	Service-Oriented Architecture
SONET	Synchronous Optical Network
SP	Special Publication
SSE	System Security Engineer
SSO	System Security Officer
SSP	System Security Plan TCB Trusted Computing Base
ST&E	Security Test & Evaluation
STIX	Structured Threat Information eXpression[17]
TAXII	Trusted Automated eXchange of Indicator Information[18]
TDM	Time Division Multiplexing
TT&E	Test, Training, & Exercise
TTP	Tactic Technique Procedure
TTX	Tabletop Exercise
UPS	Uninterruptible Power Supply
URL	Uniform Resource Locator
vIMS	virtual IMS
VLAN	Virtual Local Area Network
VMM	Virtual Machine Monitor

---

[17] https://stix.mitre.org/
[18] http://taxii.mitre.org/

VPLS	Virtual Private LAN Service
VPN	Virtual Private Network
VTL	Virtual Tape Library
WAN	Wide Area Network
WDM	Wavelength Division Multiplexing
W3C	World-Wide Web Consortium
WiFi	Wireless

Glossary

Activity – a set of one or more tasks with a defined output [ISO 22301:2019]

Affected area – location that has been impacted by a *disaster* [ISO 22300:2018]

After-action report – *document* which records, describes, and analyzes the *exercise*, drawing on debriefs and reports from *observers*, and derives lessons from it

Alert – part of *public warning* that captures attention of first responders and *people at risk* in a developing *emergency* situation [ISO 22300:2018]

All clear – message or signal that the danger is over

All-hazards – naturally occurring *event*, human induced event (both intentional and unintentional), and technology caused event with potential *impact* on an *organization*, *community*, or society and the environment on which it depends

Alternate worksite – work location, other than the primary location, to be used when the primary location is not accessible

Appropriate law enforcement and other government officials – government and law enforcement *personnel* that have specific legal jurisdiction over the *international supply chain* or portions of it

Area at risk – location that could be affected by a *disaster* [ISO 22300:2018]

Asset – anything that has value to an *organization*

Attack – successful or unsuccessful attempt(s) to circumvent an *authentication solution*, including attempts to imitate, produce, or reproduce the *authentication elements* [ISO 22300:2018]

Attribute data management system, ADMS – system that stores, manages, and controls access of data pertaining to *objects*

Audit – systematic, independent, and documented *process* for obtaining audit evidence and evaluating it objectively to determine the extent to which the audit criteria are fulfilled [ISO 22301:2019]

Auditor – person who conducts an *audit* [ISO 22300:2018]

Authentic material good – *material good* produced under the control
of the legitimate manufacturer, originator of the *goods*, or *rights
holder*

Authentication – *process* of corroborating an *entity* or attributes with a
specified or understood level of assurance

Authentication element – tangible *object*, visual feature, or *information*
associated with a *material good* or its packaging that is used as
part of an *authentication solution*

Authentication function – function performing *authentication*

Authentication solution – complete set of means and *procedures* that
allow the *authentication* of a *material good* to be performed

Authentication tool – set of hardware and/or software system(s) that are
part of an anti-counterfeiting solution and is used to control the
authentication element

Authoritative source – official origination of an attribute which is also
responsible for maintaining that attribute

Authorized economic operator – party involved in the international
movement of *goods* in whatever function that has been approved
by or on behalf of a national customs administration as conform-
ing to relevant *supply chain* security standards

Automated interpretation – *process* that automatically evaluates authenticity by one or more components of the *authentication solution*

Business continuity – capability of an *organization* to continue delivery
of *products and services* within acceptable time frames at pre-
defined capacity relating to a *disruption* [ISO 22301:2019]

Business continuity management – holistic *management process* that
identifies potential *threats* to an *organization* and the *impact* those
threats, if realized, can cause on business operations and provides
a framework for building organizational *resilience* with the capability of an effective response that safeguards the interests of key
interested parties, reputation, brand and value-creating *activities*
[ISO 22300:2018]

Business continuity management system, BCMS – *management system*
for *business continuity* [ISO 22301:2019]

Business continuity plan – *documented information* that guides an *organization* to respond to a *disruption* and resume, recover, and restore
the delivery of products and services consistent with its business
continuity objectives [ISO 22301:2019]

Business continuity program – ongoing *management* and governance
process supported by *top management* and appropriately resourced

to implement and maintain *business continuity management* [ISO 22300:2018]

Business impact analysis – *process* of analyzing the impact of a *disruption* on the *organization* [ISO 22301:2019]

Business partner – contractor, supplier, or service provider with whom an *organization* contracts to assist the organization in its function as an *organization in the supply chain*

Capacity – combination of all the strengths and *resources* available within an *organization, community,* or society that can reduce the level of *risk* or the effects of a *crisis*

Cargo transport unit – road freight vehicle, railway freight wagon, freight container, road tank vehicle, railway tank wagon, or portable tank

Certified client – *organization* whose *supply chain security management* system has been certified/registered by a qualified third party

Civil protection – measures taken and systems implemented to preserve the lives and health of citizens, their properties, and their environment from undesired *events*

Client – *entity* that hires, has formerly hired, or intends to hire an *organization* to perform *security operations* on its behalf, including, as appropriate, where such an organization subcontracts with another company or local forces

Closed-circuit television system, CCTV system – surveillance system composed of cameras, recorders, interconnections, and displays that are used to monitor activities in a store, a company, or, more generally, a specific *infrastructure* and/or a public place

Color blindness – total or partial inability of a person to differentiate between certain *hues*

Color-code – set of colors used symbolically to represent particular meanings

Command and control – *activities* of target-orientated decision-making, including assessing the situation, *planning*, implementing decisions, and controlling the effects of implementation on the *incident* [ISO 22300:2018]

Command and control system – system that supports effective *emergency management* of all available *assets* in a preparation, *incident response, continuity,* and/or *recovery process* [ISO 22300:2018]

Communication and consultation – continual and iterative *processes* that an *organization* conducts to provide, share, or obtain *information* and to engage in dialog with *interested parties* and others regarding the *management* of *risk*

Community – group of associated *organizations*, individuals, and groups sharing common interests

Community-based warning system – method to communicate *information* to the public through established networks

Competence – ability to apply knowledge and skills to achieve intended results [ISO 22301:2019]

Conformity – fulfillment of a *requirement* [ISO 22301:2019]

Consequence – outcome of an *event* affecting *objectives* [ISO 22301: 2019]

Contingency – possible future *event*, condition, or eventuality [ISO 22300:2018]

Continual improvement – recurring *activity* to enhance *performance*

Continuity – strategic and tactical capability, pre-approved by *management*, of an *organization* to plan for and respond to conditions, situations, and *events* in order to continue operations at an acceptable predefined level [ISO 22300:2018]

Continual improvement – recurring *activity* to enhance *performance* [ISO 22301:2019]

Conveyance – physical instrument of international trade that transports *goods* from one location to another

Cooperation – process of working or acting together for common interests and values based on agreement

Coordination – way in which different *organizations* (public or private) or parts of the same organization work or act together in order to achieve a common *objective*

Correction – action to eliminate a detected *non-conformity* [ISO 22300:2018]

Corrective action – action to eliminate the cause of a *non-conformity* and to prevent recurrence [ISO 22301:2019]

Counterfeit – simulate, reproduce, or modify a *material good* or its packaging without authorization

Counterfeit good – *material good* imitating or copying an *authentic material good*

Countermeasure – action taken to lower the *likelihood* of a *security threat scenario* succeeding in its *objectives* or to reduce the likelihood of *consequences* of a security threat scenario

Covert authentication element – *authentication element* that is generally hidden from the human senses and can be revealed by an informed person using a tool or by *automated interpretation*

Crisis – unstable condition involving an impending abrupt or significant change that requires urgent attention and action to protect life, *assets*, property, or the environment [ISO 22300:2018]

Crisis management – holistic *management process* that identifies potential *impacts* that threaten an *organization* and provides a framework for building *resilience*, with the capability for an effective response that safeguards the interests of the organization's key *interested parties*, reputation, brand, and value-creating *activities*, as well as effectively restoring operational capabilities [ISO 22300:2018]

Crisis management team – group of individuals functionally responsible for directing the development and execution of the response and operational *continuity* plan, declaring an operational *disruption* or *emergency/crisis* situation, and providing direction during the *recovery process*, both pre- and post-disruptive *incident* [ISO 22300:2018]

Critical control point, CCP – point, step, or *process* at which controls can be applied and a *threat* or *hazard* can be prevented, eliminated, or reduced to acceptable levels [ISO 22300:2018]

Critical customer – *entity*, the loss of whose business would threaten the survival of an *organization*

Critical product or service – *resource* obtained from a supplier which, if unavailable, would disrupt an *organization's* critical *activities* and threaten its survival

Critical supplier – provider of *critical products or services*

Criticality analysis – *process* designed to systematically identify and evaluate an *organization's assets* based on the importance of its mission or function, the group of *people at risk*, or the significance of an *undesirable event* or *disruption* on its ability to meet expectations

Custodian copy – duplicate that is subordinate to the *authoritative source*

Custody – period of time where an *organization in the supply chain* is directly controlling the manufacturing, handling, processing, and transportation of *goods* and their related shipping *information* within the *supply chain*

Disaster – situation where widespread human, material, economic, or environmental losses have occurred which exceeded the ability of the affected *organization, community, or society* to respond and recover using its own *resources* [ISO 22300:2018]

Disruption – *incident*, whether anticipated or unanticipated, that causes an unplanned, negative deviation from the expected delivery of products and services according to an *organization's* objectives [ISO 22301:2019]

Document – *information* and the medium on which it is contained

Documented information – *information* required to be controlled and maintained by an *organization* and the medium on which it is contained [ISO 22301:2019]

Downstream – handling, processing, and movement of *goods* when they are no longer in the *custody* of the *organization* in the *supply chain*

Drill – *activity* which practices a particular skill and often involves repeating the same thing several times

Dynamic metadata – *information* associated with a digital image aside from the pixel values that can change for each frame of a video sequence

Effectiveness – extent to which planned *activities* are realized and planned results achieved [ISO 22301:2019]

Emergency – sudden, urgent, usually unexpected occurrence, or *event* requiring immediate action [ISO 22301:2019]

Emergency management – overall approach for preventing *emergencies* and managing those that occur [ISO 22300:2018]

Entity – something that has a separate and distinct existence and that can be identified within context

Evacuation – organized, phased, and supervised dispersal of people from dangerous or potentially dangerous areas to places of safety [ISO 22300:2018]

Evaluation – systematic *process* that compares the result of *measurement* to recognized criteria to determine the discrepancies between intended and actual *performance* [ISO 22300:2018]

Event – occurrence or change of a particular set of circumstances [ISO 22301:2019]

Exercise – *process* to train for assessing, practicing, and improving *performance* in an *organization* [ISO 22301:2019]

Exercise annual plan – *document* in which the *exercise policy* plan has been translated to exercise goals and exercises, and in which an *exercise program* for a certain year is reflected [ISO 22300:2018]

Exercise coordinator – person responsible for *planning,* conducting, and evaluating *exercise* activities

Exercise program – series of *exercise* activities designed to meet an overall *objective* or goal

Exercise program manager – person responsible for *planning* and improving the *exercise program*

Exercise project team – group of individuals responsible for *planning*, conducting, and evaluating an *exercise* project

Exercise safety officer – person tasked with ensuring that any actions during the *exercise* are performed safely

Facility – plant, machinery, property, buildings, transportation units, sea/land/airports, and other items of *infrastructure* or plant and related systems that have a distinct and quantifiable business function or service

False acceptance rate – proportion of *authentications* wrongly declared true

False rejection rate – proportion of *authentications* wrongly declared false

Forensic – related to, or used in, courts of law [ISO 22300:2018]

Forensic analysis – scientific methodology for authenticating *material goods* by confirming an *authentication element* or an intrinsic attribute through the use of specialized equipment by a skilled expert with special knowledge [ISO 22300:2018]

Full-scale exercise – *exercise* that involves multiple *organizations* or functions and includes actual *activities*

Functional exercise – *exercise* to train for assessing, practicing, and improving the *performance* of single functions designed to respond to and recover from an unwanted *event*

Geo-location – specific location defined by one of several means to represent latitude, longitude, elevation above sea level, and coordinate system

Goods – items or materials that, upon the placement of a purchase order, are manufactured, handled, processed, or transported within the *supply chain* for usage or consumption by the purchaser

Hazard – source of potential harm [ISO 22300:2018]

Hazard monitoring function – *activities* to obtain evidence-based *information* on *hazards* in a defined area used to make decisions about the need for *public warning*

Hue – attribute of a visual sensation where an area appears to be similar to one of the perceived colors, red, yellow, green, and blue, or to a combination of two of them

Human interpretation – authenticity as evaluated by an *inspector*

Human rights risk analysis (HRRA) – *process* to identify, analyze, evaluate, and document human rights related *risks* and their *impacts*

in order to manage risk and to mitigate or prevent adverse human rights impacts and legal infractions

Identification – *process* of recognizing the attributes that identify an *entity*

Identifier – specified set of attributes assigned to an *entity* for the purpose of *identification*

Identity – set of attributes that are related to an *entity*

Impact – outcome of a disruption affecting objectives [ISO 22301:2019]

Impact analysis – consequence analysis *process* of analyzing all operational functions and the effect that an operational interruption can have upon them

Impartiality – actual or perceived presence of objectivity

Improvisation – act of inventing, composing, or performing, with little or no preparation, a reaction to the unexpected

Incident – event that can be, or could lead to, a *disruption*, loss, *emergency*, or crisis [ISO 22301:2019]

Incident command – process that is conducted as part of an incident *management system* and which evolves during the *management* of an *incident* [ISO 22300:2018]

Incident management system – system that defines the roles and responsibilities of *personnel* and the operating *procedures* to be used in the management of incidents [ISO 22300:2018]

Incident preparedness – *activities* taken to prepare for *incident response*

Incident response – actions taken in order to stop the causes of an imminent *hazard* and/or mitigate the *consequences* of potentially destabilizing *events* or *disruptions* and to recover to a normal situation

Information – data processed, organized, and correlated to produce meaning [ISO 22301:2019]

Infrastructure – system of *facilities*, equipment, and services needed for the operation of an *organization* [ISO 9000:2015, 3.5.2]

Inherently dangerous property – property that, if in the hands of an unauthorized individual, would create an imminent *threat* of death or serious bodily harm

Inject – scripted piece of *information* inserted into an *exercise* that is designed to elicit a response or decision and facilitate the flow of the exercise [ISO 22300:2018]

Inspector – person who uses the *object examination function* with the aim of evaluating an *object* [ISO 22300:2018]

Inspector access history – access logs detailing when *unique identifiers* (UID) were checked, optionally by which (privileged) *inspector*, and optionally from what specific location

Integrated authentication element – *authentication element* that is added to the *material good*

Integrity – property of safeguarding the accuracy and completeness of *assets*

Interested party – stakeholder person or *organization* that can affect, be affected by, or perceive itself to be affected by a decision or *activity* [ISO 9000:2015]

Internal attack – *attack* perpetrated by people or entities directly or indirectly linked with the legitimate manufacturer, originator of the *goods*, or *rights holder* (staff of the rights holder, subcontractor, supplier, etc.)

Internal audit – *audit* conducted by, or on behalf of, an *organization* itself for *management review* and other internal purposes, and which can form the basis for an organization's self-declaration of *conformity* [ISO 22301:2019]

International supply chain – *supply chain* that, at some point, crosses an international or economic border

Interoperability – ability of diverse systems and *organizations* to work together

Intrinsic authentication element – *authentication element* which is inherent to the *material good*

Invocation – act of declaring that an *organization's business continuity* arrangements need to be put into effect in order to continue delivery of key *products or services*

Key performance indicator (KPI) – quantifiable measure that an *organization* uses to gauge or compare *performance* in terms of meeting its strategic and operational *objectives* [ISO 22300:2018]

Less-lethal force – degree of force used that is less likely to cause death or serious injury to overcome violent encounters and appropriately meet the levels of resistance encountered

Likelihood – chance of something happening [ISO 22301:2019]

Logical structure – arrangement of data to optimize their access or processing by given user (human or machine)

Management – coordinated *activities* to direct and control an *organization* [ISO 22301:2019]

Management plan – clearly defined and documented plan of action, typically covering the key *personnel, resources*, services, and actions needed to implement the *management* process

Management system – set of interrelated or interacting elements of an *organization* to establish policies, *objectives*, and *processes* to achieve those objectives [ISO 9000:2015]

Management system consultancy and/or associated risk assessment – participation in designing, implementing, or maintaining a *supply chain security management* system and in conducting *risk assessments*

Material good – manufactured, grown product or one secured from nature

Material good lifecycle – stages in the life of a *material good* including conception, design, manufacture, storage, service, resell, and disposal

Maximum acceptable outage (MAO) – the time it would take for adverse *impacts*, which can arise as a result of not providing a product/ service or performing an *activity*, to become unacceptable [ISO 22300:2018]

Maximum tolerable period of disruption (MTPD) – the time it would take for adverse *impacts*, which can arise as a result of not providing a product/service or performing an *activity*, to become unacceptable [ISO 22300:2018]

Measurement – *process* to determine a value [ISO 9000:2015]

Metadata – *information* to describe audiovisual content and data essence in a defined format

Minimum business continuity objective (MBCO) – minimum level of services and/or products which is acceptable to an *organization* to achieve its business *objectives* during a *disruption*

Mitigation – limitation of any negative *consequence* of a particular *incident*

Monitoring – determining the status of a system, a *process*, a product, a service, or an *activity* [ISO 9000:2015]

Mutual aid agreement – pre-arranged understanding between two or more entities to render assistance to each other

Non-conformity – non-fulfillment of a *requirement* [ISO 9000:2015]

Notification – part of *public warning* that provides essential *information* to *people at risk* regarding the decisions and actions necessary to cope with an *emergency* situation [ISO 22300:2018]

Object – single and distinct entity that can be identified [ISO 22301:2019]

Object examination function (OEF) – *process* of finding or determining the *unique identifier (UID)* or other attributes intended to authenticate

Objective – result to be achieved [ISO 9000:2015]

Observer – *participant* who witnesses the *exercise* while remaining separate from exercise activities

Off-the-shelf authentication tool – *authentication tool* that can be purchased through open sales networks

Online authentication tool – *authentication tool* that requires a real-time online connection to be able to locally interpret the *authentication element*

Operational information – *information* that has been contextualized and analyzed to provide an understanding of the situation and its possible evolution [ISO 22300:2018]

Organization – person or group of people that has its own functions with responsibilities, authorities, and relationships to achieve its *objectives* [ISO 9000:2015]

Organization in the supply chain – *entity* that manufactures, handles, processes, loads, consolidates, unloads, or receives *goods* upon placement of a purchase order that, at some point, crosses an international or economy border, transports goods by any mode in the *international supply chain* regardless of whether their particular segment of the *supply chain* crosses national (or economy) boundaries, or provides, manages, or conducts the generation, distribution, or flow of shipping *information* used by customs agencies or in business practices.

Outsource – make an arrangement where an external *organization* performs part of an organization's function *process* [ISO 9000:2015]

Overt authentication element – *authentication element* that is detectable and verifiable by one or more of the human senses without resource to a tool (other than everyday tools that correct imperfect human senses, such as spectacles or hearing aids)

Owner – *entity* that legally controls the licensing and user rights and distribution of the *object* associated with the *unique identifier (UID)*

Participant – person or *organization* who performs a function related to an *exercise*

Partnering – associating with others in an *activity* or area of common interest in order to achieve individual and collective *objectives*

Partnership – organized relationship between two bodies (public–public, private–public, and private–private) which establishes the scope, roles, *procedures*, and tools to prevent and manage any *incident* impacting on *security* and *resilience* with respect to related laws

People at risk – individuals in the area who may be affected by an *incident*

Performance – measurable result [ISO 22301:2019]

Performance evaluation – *process* to determine measurable results against the set criteria [ISO 22301:2019]

Personnel – people working for and under the control of the *organization* [ISO 22301:2019]

Planning – part of *management* focused on setting *business continuity objectives* and specifying necessary operational *processes* and related resources to fulfill the business continuity objectives [ISO 22301:2019]

Policy – intentions and direction of an *organization*, as formally expressed by its *top management* [ISO 9000:2015]

Preparedness – readiness *activities*, programs, and systems developed and implemented prior to an *incident* that can be used to support and enhance prevention, protection from, mitigation of, response to, and recovery from *disruptions, emergencies,* or *disasters*

Prevention – measures that enable an *organization* to avoid, preclude, or limit the *impact* of an *undesirable event* or potential *disruption* [ISO 22300:2018]

Prevention of hazards and threats – *process*, practices, techniques, materials, products, services, or *resources* used to avoid, reduce, or control *hazards* and *threats* and their associated *risks* of any type in order to reduce their potential *likelihood* or *consequences*

Preventive action – action to eliminate the cause of a potential *non-conformity* or other undesirable potential situation [ISO 9000:2015, 3.12.1]

Prioritized activity – *activity* to which urgency is given in order to avoid unacceptable impacts to the business during a *disruption* [ISO 22301:2019]

Private security service provider – private security company (PSC) *organization* that conducts or contracts *security operations* and whose business activities include the provision of *security* services either on its own behalf or on behalf of another [ISO 22300:2018]

Probability – measure of the chance of occurrence expressed as a number between 0 and 1, where 0 is impossibility and 1 is absolute certainty [ISO/Guide 73:2009, 3.6.1.4]

Procedure – specified way to carry out an *activity* or a *process* [ISO 22301:2019]

Process – set of interrelated or interacting *activities* which transforms inputs into outputs [ISO 22301:2019]

Product or service – output or outcome provided by an *organization* to *interested parties* [ISO 22301:2019]

Protection – measures that safeguard and enable an *organization* to prevent or reduce the *impact* of a potential *disruption* [ISO 22301:2019]

Public warning – *notification* and *alert* messages disseminated as an *incident response* measure to enable responders and *people at risk* to take safety measures

Public warning system – set of protocols, *processes*, and technologies based on the *public warning policy* to deliver *notification* and *alert* messages in a developing *emergency* situation to *people at risk* and to first responders

Purpose-built authentication tool – *authentication tool* dedicated to a specific *authentication solution*

Record – *document* stating results achieved or providing evidence of *activities* performed [ISO 9000:2015]

Recovery – restoration and improvement, where appropriate, of operations, facilities, livelihoods, or living conditions of affected *organizations*, including efforts to reduce *risk* factors [ISO 22301:2019]

Recovery point objective (RPO) – point to which *information* used by an *activity* is restored to enable the activity to operate on resumption [ISO 22300:2018]

Recovery time objective (RTO) – period of time following an *incident* within which a *product or service* or an *activity* is resumed or *resources* are recovered [ISO 22300:2018]

Requirement – need or expectation that is stated, generally implied or obligatory [ISO 9000:2015]

Residual risk – *risk* remaining after *risk treatment* [ISO/Guide 73: 2009]

Resilience – ability to absorb and adapt in a changing environment [ISO 22301:2019]

Resource – asset, *facility*, equipment, material, product, or waste that has potential value and can be used

Response plan – documented collection of *procedures* and *information* that is developed, compiled, and maintained in readiness for use in an *incident* [ISO 22300:2018]

Response program – plan, *processes*, and *resources* to perform the *activities* and services necessary to preserve and protect life, property, operations, and critical *assets* [ISO 22300:2018]

Response team – group of individuals responsible for developing, executing, rehearsing, and maintaining the *response plan*, including the *processes* and *procedures* [ISO 22300:2018]

Review – *activity* undertaken to determine the suitability, adequacy, and effectiveness of the *management system* and its component elements to achieve established *objectives* [ISO/Guide 73:2009]

Rights holder – legal *entity* either holding or authorized to use one or more intellectual property rights

Risk – effect of uncertainty on *objectives* [ISO/Guide 73:2009]

Risk acceptance – informed decision to take a particular *risk* [ISO/Guide 73:2009]

Risk analysis – *process* to comprehend the nature of *risk* and to determine the level of risk [ISO/Guide 73:2009]

Risk appetite – amount and type of *risk* that an *organization* is willing to pursue or retain [ISO/Guide 73:2009]

Risk assessment – overall *process* of *risk identification, risk analysis*, and *risk evaluation* [ISO/Guide 73:2009]

Risk communication – exchange or sharing of *information* about *risk* between the decision maker and other *interested parties*

Risk criteria – terms of reference against which the significance of a *risk* is evaluated [ISO/Guide 73:2009]

Risk evaluation – *process* of comparing the results of *risk analysis* with *risk criteria* to determine whether the *risk* and/or its magnitude is acceptable or tolerable [ISO/Guide 73:2009]

Risk identification – *process* of finding, recognizing, and describing *risks* [ISO/Guide 73:2009]

Risk management – coordinated *activities* to direct and control an *organization* with regard to *risk* [ISO 22301:2019]

Risk owner – *entity* with the accountability and authority to manage a *risk* [ISO/Guide 73:2009]

Risk reduction – actions taken to lessen the *probability* or negative *consequences*, or both, associated with a *risk*

Risk register – *record* of *information* about identified *risks* [ISO/Guide 73:2009]

Risk sharing – form of *risk treatment* involving the agreed distribution of *risk* with other parties [ISO/Guide 73:2009]

Risk source – element which, alone or in combination, has the intrinsic potential to give rise to *risk* [ISO/Guide 73:2009]

Risk tolerance – or interested party's readiness to bear the *risk* after *risk treatment* in order to achieve its *objectives* [ISO/Guide 73:2009]

Risk treatment – *process* to modify *risk* [ISO/Guide 73:2009]

Robustness – ability of a system to resist virtual or physical, internal, or external *attacks*

Scenario – pre-planned storyline that drives an *exercise* as well as the stimuli used to achieve exercise project *performance objectives* [ISO 22300:2018]

Scene location – collection of *geo-locations* that define the perimeter of the viewable scene of a camera

Scope of exercise – magnitude, *resources*, and extent that reflect the needs and *objectives*

Scope of service – function(s) that an *organization in the supply chain* performs, and where it performs this/these functions

Script – story of the *exercise* as it develops which allows directing staff to understand how *events* should develop during exercise play as the various elements of the master events list are introduced

Secret – data and/or knowledge that are protected against disclosure to unauthorized entities

Security – state of being free from danger or *threat*

Security aspect – characteristic, element, or property that reduces the *risk* of unintentionally, intentionally, and naturally caused *crises* and *disasters* which disrupt and have *consequences* on the *products or services*, operation, critical *assets*, and *continuity* of an *organization* and its *interested parties* [ISO 22300:2018]

Security cleared – *process* of verifying the trustworthiness of people who will have access to *security sensitive information*

Security declaration – documented commitment by a *business partner*, which specifies *security* measures implemented by that business partner, including, at a minimum, how *goods* and physical instruments of international trade are safeguarded, associated *information* is protected, and security measures are demonstrated and verified

Security management – systematic and coordinated *activities* and practices through which an *organization* optimally manages its *risks* and the associated potential *threats* and *impacts* [ISO 22300:2018]

Security management objective – specific outcome or achievement required of *security* in order to meet the *security management policy* [ISO 22300:2018]

Security management policy – overall intentions and direction of an *organization*, related to the *security* and the framework for the control of security-related *processes* and *activities* that are derived from and consistent with its *policy* and regulatory *requirements* [ISO 22300:2018]

Security management program – *process* by which a *security management objective* is achieved [ISO 22300:2018]

Security management target – specific level of *performance* required to achieve a *security management objective*

Security operation – *activity* and function related to the *protection* of people, and tangible and intangible *assets*

Security operations management – coordinated *activities* to direct and control an *organization* with regard to *security operations*

Security operations objective – *objective* sought, or aimed for, related to *security operations*

Security operations personnel – people working on behalf of an *organization* who are engaged directly or indirectly in *security operations*

Security operations policy – overall intentions and direction of an *organization* related to *security operations* as formally expressed by *top management*

Security operations program – ongoing *management* and governance *process* supported by *top management* and resourced to ensure that the necessary steps are taken to coordinate the efforts to achieve the *objectives* of the *security operations management* system [ISO 22300:2018]

Security personnel – people in an *organization in the supply chain* who have been assigned *security*

Security plan – planned arrangements for ensuring that *security* is adequately managed [ISO 22300:2018]

Security sensitive information – security sensitive material *information* or material, produced by or incorporated into the *supply chain* security *process*, that contains information about the *security* processes, shipments, or government directives that would not be readily available to the public and would be useful to someone wishing to initiate a security *incident*

Security threat scenario – means by which a potential *security incident* can occur [ISO 22300:2018]

Self-defense – *protection* of one's person or property against some injury attempted by another

Semantic interoperability – ability of two or more systems or services to automatically interpret and use *information* that has been exchanged accurately [ISO 22300:2018]

Sensitive information – *information* that is protected from public disclosure only because it would have an adverse effect on an *organization*, national *security*, or public safety

Shelter in place – remain or take immediate refuge in a protected location relevant to the *risk*

Specifier – *entity* who defines the *requirements* for an *authentication solution* to be applied to a particular *material good*

Stand-alone authentication tool – *authentication tool* that is either used to reveal a covert *authentication element* to the human senses for human *verification* or that integrates the functions required to be able to verify the authentication element independently

Static metadata – *information* associated with a digital image aside from the pixel values that does not change over time (or at least does not change over the addressed sequence)

Strategic exercise – *exercise* involving *top management* at a strategic level [ISO 22300:2018]

Subcontracting – contracting with an external party to fulfill an obligation arising out of an existing contract

Supply chain – two-way relationship of *organizations*, people, *processes*, logistics, *information*, technology, and resources engaged in *activities* and creating value from the sourcing of materials through the delivery of *products and services* [ISO 22301:2019]

Supply chain continuity management (SCCM) – application of *business continuity management* to a *supply chain*

Syntactic interoperability – ability of two or more systems or services to exchange structured *information*

Tamper evidence – ability of the *authentication element* to show that the *material good* has been compromised

Target – detailed *performance requirement*, applicable to an *organization* or parts thereof, that arises from the *objectives* and that needs to be set and met in order to achieve those objectives [ISO 14050: 2009]

Target group – individuals or *organizations* subject to *exercise*

Test – unique and particular type of *exercise*, which incorporates an expectation of a pass or fail element within the aim or *objectives* of the exercise being planned [ISO 22300:2018]

Testing – *procedure* for *evaluation*; a means of determining the presence, quality, or veracity of something [ISO 22300:2018]

Threat – potential cause of an unwanted *incident*, which may result in harm to individuals, *assets*, a system or *organization*, the environment, or the *community*

Threat analysis – *process* of identifying, qualifying, and quantifying the potential cause of an unwanted *event*, which may result in harm to individuals, *assets*, a system or *organization*, the environment, or the *community*

Tier 1 supplier – provider of *products or services* directly to an *organization* usually through a contractual arrangement

Tier 2 supplier – provider of *products or services* indirectly to an *organization* through a *tier 1 supplier*

Top management – person or group of people who directs and controls an *organization* at the highest level [ISO 9000:2015]

Track and trace – means of identifying every individual *material good* or lot(s) or batch in order to know where it has been (track) and where it is (trace) in the supply chain

Training – *activities* designed to facilitate the learning and development of knowledge, skills, and abilities, and to improve the *performance* of specific tasks or roles [ISO 22301:2019]

Trusted query processing function (TQPF) – function that provides a gateway to *trusted verification function* (TVF) and attribute management data system (ADMS)

Trusted verification function (TVF) – function that verifies whether a *unique identifier* (UID) received is valid or not and manages a response according to rules and access privileges [ISO 22300:2018]

Undesirable event – occurrence or change that has the potential to cause loss of life, harm to tangible or intangible *assets*, or negatively *impact* the human rights and fundamental freedoms of internal or external *interested parties*

Unique identifier (UID) – code that represents a single and specific set of attributes that are related to an *object* or class of objects during its life within a particular domain and scope of an object *identification* system

Upstream – handling, processing, and movement of *goods* that occur before the *organization in the supply chain* takes *custody* of the goods

Use of force continuum – increasing or decreasing the level of force applied as a continuum relative to the response of the adversary, using the amount of force reasonable and necessary

Verification – confirmation, through the provision of objective evidence, that specified *requirements* have been fulfilled [ISO 9000:2015]

Vulnerability – vulnerability analysis and vulnerability assessment *process* of identifying and quantifying something that creates susceptibility to a source of *risk* that can lead to a *consequence*

Vulnerable group – individuals who share one or several characteristics that are the basis of discrimination or adverse social, economic, cultural, political, or health circumstances and that cause them to lack the means to achieve their rights or, otherwise, enjoy equal opportunities

Warning dissemination function – *activities* to issue appropriate messages for *people at risk* based on evidence-based *information* received from the *hazard monitoring function* [ISO 22300:2018]

Work environment – set of conditions under which work is performed [ISO 22301:2019]

World Customs Organization (WCO) – independent intergovernmental body whose mission is to enhance the *effectiveness* and efficiency of customs administrations

Introduction

The term *business continuity management (BCM)* appeared recently and, today, attracts constant interest from top managers of international companies. Since approximately 1988, a number of high-tech countries around the world, mainly in the *United Kingdom, the United States, Canada, the European Union, Russia, Australia, China, Singapore,* and *Japan,* have held annual hearings and meetings of specially created committees and commissions on *business continuity management.* Over a dozen different international and national standards and specifications on *business continuity management* were prepared, including the most famous: *ISO 22301:2019 (replaced part 2 of the standard BS 25999 (PAS 56)), ISO/IEC 27001:2013(A. 17) and ISO/IEC 27031:2011, ASIS ORM.1-2017, NIST SP800-34, NFPA 1600:2019, CSA Z1600, AS/NZS 5050 (HB 292), SS540:2009 (TR19:2004), SI 24001:2007, High Level Principles for Business Continuity (2006), COBIT ®2019, RESILIA 2015, V4 ITIL, MOF 4.0 in the BCM part,* etc. For example, the ISO 22301:2019 standard *"Security and resilience – Business continuity management systems – Requirements"* is intended for certification of *BCMS – business continuity management systems* of organizations operating internationally. *ISO 22301:2019* is coordinated with other well-known international standards *ISO 9001 "Quality management systems," ISO 14001 "Environmental management systems," ISO 31000 "Risk management," ISO/IEC 20000-1 "Information technology – Service management," ISO/IEC 27001:2013 "Information security management systems," ISO 28000 "Specification for security management systems for the supply chain,"* etc.

Currently, *business continuity management* is one of the most relevant and dynamically developing areas of strategic and operational management of modern enterprises. The relevance of this trend for each company is explained by the need to ensure the survival and preservation of their business in emergency situations. The term business continuity management usually refers to the systematic process of assessing the consequences of emergencies and making appropriate decisions to preserve the company's business. Therefore, the main goal of the relevant *enterprise continuity program (ECP)* is minimizing the risk of business loss in case of its interruption and to continue the company's activities in emergency situations.

In a number of countries, including *Russia*, the practice of developing and implementing corporate ECP programs is just beginning. One of the best initiatives of the *Bank of Russia* prepared the corresponding section 8.11 of the STO BR IBBS-1.0-2008, on the grounds of recommendations of *ISO/IEC 27001:2005 (A. 14)*, and then based on the document of the Basel Committee on the Banking supervision (*High Level Principles for Business Continuity*) developed Paragraph 3.7 of the Bank of Russia regulations dated December 16, 2003. N 242-P "*On the organization of internal control in credit organizations and banking groups*" (updated in accordance with the instruction dated March 5, 2009 No. 2194-U "*On amendments to the regulations of the Bank of Russia*" dated December 16, 2003 N 242-P).

At the same time, in Europe and the United States, the implementation and support of these corporate programs is on fast-forward, and in some government and commercial structures, business continuity management issues are given the closest attention. For example, US Federal departments carry out business continuity planning in accordance with approved continuity of operations (COOP) directives. In the financial field, business continuity issues for American companies are regulated by the recommendations of the *Gramm-Leach-Bliley* and the *Expedited Funds Availability law*s, as well as the recommendations of the *SAS 78/94* standard. In the field of health, the guiding document in the BCM part is *HIPAA*. In addition, for most companies that provide essential services (electricity, water, gas, communications, etc.), certain benefits are provided by the state when using business continuity procedures. The fact is that the continuity of these companies plays an important role in ensuring the continuity of various Federal organizations and structures (hospitals, police, fire departments, schools, and government agencies), as well as large commercial structures (banks, financial organizations, insurance companies, Internet service providers, and so on). In the *USA*, *Canada*, and the *EU*, the most active users of *business continuity plans* (BCP) are various financial institutions and organizations, enterprises of the raw materials and oil refining industry, airlines, telecommunications companies, etc.

The recent tragic events, such as the terrorist attacks in September 2001 in New York at the World Trade Center, the blackout in North-Eastern USA and South-Eastern Canada in 2003–2009, volcanoes in Guatemala, New Zealand, Indonesia and Iceland in 2010–2018, natural disasters in India, Philippines, and China in 2018, traffic collapses in the European Union in 2019, and, finally, the pandemic threat of the virus *COVID-19* in early 2020 that we have only begun to fight and that has already claimed hundreds of thousands of lives, clearly showed that only those companies that took timely advantage of the recommendations for business continuity were able to avoid major financial losses and maintain their business. The rest of the companies suffered significant financial losses and some even lost their business. Therefore, companies are constantly improving their business continuity plan and its various derivatives: the *Business Crash Plan, the Business Disaster Plan, the Anti-Terrorist plan, the Anti-Bomb Plan, the Business Continuity Plan, the Business Recovery Plan, the Anti-Crisis Plan, and so on.*

Emergencies occur almost every day; therefore, every company probably raises the following questions:

1. What are the legal guidelines and requirements for ensuring business continuity? How should we organize work within this scope?
2. How to create and implement a cost-effective corporate business continuity management program?
3. What kind of BCM solutions or services best meet our company's needs?
4. Should our company itself create and maintain business continuity and recovery plans or is it sufficient to enter into an appropriate contract with a consulting company?
5. What tools exist for automating business continuity planning and management?
6. How to control business continuity management?
7. How to evaluate and manage the costs of support and maintaining an enterprise continuity program?

The answers to these and many other questions will create and implement a truly effective and cost-effective *enterprise continuity program* and, at the same time, make the aforementioned program "transparent" and understandable both for the management and ordinary employees, as well as for business partners and clients of the company.

Table.   Stages of the enterprise continuity program lifecycle.

Stages of the *enterprise continuity program lifecycle*	Possible outcome
Stage 1: Analysis of business continuity requirements	
Necessary:	The results:
• analyze the company's critical business processes and supporting infrastructure; • identify and verify current threats and vulnerabilities of business processes; • assess the main risks (RA) of business processes; • assess potential financial losses in the event of an emergency; • conduct a full business impact analysis (BIA) for the company's business units.	• methods of assessment and ranking of company critical business processes; • methods of verifying threats and vulnerabilities of company business processes; • methods of risk assessment; • RA report with the analysis of risks and priorities, and priority tasks to ensure business continuity; • methodology of damage assessment in case of emergency situations; • BIA report with estimates of company assets and possible damage as a result of emergency situations.
Stage 2: Business continuity planning	
Necessary:	The results:
• form and approve the BCPM business continuity planning and management group; • develop strategies and continuity plans for each business unit of the company; • identify priority measures to ensure business continuity; • develop alternative solutions; • choose the best solution from the available alternatives; • determine the necessary resources for business continuity planning and management; • form and approve the BCPM business continuity planning and management group.	• membership of the business continuity planning and management group; • business continuity strategies; • business continuity plans for each business unit of the company; • list of priority measures to ensure business continuity; • list of alternatives and criteria for choosing the optimal solution; • official instructions of the company's employees on business continuity provision with definition of the role, responsibilities and degree of responsibility of each employee; • formalized requirements for business continuity planning and management; • estimates of the cost of possible solutions for business continuity management; • criteria for selecting BCP solution providers; • extracts from the company budget for business continuity planning and management.

Stage 3: Support and maintenance of the ECP program

Necessary:

- train company employees on business continuity and management issues;
- develop regulations for maintaining and supporting business continuity plans (BCP);
- purchase the necessary BCP support tools;
- install and configure BCP support tools;
- develop a notification system for adjustments and changes to the BCP;
- develop control tests of the effectiveness of business continuity plans and a schedule of control checks;
- develop formal criteria for evaluating BCP audits;
- develop a procedure for making changes to the BCP.

Results:

- employee certificates in the field of BCM;
- methods and guidelines for installing, configuring, and servicing BCP tools;
- specifications of BCP support regulations;
- annunciation scheme introducing the changes that are being made.
- BCP testing and verification methods;
- formal BCP evaluation criteria for presenting test results;
- reports on testing BCP plans;
- instructions on how to make changes to the BCP;
- guidelines for maintaining and supporting business continuity.

However, the first experience of developing and implementing corporate ECP programs revealed the following problems:

- a number of organizations do not have any kind of policies, strategies, plans, or procedures for business continuity and recovery in emergency situations;
- insufficient system development of the subject area and, as a result, the focus on disaster recovery of IT services and poor coverage of critical business processes, including services provided to customers and partners;
- the lack of a formal description of business processes with the names of responsible persons and, as a result, difficulties in determining the acceptable recovery time and optimal recovery point;
- irregular and/or incomplete analysis of external and internal impacts on critically important business processes of the company, which leads to the fact that business continuity plans do not always meet the goals and objectives of the business, not to mention inadequate expenses for business continuity;
- outdated methods and approaches to business continuity planning and management that are poorly adapted to the requirements of international legislation and relevant regulatory documents of state bodies and regulators;

- insufficient training of employees of organizations in business continuity management, lack of knowledge, and practical skills in emergency situations.

This monograph contains possible solutions to these problems. According to the author, the monograph compares favorably with other books previously published because it consistently outlines all the main issues of development and implementation of ECP programs. In particular, the possible methods of *BIA* and *RA*, examples of developing *BCP/DRP* plans and appropriate *BCP/DRP* testing plans, and the practice of successfully completed BCM projects in government and commercial organizations are considered. In addition, the book discusses for the first time a number of problematic issues in the development of quantitative metrics and measures to ensure business continuity management, *BCM*. The article presents well-known and author's methods of ensuring business continuity and stability that meet the goals and objectives of the well-known technical committee ISO/TC 292 *"Security and resilience"* of the international organization ISO[19], responsible for the development and improvement of the ISO 22300 family of standards. It is important that the book contains the results of both qualitative and quantitative studies of cyber resilience, which allows us to discover for the first time the ultimate law of the effectiveness of protecting critical information infrastructure.

The monograph is aimed at the following main groups of readers:

1. **Managers of the highest level of company management (top managers)** who want to get answers to the following questions: How to effectively manage business continuity in the interests of the company? What is the *enterprise continuity program*? What is its essence? Why and who needs it? How relevant is it for the company and its business activities? What is its cost and subsequent costs? Who is responsible for implementing and maintaining the business continuity management program? What are the legal restrictions? What are the consequences for the company? What role do business continuity planning and management tools play? What national and international business continuity management techniques and technologies are preferable to use? How should your company be prepared for audit and accreditation in accordance with the requirements of the international standard *ISO 22301:2019*?

    The answers to these questions are given in Chapters 1, 2, and 4.

---

[19] https://www.isotc292online.org

2. **Chief Information Officer (CIO) and Chief Information Security Officer (CISO)** who wish to obtain an objective and independent assessment of the current state of continuity of critical business processes and it services, to assess in a timely manner the consequences of emergencies and develop requirements for business resumption, check the adequacy and effectiveness of the enterprise continuity management program, calculate the necessary costs for improving the aforementioned ECP program, and take all necessary organizational and technical measures to increase (adequately ensure) the level of maturity of the *business continuity management system (BCMS)* of the organization.

   The answers to these questions are given in Chapters 1, 2, and 4.

3. **BCM professionals, business continuity and security managers, and internal control specialists** who want to get a detailed understanding of *business continuity management (BCM)*, sufficient to understand these issues correctly and possibly lead the work related to the development and implementation of the *enterprise continuity program (ECP)* in their company. Chapters 1, 2, and 3 are addressed to this category of readers.

4. **BCM internal and external auditors, consultants, and trainers.** This category of readers will be able to get answers to their questions in Chapters 1, 2, and 4.

The monograph can also be used as a textbook by students and postgraduates of relevant technical specialties, especially since the materials of the chapters of the monograph are based, among other things, on the author's teaching experience at the *Moscow Institute of physics and Technology (MIPT) and Innopolis University.*

   The monograph contains four chapters dedicated to:

- The role of the *enterprise continuity program (ECP)* in the organization's management system in the context of an unprecedented growth of threats to business interruption. Problematic issues of business continuity management (BCM) and possible ways to resolve them. Author's models and methods for ensuring business continuity and sustainability.

- Best practices in business continuity management (BCM) *ISO 22301:2019, ISO 22313:2020, ISO/TS 22317:2015, ISO 31000:2018, ISO/IEC 27001:2013 (A.17) and ISO/IEC 27031:2011, Good Practice Guidelines 2018 Edition (BCI), The Professional Practices for Business Continuity Management 2017 Edition (DRII), ASIS*

ORM.1-2017, NIST SP800-34, NFPA 1600:2019, COBIT ®2019, ITIL V4 and MOF 4.0, RESILIA 2015 etc.

- BCM project management on the example of well-known *"big four"* companies *Deloitte Touche Tohmatsu, PricewaterhouseCoopers, Ernst & Young, and KPMG*, as well as leading manufacturers of fault-tolerant solutions *IBM, HP, EMC, Microsoft*, etc.

- Examples of *enterprise continuity program (ECP)* development for a number of leading technologically advanced national and international companies. Here are the author's recommendations for creating and implementing ECP.

The "**Developing an Enterprise Continuity Program**" monograph was written by *Sergei Petrenko, Prof. Dr.-Ing., Head of the Information Security Center at Innopolis University* based on the author's many years of practical experience in the field of business continuity. The author personally managed a number of major projects that were successfully completed together with leading auditors and consultants of the *"big four"* Deloitte Touche Tohmatsu, PricewaterhouseCoopers, Ernst & Young, and KPMG, as well as with consultants of *IBM, HP, EMC, and Microsoft* in the BCM part. In particular, the following unique projects:

- *"Development and implementation of procedures for ensuring the continuity of information systems of the Federal Customs Service of Russia"*;
- *"Development of a strategy for managing the continuity of critical technological processes of the leading Russian telecommunications company MTS PJSC (Mobile TeleSystems)"*;
- *"Development of a program ensuring the continuity of the largest Bank in Russia, Central and Eastern Europe, Sberbank PJSC, in the context of local and global disasters"*;
- *"Analysis of the business impact on the activities of a large commercial Bank of Russia, PJSC Gazprombank"*;
- *"Development of ECP for LLC Home Credit and Finance Bank – a Russian commercial Bank, one of the leaders of the Russian consumer lending market"*;
- *"Development of a business continuity plan (BCP) for the world's largest aluminum company "Russian aluminum" (United Company Rusal)"*;
- *"Development of a Business Continuity Concept for LUKOIL-inform LLC, a subsidiary of LUKOIL, the world's largest oil company"*;

- *"Development of ECP for a major Russian manufacturer of soft drinks and dairy products "Wimm Bill Dann" (WBD) – acquired by PepsiCo, Inc.,"* and others.

The author is grateful in advance to all readers who are ready to share their opinions on this book. You can send your emails to the author at *S.Petrenko@innopolis.ru.*

Professor Sergei A. Petrenko
S.Petrenko@innopolis.ru
Russia-Germany
January 2021

1

ECP Relevance

In the terms of business as well as IT specialists, the concept of *business continuity management* is often identified as *disaster recovery* after. It should be clearly understood that the main goal of *BCM* is to maintain up-to-date a sufficient number of *structures, operations, and resources* (*assets*) necessary for the stable functioning of the organization in emergency situations. This representation of *BCM* differs significantly from the concept of disaster recovery, which is closely, if not exclusively, associated with information technology [1, 2, 3]. Today, the focus of the continuity concept is shifting to the organization in general (Figure 1.1), to critical business processes (*main and supporting*), expanding the horizons of previous consideration of the problem beyond only information systems and IT services, despite their importance for modern companies.

1.1 Motivation and Achievable Benefits

The enterprise business continuity management programs are relevant primarily for medium and large companies. At the same time, companies, as a rule, must already have a certain level of business culture: a clearly defined business development strategy, an appropriate IT, and a security strategy [4, 5, 6 , 7, 8, 9, 10, 11]. It is important that strategies appear not only as conceptual normative documents but also as specific metrics and measures (preferably quantitative) that allow assessing the achievement of goals related to the business continuity management (Figure 1.2).

It is obvious that the case begins to take real shape only when the responsibility for IT will be delegated to a specific official. Therefore, the starting point is to determine the owner of the business continuity management process. While creating an appropriate business continuity program in an organization, a responsible person – an employee interested in ensuring business continuity that has sufficient authority to implement it – must be identified [12, 13, 14, 15, 16].

1

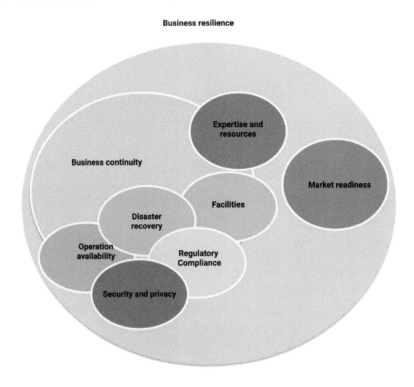

Figure 1.1   Basic concepts and definitions of business resilience.

Preferably, the question of business continuity management should be raised by the Head of the organization. However, this is not always the case, especially in national practices. If CFO in the organization has a large authority, then, as a rule, he first draws attention to the need of ensuring the business continuity. Most of the surge in interest occurs after audits (for example, for compliance with *SOX* requirements), which require the company to develop and maintain up-to-date continuity plans.

The security service can also initiate the business continuity management process. But security services, with rare exceptions, rarely consider business continuity as an integral part of the security system.

IT-dependent businesses are normally the first to pay attention to business continuity issues as they are more mature in terms of organization and business management. It is clear that ensuring business continuity is not exclusively an IT task. However, it is often the IT professionals who become the engines of business continuity processes in the organization. Partially, this situation happens due to the fact that they are eternal

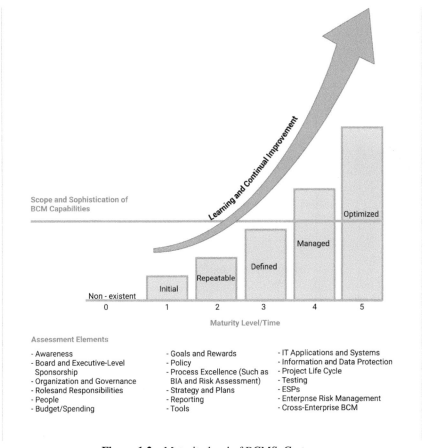

Figure 1.2 Maturity level of BCMS, Gartner.

"switchers": the fault for any failure in the organization's work related to information systems automatically falls on the IT service. In many companies, most of the risks are transferred by management to the IT sphere, which is not always legal, but, unfortunately, this is currently a fairly common practice. Therefore, the CIO, as no one else, is interested in building and optimizing business processes correctly.

However, the Head of the organization should be aware that this solves only the part of the overall complex task. The systems that are critical to the organization can be duplicated, fault tolerance and disaster tolerance can be implemented, as well as *Firewall* and virtual private networks (*VPNs*), antivirus protection, and intrusion and anomaly detection (*IDS/IPS*), but without other measures that are key to the business, IT efforts will not have the desired effect [17, 18, 19, 20, 21, 22, 23].

Trying to establish the continuity of the computing environment, CIOs inevitably come to the questions of business organization. However, they come to these questions in different ways. If a CIO comes from an IT environment, he is usually focused on ensuring the continuity of IT services. For instance, by provision of support services based on *artificial intelligence (AI), cloud and foggy computing, 5G+, Internet of Things (IoT)/industrial Internet of Things (IIoT), Big Data and ETL, Q-computing, blockchain, virtual and augmented reality (VR/AR)*. The practice is different if the CIO is a representative of a business and an international business. These specialists initially consider the tasks of supporting the organization's business processes as well as the tasks of combining data into unified reporting forms and statements to be one of the key tasks for the business.

1.1.1 Examples of Incidents

What can a successful business face (Figures 1.3 and 1.4) be, regardless of its size, scope, and distribution structure?

Incident 1

Employees of the telephone service provider, while updating the software on the telephone exchange (PBX), made an error, as a result of which all phones in the quarter served by this PBX were unavailable during the day. The vast majority of your clients only knew the official landline phone numbers set up at your organization's head office. You have

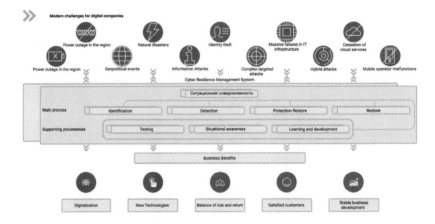

Figure 1.3   Modern challenges of the digital organizations.

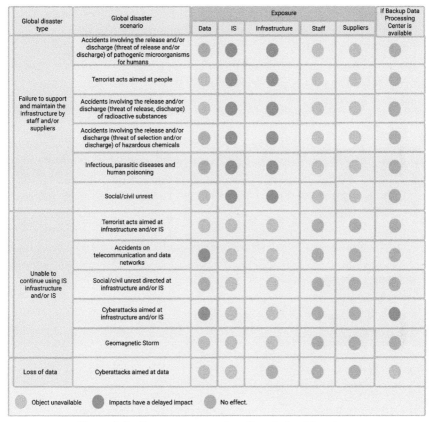

Global disaster type	Global disaster scenario	Exposure						If Backup Data Processing Center is available
		Data	IS	Infrastructure	Staff	Suppliers		
Failure to support and maintain the infrastructure by staff and/or suppliers	Accidents involving the release and/or discharge (threat of release and/or discharge) of pathogenic microorganisms for humans	◐	●	●	◌	◌		●
	Terrorist acts aimed at people	◌	●	●	◌	◌		●
	Accidents involving the release and/or discharge (threat of release, discharge) of radioactive substances	◌	●	●	◌	◌		●
	Accidents involving the release and/or discharge (threat of selection and/or discharge) of hazardous chemicals	◐	●	●	◌	◌		●
	Infectious, parasitic diseases and human poisoning	◐	●	●	◌	◌		●
	Social/civil unrest	◌	●	●	◌	◌		●
Unable to continue using IS infrastructure and/or IS	Terrorist acts aimed at infrastructure and/or IS	◐	◌	◌	◌	◌		◌
	Accidents on telecommunication and data networks	●	◌	◌	◌	◌		◌
	Social/civil unrest directed at infrastructure and/or IS	◐	◌	◌	◌	◌		◌
	Cyberattacks aimed at infrastructure and/or IS	●	◌	◌	◌	◌		●
	Geomagnetic Storm	◌	◌	◌	◌	●		◌
Loss of data	Cyberattacks aimed at data	◌	◌	●	●	●		◌

◌ Object unavailable   ● Impacts have a delayed impact   ◌ No effect.

Figure 1.4 Emergency scenarios of a catastrophic nature.

managed to call on the cellular network and report problems and backup cell numbers only to the largest customers. As a result, some of the customers for whom the unavailability of your company turned into real losses switch to competitors. Most of your clients have left an unpleasant impression that you cannot be reached for a whole working day.

Incident 2

The virus, which uses the vulnerability of the operating system "0 day," i.e., not yet fixed by the OS developers, quickly spread through the company's divisions. Each new copy of malicious software began actively searching for new victims in Your Enterprise network, and the communication channels were filled with malicious requests for 100%. As a result, the active network equipment of the enterprise network, unable to cope with the flow of virus attacks on MAC address tables in the local network, began to switch to the "denial of service" mode. IT specialists spent 6 hours to find a solution, treat the virus by removing it, and write and run a script to disable vulnerable services in Your Enterprise network. During this period, the company could not perform any operations requiring IT support.

Incident 3

A fistula in the water heating pipe that opened at night, in the room next to the server room, and connected to it by a common ventilation pipe, irrevocably destroyed all printed documentation and furniture in the affected room. However, even worse, that the saturated steam that spread unchecked into the server room overnight permanently disabled most of the network equipment and some of the servers. Fortunately, the fire did not occur, and the information on the media itself was undamaged. However, the search for temporary computing capacity to transfer the most important data carriers and backup recovery of the operating environment of the organization took 2 hours, the restoration of the main communication channel took 2 hours, and the server room was fully operational only a day later.

Incident 4

At the height of the season of maximum demand for the company's services, the hard disk controller of one of the two key corporate servers failed. As a result, electronic records of services rendered since the beginning of the day were lost (about 600 requests processed by operators), which subsequently required their re-entry from printed copies during non-working hours (from 20 to 24 PM). Its specialists who tried to transfer data storage and software to the remaining functioning server failed, and, moreover, because of the transfer work, they kept the second server in an inoperable state for about an hour, until it was decided to stop trying to run both systems on the same hardware. The new controller was delivered to the company only by 18 PM, and all employees involved in the accident response, including many operators, went home only at 2 AM.

Incident 5

Local fire that occurred in the mine, which housed the power shield, power cables, distribution Ethernet switch floor, and the LAN cable was *seen by staff from the smoke and quickly extinguished using powder extinguishers that were on the floor. The most likely cause of the fire was recognized as a spark that fell on a pile of dust and sawdust accumulated in the rarely opened mine. For reasons of electrical and fire safety, the shield was de-energized prior to replacement. In addition, it turned out that the fire caused damage to two of the three cables of the interstorey communication of the local network, and the distribution switch itself refused to work after running two extreme blocks of eight Ethernet ports in each. Work on laying a temporary power cable for de-energized office premises bypassing the damaged mine, reconnecting to the remaining Ethernet ports of the most important employees took 2 hours, and purchasing equipment to change the scheme of interfloor switching in order to restore the local network on the upper floors another 2 hours.*

Incident 6

An hour before the end of the working day, after the next application software update for receiving and processing requests from customers, developed locally for the needs of the company by a team of 18 programmers in the state, at the stage of forming a general order, an employee of the company found incorrect entries in the databases. The software developers did not take the message seriously because the night changes were related to a completely different functionality of the enterprise software system. However, a review of all entries made during the working day, initiated by the head of the operator's division, found that erroneous information occurs quite randomly in about 15% of cases. The final reason was an error in processing the decimal separator character that was changed by the night update, which was only shown in some orders as a result. Almost the entire day following the accident was spent on data recovery. Some of the information was restored by a specially formed query to the database. In about a third of cases, specialists had to make calls to customers with apologies for the day's delay and clarification of order volumes.

No matter how similar the above episodes are to "intentional panic-mongering," all this is only a small part of the incidents from actual practices (Figure **1.3** and Figure **1.4**), but each of them almost completely paralyzed the functioning of the organization, at least for half a working day, and often for much longer.

1.1.2 The Main Reasons

In many areas of activity, business continuity is, unfortunately, not always regulated by the state or business. Where there is no such control, the situation usually develops according to the scheme "lock the barn door after the horse is stolen." Like any other complication of business, ensuring business continuity entails additional financial costs. Meanwhile, we have to spend money today, and the negative impact may occur only tomorrow, the day after tomorrow, or even in ten years. Therefore, the return on investment in business continuity is postponed indefinitely. Therefore, in most cases, the leadership began to seriously think about the need to support the processes of subordinate structures in a state of continuity and stability only when the organization is disrupted due to a failure in the information system when it suffers losses (in case of business) or receives a suggestion from the parent organization (for example, in the case of audits of state agencies) [24, 25, 26, 27, 28, 29, 30, 31, 32].

What can motivate companies to develop and implement an ECP program?

Motive 1. Awareness through analysis:
The company's management's awareness of the need to ensure business continuity as a commitment to its partners and customers is the most "correct," evolutionary path, which is, unfortunately, less common in modern practice than we would like it to be. At the same time, we understand the importance of implementing best business continuity practices and developing and implementing the *enterprise continuity program.* The company's business units, in compliance with the management requirements, take responsibility for creating and updating appropriate business continuity plans and procedures in emergency situations with the appropriate level of detail.

Being the most correct, this path is actually quite demanding (Figure 1.5). Only management with a high managerial culture and the ability to make adequate forecasts in the medium and long term can lead to an awareness of the need to implement the entire range of measures to ensure continuity without any external influences. In general, the quality of development and implementation of the enterprise business continuity program is inextricably linked to the level of maturity of management activities in the field of risk management.

Motive 2. Awareness through the incident:
Unfortunately, the less successful but, as practice shows, more driving potential way for management to realize the need for business continuity

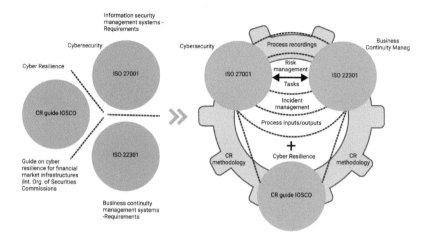

Figure 1.5 Improving the organization's BCMS maturity.

measures is to go through the incident of continuity violation itself. At best, this incident may be the sad experience of partner companies, neighbors, competitors, and, at worst, the company itself.

The assessment of the reasons (Figure 1.6) that can lead to various kinds of unforeseen circumstances is adequately disclosed by the results of the next survey conducted by Innopolis University in early 2020. For example, it turned out that the most common causes of operational downtime are:

- power outages – 14%;
- hardware and software failures and faults – 13%;
- problems in communication – 11.5%;
- human factor – 11%;
- the impact of the external environment (flooding, fire, natural phenomena, etc.) – 10.5%.

The only subtlety, perhaps, in cases where the driving force of the continuity program is the actual fact of a fairly long disruption of the production process, is the need to necessarily look at the problem as a whole. Unfortunately, we often have to deal with situations when after the failure of network equipment, measures are taken only to ensure the reserve of the network infrastructure, and after the failure of storage devices, to ensure the reserve of the storage system. In practice, this "band-aid approach" is very rarely justified – when using technologies

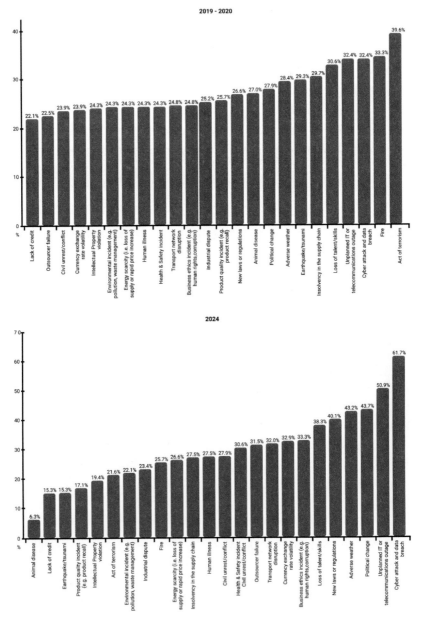

Figure 1.6 Main threats to business interruption[20].

[20] https://insider.zurich.co.uk/app/uploads/2019/11/BCISupplyChainResilience
ReportOctober2019SingleLow1.pdf.

of approximately the same level, the failure of a particular subsystem is approximately equal. As a result, the real impact of proactive measures can only be felt when building a comprehensive system for warning and responding to serious accidents in all subsystems of the enterprise that are critical for business.

Motive 3. Compliance with regulatory requirements:
Ensuring business continuity is dictated not only by internal necessity but also by a number of legal norms. For example, in the United States and Canada, these norms are industry regulations, Sarbanes–Oxley, Basel II, Federal Information Security Management Act (FISMA), Federal Preparedness Circular 65, or Health Insurance Portability and Accountability Act (HIPAA). And companies whose shares are listed on the London stock exchange are required by the Turnbull Committee to include sections on risk management in their annual financial and economic reports. In April 2004, the US Securities Commission approved a requirement asking each National Association of Securities Dealers (NASD) member organization to develop and implement a business continuity plan. Various regulatory agencies monitor compliance with the relevant requirements of regulatory documents, imposing significant fines if the company does not properly comply with business continuity requirements.

International business continuity management standards of the ISO 22300 family continue to develop. Including ISO 22301: 2019 *"Security and resilience — Business continuity management systems — Requirements,"* which is intended for certification of business continuity systems, BCMS. ISO 22301:2019 is harmonized with other well-known international standards ISO 9001 *"Quality management systems,"* ISO 14001 *"Environmental management systems,"* ISO 31000 *"Risk management,"* ISO/IEC 20000-1 *"Information technology-Service management,"* ISO/IEC 27001:2013 *"Information security management systems,"* ISO 28000 *"Specification for security management systems for the supply chain,"* etc. The *Good Practice Guidelines 2018 Edition* of the British Business Continuity Institute (BCI)[21], *the Professional Practices for Business Continuity Management* Disaster Recovery Institute International (DRI)[22], SANS[23], and others are also being developed. Based on these standards, *business continuity management industry practices are also*

---

[21] *www.thebci.org*
[22] *www.drii.org*
[23] *www.sans.org*

formed, including those for the credit and financial sector, retail, engineering, telecommunications, and so on.

Motive 4. Preparing for certification for compliance with best practices: Certification of the organization's management system for compliance with the best practices of BCM is one of the proofs that the company will be able to adequately ensure the stability and stability of critical business processes in practice. For example, certification for compliance with *ISO 22301:2019* confirms the necessary level of maturity of the *business continuity management system (BCMS)* of the organization. *ISO 22301:2019* contains requirements and allows for the certification audit of an organization's BCMS (Figures 1.7 and 1.8).

Motive 5. Fulfilling the requirements of customers and partners: Clients and partners of the company often want some assurance that their critical business processes are properly protected when interacting with the company and may require legal confirmation of this in contracts. In this case, the ECP program, BC policies, standards, and regulations, as well as BCP/DRP plans and their testing plans are proof of the provision of such guarantees since enterprise documents declare the company's intentions regarding the quality of business continuity and contingency plans. It is interesting that business partners and clients of the company are usually interested in these "intentions," i.e., high-level, final obligations, and not the technical means by which these intentions can be achieved.

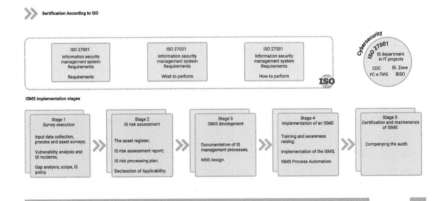

Figure 1.7 Certification according to ISO 27001:2013.

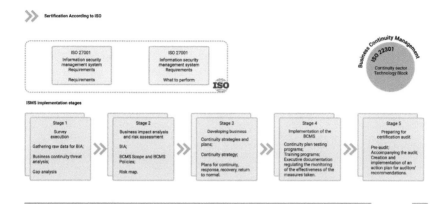

Figure 1.8 Certification according to ISO 22301:2013.

Motive 6. Elimination of auditors' comments:
Any external audit draws attention to the need to develop and implement the *enterprise continuity program*. Special attention is paid to the business continuity program (BCP), crisis management plan (CMP), and disaster recovery plans (DRP). In practice, it is recommended to use the requirements and recommendations of *ISO 22301:2019, COSO Internal Control Integrated Framework and the Combined Code on Corporate Governance*, the *COSO Enterprise Risk Management Integrated Framework, CobIT 2019*, and *ITIL v. 4* recommendations, and the requirements of *articles 404 and 302 of the Sarbanes–Oxley act (SOA)* and the *Public Company Accounting Oversight Board (PCAOB)*. In particular, in the comprehensive business continuity plan, auditors recommend to reflect the following issues:

- ranking of critical business processes and supporting IT services by the degree of importance and the amount of risk (the number of financial losses in case of unavailability);
- composition and structure of the group responsible for business management in crisis situations, crisis management team (CMT);
- rights and responsibilities of the disaster recovery team supporting business infrastructure;
- issues of mobilization of the required personnel;
- contact details of responsible persons (full name, position, office, mobile and home phone numbers, etc.) and ways to notify them;

- issues of information interaction both inside and outside the company (with suppliers, partners, customers, regulatory authorities, mass media, etc.);
- issues of transferring critical business processes to alternative sites;
- a list of resources, structures, and operations that are minimally necessary for disaster recovery, including archived information, primary documents, forms, and business stationery;
- requirements for restoring core business processes and IT services, including priorities and recovery time;
- procedures for restoring the supporting infrastructure, etc.

Attention is drawn to the fact that all employees responsible for its implementation must be familiar with the BCP plan, and the plan itself must be approved by the company's management. Copies of the business continuity plan (both printed and electronic) should be available to all employees involved in the business continuity process. The effectiveness of the activities included in the plan must be tested periodically (at least once a year). The plan should be reviewed and modified for all changes in operations, organizational structure, business processes, and IT systems that affect business recovery in emergency situations.

Inset 1 Examples of auditors' comments on the "ensuring business continuity and IT services" process for a certain telecommunications company.

Risk P01. Interrupting critical business processes (production, sales, execution, payment processing, financial reporting, marketing research, etc.) in the event of an emergency

KR01. 1. Develop a "matrix" of business processes and supporting IT services, which describes the areas of responsibility of competent persons in terms of ensuring business continuity.

KR01. 2. Use a redundant data storage and processing center connected via two independent Internet service providers.

KR01. 3. Apply fault-tolerant systems (geographically distributed clusters, backup systems).

KR01. 4. Develop disaster recovery plans, DRP.

KR01. 5. Use a centralized backup system.

Risk P02. Alternative options for performing critical business processes of the company in the event of failure of supporting IT services have not been developed.

KR02. 1. Use wireless communication channels to reserve wired communication channels.

> **KR02. 2.** To apply the technology of "thin client" to preserve the possibility of remote connection and perform sensitive operations.
> **KR02. 3.** Provide an independent backup pool of email addresses for company management.
> **KR02. 4.** Use the "Right Fax" system as an alternative to the existing email system.
> **KR02. 5.** Use a backup data center connected via two independent Internet service providers.

1.1.3 Economic Feasibility

A properly designed and implemented ECP program allows increasing the availability time and availability factor of business processes and IT services. This increases the overall business resilience and ensures the organization's competitive advantage in the business community.

Note that from 60% to 80% of the organization's efforts to ensure business continuity should be directed specifically to organizational measures and the development of an appropriate ECP program. As one can see from the diagram (Figure 1.9), an ECP program and BCP/DRP plans can be both the cheapest and most effective way to ensure business continuity in an organization.

The remaining 20%–40% of the organization's efforts to ensure business continuity are directed to the selection and implementation of

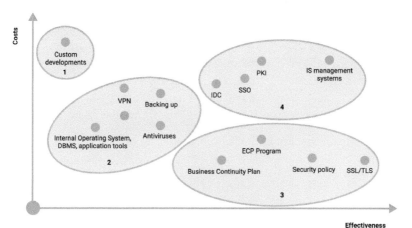

Figure 1.9 Evaluation of performance/price solutions in the field of BCM.

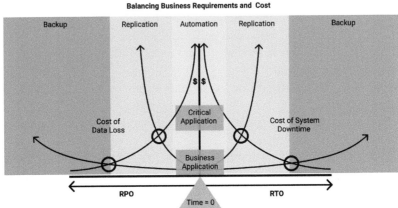

Figure 1.10   Cost effectiveness of BCM technical measures.

Figure 1.10 Cost effectiveness of BCM technical measures.

adequate technical measures (solutions) that will have to be economically justified (Figure 1.10).

What are the economic benefits of the ECP program?

Advantage 1. Directly reducing the impact of an incident on the organization's operations.

The main reasons for reducing the damage caused by business interruption in the presence of a developed and implemented ECP program are:

- minimization of time for making decisions about the required actions in the event of an emergency;
- reducing the risk of human error due to stressful situations;
- provision of personnel with means (including communications) both for accident elimination and for performing some part of their official duties;
- the staff has experience and skills in emergency situations, including those obtained during regular cyber training (Figure 1.11).

It is significant that the ECP program can bring returns that exceed even the sum of individual disaster recovery plan for each specific area.

The ECP program is aimed at restoring the organization's performance in cases of major negative impacts that affect most often several objects and areas of the infrastructure that supports the business. With such impacts, individual DRP plans, often developed by structural divisions

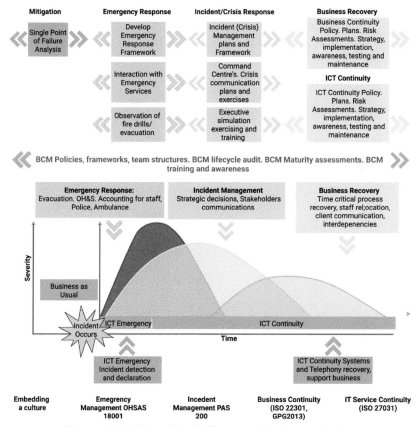

Figure 1.11   Value of the ECP program for the organization.

in isolation from each other and based on the assumption of the health of other services, are not able to adequately reflect the really required measures to restore the business. Only a comprehensive program that takes into account various scenarios and covers many areas can form an optimal strategy for behavior in such situations.

Advantage 2. Reducing the risk of financial liability to customers.

Implementation of continuity measures allows avoiding or significantly reducing the risk of legal liability to clients and customers which is significant for many organizations, and, as a result, reducing the volume of insurance payments, penalties, etc.

Advantage 3. Reducing the risk of damage to reputation.

The risk of damage to the company's reputation is closely related to incidents of disruption of the organization's performance and is one of the uninsured risks. In this regard, the implementation of the BCM plan and its corresponding technical support is one of the few tools for managing this type of risk.

1.1.4   Additional Advantages

Considering the issues of management motivation when implementing the ECP program, it should also be noted that such a process always brings the organization a number of additional advantages that are not directly related to the goals of ensuring business continuity.

Advantage 4. The organization's awareness of "itself."

A well-developed ECP program allows the company's key people to understand the structure of the company's supporting processes and internal services, their importance to the business, the significance of existing relationships with third-party organizations, and the degree of dependence of the business on them.

Advantage 5. Documenting processes.

In organizations that have not previously conducted a uniform inventory of existing business processes and its services in terms of impact on the performance of the company, the process of developing the ECP program forms a detailed, and, importantly, "business-oriented" package of documentation about the company as an additional result. This material will be useful for top managers of the company, middle managers, and even newly hired specialists (for example, for the purpose of quick and targeted acquaintance with the principles of functioning of the company's business processes and systems that serve them).

Advantage 6. Reducing dependence on "irreplaceable" people and functions.

The degree of dependence of the company on key figures in their company who have exclusive knowledge of the subject area or business contacts has always been and is one of the major risks. The special attention was always paid to reduce the mentioned risks by the management services and external auditors.

1.2   ECP Content and Structure

With the *enterprise* continuity *program*, each organization monitors its business processes, determines the most critical of them, and makes decisions about acceptable levels of risk, then regularly adjusts them to the current situation. Using the term *business continuity*, it is customary to talk about the processes of activity of any organization, regardless of whether they are engaged in commercial activities or not.

The mentioned business continuity program *ECP* (Figure 1.12) must explicitly include the following steps:

- *Business environment analysis*, which involves identifying and ranking business-relevant processes and defining continuity requirements for them.

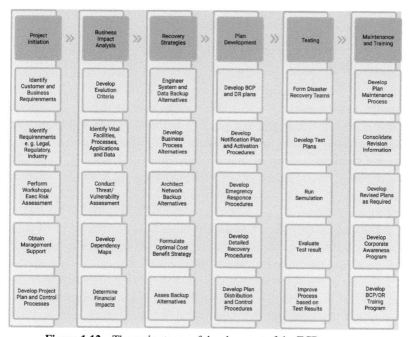

Figure 1.12   The main stages of development of the ECP program.

- *Risk analysis* – evaluating and ranking significant threats and vulnerabilities to business continuity, as well as evaluating the adequacy of existing organizational and technical measures to prevent business interruptions.
- *Business impact analysis* – analyze the impact on business processes and determine the recovery goals for each mentioned business process and its supporting infrastructure.
- *Business continuity strategy definition* – defining key parameters (such as *RPO* and *RTO)* for each business process under consideration and selecting appropriate organizational and technical solutions to ensure business continuity.
- Development and support of *business continuity plan* and *disaster recovery plan* for documenting appropriate solutions to ensure business continuity and restore business infrastructure in emergencies.
- Information support for the company's employees (awareness program) on current *BCM* issues.
- Support and maintenance of the *enterprise continuity program*; however, it is essential that the process of creating and maintaining the *ECP* business continuity management program is a constant process: the proper program must change along with changes in the organization and the external environment.

1.2.1 Background

Historically, *business continuity* was primarily provided with protection measures against unlikely but serious events, such as fire, flood, and other natural disasters. However, for modern technologically equipped enterprises operating in real time, the so-called *real-time enterprises (RTE)*, even the smallest violations in the continuity of business processes can have the most serious problems for the business as a whole. Therefore, ECP programs have undergone significant changes over the past years (Table 1.1 and Figures 1.13–1.16) [33, 34, 35, 36, 37, 38, 39, 40, 41].

In the early 1990s, BCM programs meant disaster recovery as well as protection from natural disasters and large-scale failures by switching to a different computing center within 72 hours. In the mid-1990s, recovery plans similar to the recovery plans for telephone customer support centers were introduced. In the run-up to the 2000 problem, BCM programs were revised, and it turned out that traditional recovery plans with a 72-hour recovery period were no longer meeting business requirements. It took a reduction in recovery periods to 2–4 hours.

Table 1.1 Example of calculating RTO and RPO for BC 2000 solutions.

	Alignment Scheme	Alignment Specification	Tier 1	Tier 2	Tier 3	Tier 4
Primary Storage	Guranted Performance	Performance Throughput Per Port (I/O sec)	5,000+	up to 5,000	up to 3,500	up to 1,500
		Response Time (ms)	< 8ms	7–14ms	12–30ms	22–30ms
	Availability	Maximum unplanned Downtime Per Year (Min)	< 26.5	< 26.5	< 52.5	< 263
	Cost	$/Usable GB	$111	$65	$33	$22
Archiving storage	Performance	Response Time	< 1 Second	< 1 Second	< 24 Hours	
		Throughput	<= 300 Mbps	<= 700 Mbps	280 Mbps	
	Availability	Downtime (Yr)	< 5.25 Min	< 52.56 Min	< 175.2 Hr	
	Retention & Disposition	Retention period	< 30 Years	< 10 Years	Years	
		Data schredding compliance	Yes	No	No	
	Accessibility	Read/Annual access frequency	< Hourly	< Hourly	Daily	
	Data Integrity	Gurarantee of authencity	Yes	No	No	
	Off-site	Recovery point objective	<1 Minute	<28 Hours	<38 Hours	
	Cost	$ Usable GB	$35	$25	$11	
Operational recovery	Recovery Classification	Recovery Classification	Complet application restore	Complete application restore		
	Operational Recovery Point Objective	Amount of dataloss	1 Hour	24 Hours	24 Hours	30 days
	Operational Recovery Time Objective	Time required for recovery	<30 Minutes	<30 Minutes	7GB/Min	5GB/Min
	Recoverability	Ability to recover backed up data	100%	100%	98%	95%
	Retention period	Length of time that data is retained	2 Hours	24 Hours	3 Weeks	15 months
	Cost	$/Usable GB	$46-$13	$46-$13	$8	$5
Disaster recovery	Disaster Recovery Point Objective (RPO)	Amount of Data Loss	0 Minutes	< 4 Hours	24 – 48 Hours	24 – 48 Hours
	Disaster Recovery Time Objective (RTO)	Time Required To Restore Data	< 2 Hours	< 12 Hours	< 48 Hours	< 72 Hours

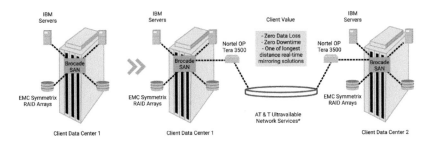

Figure 1.13   An example of a financial services client solution.

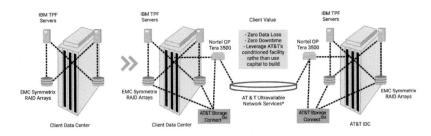

Figure 1.14   An example of a travel services client solution.

The emergence of e-commerce and businesses that operate on a real time scale required another downward revision of recovery period. The fact is that the downtime of computing power even for 15–30 minutes for many RTE enterprises would mean irreparable losses and even possibly the collapse of their successful business activities. Therefore, BCM programs were required to ensure that IT services are ready in 24 × 7 mode (24 hours a day, 7 days a week). Technologically, this means that indicators such as *recovery time objective* and *recovery point objective* have been reduced and are close to real-time indicators, i.e., close to "zero." Here, the RTO determines how quickly the IT services are able to return to a working state after an incident, and the RPO determines the latest backup that will be restored.

During the terrorist attacks of September 11, 2001 in the United States and in the fall of 2004 in Moscow, a number of major energy crises – in 1996, in the Western United States; in 2002, in Buenos Aires (Argentina); in 2003, in the United States and Canada; in 2006 in Moscow; in 2009, in Russia at the Sayano-Shushenskaya GRES – forced enterprises to review their BCM programs since most enterprises were unable to cope with such events. Among the drawbacks found in BCM's programs were, above all, inadequate management of emergency events and the lack of activities for

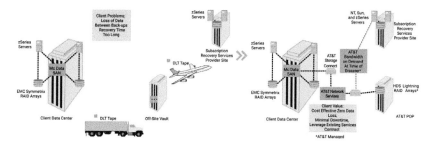

Figure 1.15   An example of a transportation services solution.

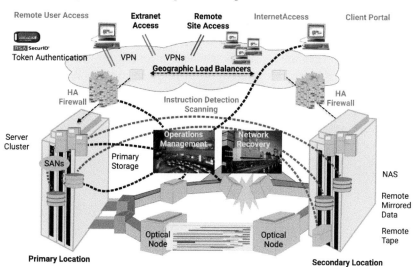

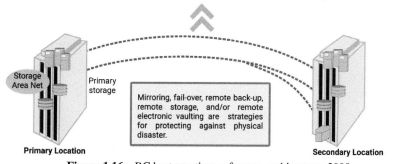

Figure 1.16   BC best practices reference architecture, 2000.

scenarios that had not previously been considered. Businesses have become more focused on the human factor, crisis management, and planning for scenarios of loss of life, lack of transport, and complete physical destruction of assets. The lessons of September 11 as well as the lessons of the subsequent financial crashes, in particular of *Enron*, have led to increased interest in such BCM programs that would guarantee the required business continuity.

A number of international studies on cybersecurity, including Ernst & Young (E&Y) 2018, have shown that more than **85%** of representatives of boards of directors and executives of leading companies around the world constantly feel *insecure* about the sufficiency of cybersecurity measures taken. At the same time, most managers strive to increase the speed of rapid response to the emergence of new challenges and threats in cyberspace. This includes significant investments in the creation and development of second- and next-generation *SOC*. However, the main question here is: *does the company have the required cyber resilience? Does it have enough capacity to minimize the risks of business interruption?* Apparently, the high cyber resilience of a critical business information infrastructure is not limited solely to the ability to quickly respond to new challenges and cyber threats (Figure 1.17). Here we need fundamentally new ideas and new approaches to ensuring the sustainability of business as a whole.

Also, the results of these studies have shown that there is a need to change the paradigm – *from* fault tolerance *to cyber resilience* [42. 43, 44, 45, 46, 47, 48, 49, 50]. Indeed, until recently, the issues of building

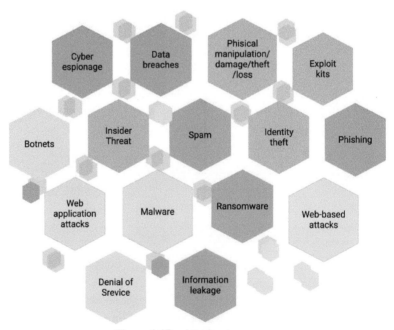

Figure 1.17   Attacker types.

fault-tolerant enterprise systems that can effectively withstand typical failures and failures in normal operating conditions were among the priorities. However, it is no longer enough to limit us to provide fault tolerance in the face of an unprecedented increase in security threats. It requires a new paradigm of building an enterprise system *to ensure cyber resilience*, which will *warn and prevent* attacks, and in the case of cyber-attacks *cushion a shock, reduce his power and nature of destructive effects, and minimizes the consequences*. Moreover, such a *"smart" organization of protection*, if necessary, *should allow sacrificing* some functions and components of the protected infrastructure to *resume business*.

According to CSIRT data from Innopolis University, the average flow of cybersecurity events in 2020 *was 57 million events per day*[24]. At the same time, the share of critical security incidents exceeded *18.7%*, i.e., every fifth incident became critical. This dynamics correlates with the results of cyberspace monitoring and cybersecurity threats monitoring by leading international CERT/CSIRT in the US and the European Union and also confirms the results of investigations of well-known cyber-attacks: *"STUXNET" (2010), "Duqu" (2011), "Flame" (2012), "Wanna Cry" (2017), "Industriyer" "TRITON/TRISIS/HATMAN" (2018)*, etc. However, there is growing concern that the number of unknown and, therefore, undetectable cyber-attacks range ***from 60% to 40%*** of all possible ones. All this together shows that the known methods of ensuring that cybersecurity and fault tolerance are no longer sufficient to provide the *required cyber resilience* and prevent the transfer of critical information infrastructure to irreversible catastrophic states.

The above is a *problematic situation*, the content of which is a contradiction between the increasing need to provide *cyber resilience* critical information infrastructure of the state and business in the context of destructive software impacts and the imperfection of methods and means of preventing, detecting, and neutralizing cyber-attacks in a timely manner. The removal of this contradiction required the resolution of an urgent *scientific and technical problem* – *the creation of secure and stable cloud computing platforms based on methods and algorithms of artificial immune systems, distributed data processing, container orchestration, software-defined data storage, and universal data bus* (Figure 1.18).

---

[24] https://university.innopolis.ru/research/tib/csirt-iu

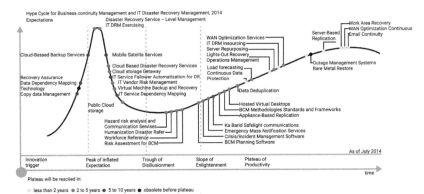

Figure 1.18   Hype cycle for business continuity management and IT resilience, Gartner.

1.2.2 Cloud Perspectives

A private cloud is a cloud computing model in which all infrastructure resources and services are dedicated to a single organization. In contrast to the public cloud, this model is characterized by an increased security level, extensive control capabilities, more flexible selection of hardware, and compliance with the requirements of legislation on data processing in the country, e.g., compliance with personal data protection requirements.

The demand for private clouds is confirmed by the dynamics of growth in the market volume of the corresponding solutions. According to PRnewswire, the market for private cloud solutions will reach $183 billion by 2025[25]. At the same time, the average annual growth rate of the CAGR will be 29.4% during the forecast period. According to the Analytics Company Grandview research, the global market for private cloud solutions was estimated at $30.24 billion in 2018, and the CAGR is expected to be 29.6% between 2019 and 2025[26].

There are two main segments in the market for private cloud solutions: platforms for building *virtual and dedicated* private clouds. Here, the known solutions are:

- *Red Hat OpenShift Container Platform* is a platform for developing, deploying, and operating classic and container applications in physical, virtual, and public cloud environments.

---

[25] https://www.prnewswire.com/news-releases/the-global-private-cloud-server-market-size-is-expected-to-reach-183-billion-by-2025–rising-at-a-market-growth-of-29-4-cagr-during-the-forecast-period-300952631.html

[26] https://www.grandviewresearch.com/industry-analysis/private-cloud-server-market

- *Amazon Virtual Private Cloud, Google Cloud Platform, Mail.ru cloud, Yandex Cloud* – these decisions represent a public cloud and are designed for a large number of clients.
- *VMware vCloud Suite, Microsoft Azure Stack, Cisco Private Cloud,* and *Oracle Private Cloud* are characterized by advanced functionality for multiple clients and the availability of standard solutions for building a private cloud for a single client.
- HPE Helion CloudSystem – a set of software solutions for deploying private and hybrid cloud environments.
- Dell Active System is a family of ready-made integrated systems for private cloud, application virtualization and deployment, etc.

Red Hat OpenShift Container Platform is a platform (Figure 1.19) for developing, deploying, and operating classic and container applications in physical, virtual, and public cloud environments[27]. The platform allows preparing, building, and deploying applications and their components. At the same time, automation tools for building source code based on S2I images (from source to image) simplify the Assembly of Docker containers based on code extracted from the version control system. And combining with DevOps (development and operations) and CI/CD (continuous delivery and integration) tools has helped to develop the solution.

Note that the development of universal platforms based on open source software is also relevant. This approach has a number of advantages in practice. Let us look at an example of developing the mentioned platform on the basis of open code, which used advanced cloud technologies. This includes models and methods for distributed data processing, container orchestration technologies, and architecture solutions for software-defined data storage and the universal data bus. Unique author's technologies for monitoring and managing cybersecurity and cyber resiliency based on models and methods of AI are also used here [51. 52, 53, 54, 55, 56, 57, 58, 59, 60, 61, 62, 63, 64, 65].

The universal platform for building a private cloud based on open source code (Figure 1.20) includes the following main components:

- engineering infrastructure of the data center;
- platform hardware;
- *Core Services* software;
- *Data Services* software;

---

[27] https://www.redhat.com/en/technologies/cloud-computing/openshift OpenShift

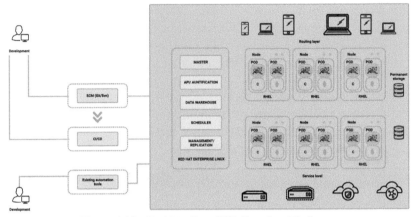

Figure 1.19   Red Hat OpenShift Container Platform.

- *Management* software;
- software-defined *Ceph* data storage;
- managed application runtime in *Kubernetes* containers;
- database software;
- Pre-installed software;
- Auxiliary software and services.

Let us look at the functionality and technical characteristics of the listed platform components in more detail.

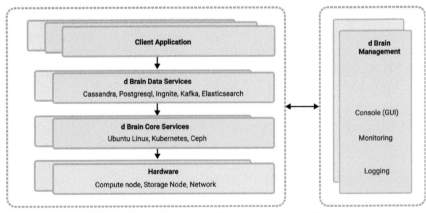

Figure 1.20   Structure of the open source platform.

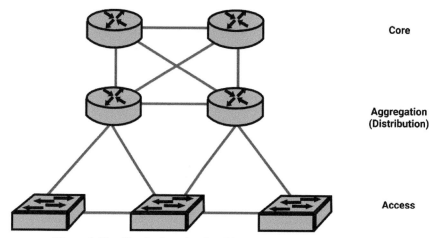

Figure 1.21   Classic three-level architecture of data center networks.

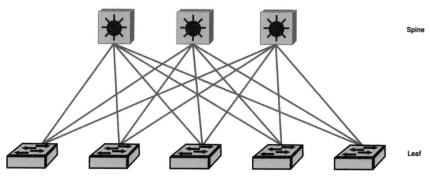

Figure 1.22   Leaf–Spine network architecture.

Network infrastructure:

The platform's network infrastructure is designed on a relatively new *Leaf–Spine* architecture[28]. This architecture replaced the classic three-level network architecture in data centers (Figure 1.21), which used the *spanning tree protocol (STP)* (including to prevent loops during data transfer).

A distinctive feature of the *Leaf–Spine* (Figures 1.22 and 1.23) architecture is the ability to adapt to the growth of big data by rapidly scaling.

The advantages of the Leaf–Spine architecture include:

- dynamic layer 3 routing based on the *equal-cost multipath protocol (ECMP)*, which improves network throughput and provides the required data transfer stability;

---

[28] https://networklessons.com/cisco/ccna-200-301/spine-and-leaf-architecture

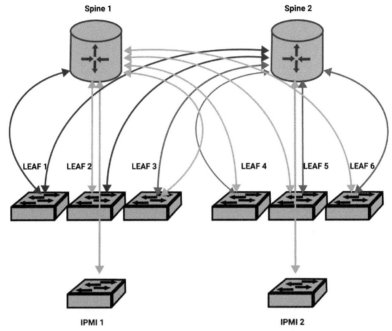

Figure 1.23 Leaf–Spine network architecture of the open source platform.

- ability to expand network bandwidth by adding additional hardware.

Disadvantages of the Leaf–Spine architecture include an increase in the number of connections in the leaf and spine switching scheme of devices, as well as certain difficulties when deploying *VLAN* in the network. To solve this problem, one can use the *software-defined networking (SDN)* functionality and create a virtual layer 2 on top of the Leaf–Spine network.

Server infrastructure:
The platform's server infrastructure (Figure 1.24) includes:

- AUX servers – responsible for collecting and processing system information, metrics, logs, etc. These servers are also responsible for quickly and conveniently deploying configurations to multiple servers using techniques used in cloud platforms. Linear scaling is applied.
- Kubernetes cluster – server nodes (from lat. nodus-node) K8S-MASTER plus several K8S-WORKER. The number of Kubernetes nodes in a cluster depends on the required computing power. K8S-LB nodes are responsible for user load balancing. Linear scaling is applied. In Figure 1.24, k8s server nodes are designated as compute nodes.

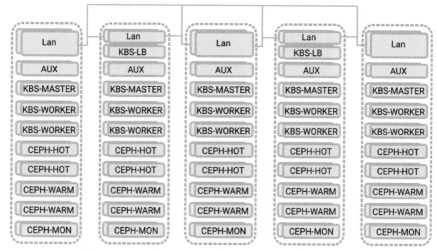

Figure 1.24   Open source platform server infrastructure.

- Ceph cluster – Ceph-HOT server nodes (ceph cluster nodes with disks that have high read and write speeds, usually SSD) and Ceph-WARM server nodes (cluster nodes with disks that have standard read/write speed parameters, usually HDD). The number of nodes in the CEPH cluster depends on the required amount of information storage. Linear scaling is applied. In Figure 1.24, server nodes are designated as compute nodes.

Core Services software:
Core Services software is designed to deploy a secure and stable platform for running applications and storing data, as well as for updating the software component of the platform.

Data Services software:
Data Services software is designed to solve the following tasks:
- the exchange of messages between different applications;
- ensuring scalability and fault tolerance;
- providing a delivery guarantee;
- support for storing a limited number of recent messages on disk for batch processing.

Management software:
Management software is designed to solve the following tasks:
- collecting and storing metrics;
- collecting and storing logs;

- detecting problems based on specified triggers with subsequent notification;
- providing information about the state of cluster components;
- routine operations are performed over the components of the cluster;
- identifying problems with the performance and stability of cluster components.

The Management software includes the following components:

- Monitoring and logging system: collect metrics from all machines and applications for further processing and analysis. The system allows configuring triggers for certain types of events that occur in the cluster, with subsequent notification via email and Slack.
- The logging system collects logs from all servers and applications.
- Cluster management system: service personnel are provided with a web interface for managing and getting information about the status of cluster components. All operations performed on the cluster are logged for subsequent resolution of incidents. A notification mechanism is provided for prompt notification of problems.

Software-defined Ceph data storage:

Here we selected the promising data storage system *Ceph*[29], which is a fault-tolerant software-defined data storage. An important feature of Ceph (Figure 1.25) is the ability to scale by tens of thousands of nodes, which ensures that the storage is scaled at the petabyte (Pb) scale. Built-in data replication mechanisms allow ensuring the required high reliability and availability of the system.

The Ceph storage system is represented by a set of several types of entities:

- OSD (object storage daemon) – the entity responsible for storing data. It is a daemon that works directly with a separate physical data store. Each cluster node can host a large number of OSD entities.
- Mon (monitor) – the ceph infrastructure coordinator that contains information about connected storage nodes (OSD), data distribution, and cluster status, and provides data addressing and replication.
- RGW (RADOS Gateway) – an auxiliary daemon that acts as a gateway for providing object storage. Ceph provides two different abstractions for working with storage, namely object storage (CEPH OBJECT GATEWAY) and block device (CEPH BLOCK DEVICE).

---

[29] https://www.redhat.com/en/technologies/storage/ceph

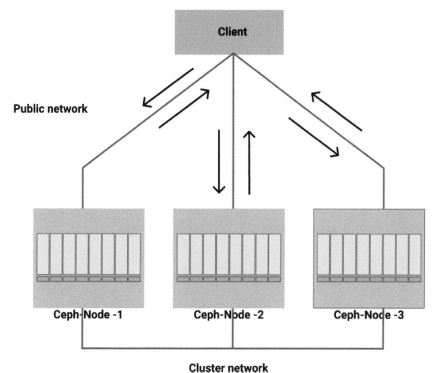

Figure 1.25   Ceph storage system operation.

Block device abstraction provides the user with the ability to create and use virtual block devices of any size. Object storage abstraction allows using the Ceph to store custom objects using the S3-compatible protocol.

Managed application runtime in Kubernetes containers:
A promising technology has been used for managing *Docker containers and Kubernetes* (Figure 1.26). *Docker* (a well-known open source project) allowed packaging applications (microservices) in containers along with all dependencies. This made easier the deployment of applications since the image was immediately launched on any *Linux* system without install-ing additional libraries and configuration files. The orchestration system was used to deploy applications packaged in *Docker*. *Kubernetes* (also an open source project) is an automatic deployment and project management system based on the microservice architecture and *Docker* technology. *Kubernetes allows* combining a large number of computing nodes into a single cluster. Here, horizontal scaling was achieved by running image

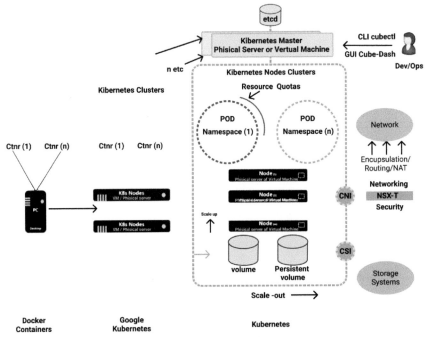

Figure 1.26   *Docker and Kubernetes* container management systems.

instances on different physical servers, and the load on the service was automatically distributed between the running instances.

Database software:
Cassandra (Figure 1.27) was chosen for database management; it belongs to the NoSQL System class and is designed to create highly scalable and reliable storage of large data arrays represented as a hash[30]. Cassandra uses a data storage model based on a family of columns. Data placed in the database is automatically replicated to multiple nodes in the distributed network. Cassandra is characterized by high scalability and reliability, high throughput for write and read operations, configurable consistency, fast writing, and flexible circuitry. An SQL-like pattern mining language supports searching for data using secondary indexes.

Monitoring and logging system:
The monitoring and logging system is designed to collect data from machines and applications for further processing and analysis. The system

---

[30] http://cassandra.apache.org/

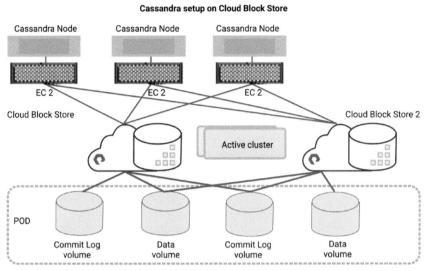

Figure 1.27 Cassandra-based large data storage architecture.

allows configuring triggers for certain types of events that occur in the cluster, with subsequent notification via email and enterprise messenger. The mentioned system also collects logs from all servers and applications and combines and systematizes the logging infrastructure.

The logging and monitoring system processes, aggregates, and outputs the following data in accessible form:

- system information that reflects the operation of systems, processes, and services, including information about the operation of a software-defined distributed file system;
- temperature, voltage, fan speed, and information from other sensors on the server and network equipment;
- indicators for monitoring and using disk partitions;
- indicators of the self-monitoring, analysis, and reporting mechanism contain information received from network devices that contain technical data about this device;

Subsystem of cybersecurity:
The platform's security is ensured by compliance with international standards (ISO/IEC 27001: 2013, ISO/IEC 27031: 2011, etc.), as well as by the use of open source software, the ability to select equipment, and the use of advanced tools for monitoring and managing cybersecurity based on models and methods of AI.

The cybersecurity tools of the mentioned platform include the following components:

- hardware and software complex for protection against unauthorized access;
- software and hardware complex for detecting, preventing, and neutralizing cyber-attacks;
- software and hardware complex of immune protection for detecting and neutralizing previously unknown cyber-attacks;
- hardware and software complex of homomorphic encryption;
- antivirus software;
- security monitoring system;
- the system is multithreaded scanning of files, etc.

At the same time, a number of unique proprietary cybersecurity technologies were applied [66, 67, 68, 69, 70, 71, 72, 73, 74, 75, 76, 77, 78, 79, 80, 81, 82, 83, 84, 85, 86, 87], including the following innovative technologies:

- *Immune protection.* A corresponding SDK has been developed that, for the first time in cybersecurity practice, allowed the accumulation and use of artificial "cyber-immunity" to protect against previously unknown types of cyber-attacks and malicious software (more than 50% of the total number of cyber-attacks). The SDK is integrated into the most common tool environments and development studios for Industry 4.0 digital platforms and applications in programming languages (*JavaScript, Java, C#, Python, C++, TypeScript, Swift, Kotlin, Rubi, Go, 1C, C, Scala, Pascal/Delphi, T-SQL, Dart, PL-SQL, Erlang, Apex*, etc.). The practical significance of the SDK is that it allowed implementing the so-called dynamic control and preventing irreversible and catastrophic transfers of critical information infrastructure of the state and business.
- *Detection of "digital bombs."* It is implemented based on deep semantic analysis and "passporting" of applications using dimensions and similarity invariants. Well-known methods for detecting software bookmarks include methods of static and dynamic software analysis. The so-called profiling methods based on monitoring the behavior of critical information infrastructure in the face of growing security threats are also widely used. However, performing static analysis in the absence of source code and software documentation required the search for new approaches to effectively detect, warn, and block software bookmarks. Therefore, an original method for solving the above problem was proposed, justified, and implemented

on the basis of detecting software defects using software research, taking into account the *structural, logical,* and *operational* properties of programs based on a set of models and methods *of graph theory* (to study the structure of software bookmarks), *RSL logic* (to study the logic of software bookmarks), zero-checking *Petri nets* (to study the actions of software bookmarks), and *similarity and dimension theory* (for dynamic control of the semantics of computationcomputations). As a result, an original tool and automatic dynamic control were developed to detect and neutralize destructive software bookmarks and "digital bombs" of critical information infrastructure.

- Innovative technology of homomorphic encryption and quantum cryptanalysis. Today, homomorphic encryption systems (*RSA, El Gamal, peye, gentry-Palevi-Smart, Gribov-Mikhalev, Burtyka-Babenko*, etc.) are widely used to protect cloud computing. A unique technology of previously unknown vulnerabilities of cryptographic primitives and asymmetric encryption algorithms is used. In particular, it is based on the quantum cryptanalysis – a modified algorithm of *Shor and Grover*. This includes a hardware and software package for quantum cryptanalysis of cloud computing in the Python programming language in the Jupiter computing environment. The results obtained made it possible to significantly improve the cryptographic protection of modern cloud computing.

1.2.3 ECP Practice

The main elements of any ECP program are:

- recovery after a disaster;
- business recovery;
- business resumption;
- planning of unforeseen circumstances;
- crisis management.

The differences between the listed BCM elements are given in Table 1.2.

Due to the continuous improvement of business continuity scenarios, the concept of "recovery" has become somewhat more meaningful than just "recovery after a disaster." The shift from disaster recovery planning to contingency planning highlights the role of its services, which have become important components of enterprise business. The implementation of the anti-crisis management in modern BCM (Table 1.3) programs has become extremely important.

Table 1.2   Business continuity management system components.

	Disaster Recovery	**Business Recovery**	**Business Resumption**	**Contingency Planning**
Objective	Mission-critical application	Mission-critical business processing (workspace)	Business process workarounds	External event
Focus	Site or component outage (external)	Site outage (external)	Application outage (internal)	External behaviour forcing change to internal
Deliverable	Disaster recovery plan	Business recovery plan	Alternate processing plan	Business contingency plan
Sample Event(s)	Fire at the data center; critical server failure	Electrical outage in the building	Credit authorization system down	Main supplier cannot ship due to its own problem
Sample Solution	Recovery site in differenct location	Recovery site in a different power grid	Manual procedure	25 percent a backup of vital products; backup supplier

The success of any BCM program is based primarily on supporting the management of enterprises and creating an enterprise culture of business continuity, which allows integrating business continuity processes into the lifecycle of each project and change management processes. For example, to determine the maturity level of an enterprise BCM system on a traditional six-level scale (from "chaos (0)" to "optimized (5)"), Gartner[31] suggests using the following evaluation elements:

- awareness;
- support at the level of the Board of Directors and senior managers;
- organization and management;
- roles and responsibilities;
- staff;
- budget/costs;
- goals and rewards;
- politics;
- quality of the process (for example, BIA-risk assessment);

---

[31] https://www.gartner.com

Table 1.3   Comparison of micro and macro continuity management.

Crisis management (macro)	Incident response (micro)
Scale enterprises	Scale the company's business unit
Site or component outage (external)	Site outage (external)
Task: warning and parry incidents	Task: neutralizing incidents
Goal: ensuring the stability of the company	Goal: ensuring the stability of the business unit
Strategy: ensuring. the company's business continuity	Strategy: ensuring business unit continuity
Decision maker (PLR): TOP management of the company	Decision maker (PLR): manager and technical specialists
Recommendations for companies	
• Know your rights under contracts and agreements to respond to incidents • Check the expiration dates of contracts and agreements • Make an inventory of business services provided by external providers • Use redundancy and duplication of business services from external providers • Check RTO and RPO in BCP and DRP plans • Include in your contract the clause "on termination of obligations under the contract without liability" in case of force majeure	
BC- business continuity **DR-** disaster recovery of IT service **BCM-** business continuity. Management **RPO-** recovery point objective **RTA-** recovery time actual	

- strategy and plans;
- reporting;
- workbench;
- IT applications and systems;
- information and data protection;
- project lifecycle;
- testing;
- external service providers (ESP);
- enterprise risk management;
- business continuity management between enterprises.

To implement an effective ECP program, one must complete the following steps (Figure 1.28):

1. Conduct a primary business impact analysis, *business impact analysis (BIA)*, the purpose of which is to rank the business processes and IT

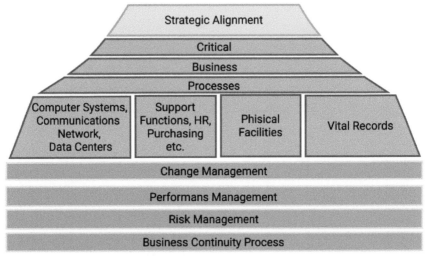

Figure 1.28   Support and development of the BCM program.

services of the enterprise by the degree of criticality and obtain a primary, fairly general, assessment of the likely losses to the business in the event of a malfunction of these processes and services.

2. Carry out a risk analysis (*risk analysis (RA)*). Identify in detail the existing threats and vulnerabilities to the business processes and IT services of the enterprise, as well as possible business losses in the event of possible threats. Clarify the results of the primary business impact analysis, *BIA*.

3. Develop a BCM strategy. The purpose of developing the BCM strategy is to determine the main directions of development of the BCM enterprise system for the next year and in the future for five years. The BCM strategy allows creating the qualitative and quantitative goals for the development of the enterprise's business continuity management system in the future for five years, sets their priority, and determines the main ways to achieve them.

4. Create a business unit of the company responsible for the BCM planning process.

5. Develop a BCP plan that meets the requirements and objectives of the business. Develop the necessary regulations and procedures for the BCM.

6. Implement a process for regularly testing the BCP plan and procedures. Testing is necessary to ensure that the required BCM requirements are met.

7. Implement a process approach to BCM. Keep the BCP plan up-to-date and make necessary changes if business processes and IT services change.

Role of business impact assessment (BIA):

BIA is an extremely important step for implementing any successful BCM program. It is the results of BIA (Table 1.4) that allow determining what, how, and to what extent to protect and preserve. BIA is recommended to be performed by a team of specialists that includes representatives of the enterprise's business divisions, information security service, and IT department.

The BIA surveys the business processes and enterprise IT services. When analyzing business processes, the focus is on assessing the possible losses that an enterprise may incur as a result of interrupting business

Table 1.4 Main stages of business impact assessment and their results.

No	Key milestones in BIA	BIA result
1	**Analysis of business process and information flows of the company** • building a map of the company's business processes and information flows; • mapping of potential violations of business process function and of the company's information flows; • choice of approach to loss assessment (qualitative, quantitative mixed); • building an analytical model to assess losses due to disruption in the functioning of business processes and information flows of the company; • aligning the model with the company's employees; • obtaining the resulting estimates of criticality of business processes and information flows of the company; • construction of the resulting estimates of criticality of various kind of violation of business processes and information flows of the company with reference to the scale of possible losses.	**The result of the analysis of business processes and information flows of the company refer to** • map of business processes and information flow of the company with necessary details; • map of possible disruptions in the functioning of business processes and information flow of the company; • analytical model for estimation of losses as a result of malfunctioning of business processes and information flow of the company (quantitative and/or qualitative estimates); • resulting estimates of criticality of business processes and information flow of the company; • resulting criticality assessment of various kinds of business process and information disruptions of company flows linked to the scale of possible losses.

Continued

Table 1.4   Continued

No	Key milestones in BIA	BIA result
2	**Analysis of IT services:** • coordination of the list of information services under consideration; • construction of a map for the use of information services by the company's business processes; • building a map of possible violations of information services functioning; • preliminary assessment of information services' criticality for business.	**The result of the IT services analysis include:** • list of information services under consideration with a brief description; • map of the use of information services by the company's business processes; • map of possible violations of information services functioning; • preliminary assessment of information services' criticality for business.
3	**Analysis of IT services impact on business:** • building an analytical model of cause and effect relationship between the functioning of information services and the functioning of business processes and information flow; • building of the analytical model for estimation of losses as a result of information services functioning disturbance; • obtaining by means of the model estimations of information services' criticality per se and various violations of their functioning; • development of preliminary estimation of economically justified expenses for increasing the level of information services accessibility.	**The result of the analysis of the impact of IT services on business include:** • analytical model for estimation of losses as a result of information services functioning disruption; • the resultant estimation of the information services' criticality per se and different kinds of their functioning disturbances; • preliminary estimates of economically justified expenses for increasing the level of information services accessibility.

processes. Various categories of losses are considered: exceeding the standard level of operating costs, penalties as a result of a violation of contractual obligations, reduced return on an investment relative to the planned level, loss of business reputation of the company, and so on up to a decrease of the enterprise in the market value.

BIA analysis begins with building a map of processes (*management processes, main and supporting business processes*, etc.) and the main information resources (*telecommunications channels, server infrastructure, application services, databases,* etc.) of the enterprise. For processes and information resources, various scenarios of malfunction are considered, which potentially lead to losses. Based on the constructed map, an analytical model is constructed that links various violations in the functioning of business processes and information flows of the enterprise with the category and scale of losses as a result of such violations. Depending on the scale, losses can be estimated quantitatively (in monetary terms) or qualitatively (according to a specially developed qualitative scale). Based on the results of possible loss evaluation, the model allows assessing both the criticality of business processes and information flows and the criticality of business interruption.

At the same time, with critical analysis of business processes and the dependence of the scale of losses from violations of the business processes functioning, the analysis of information services and their binding to business processes and information flows is performed. Information services are the services such as enterprise accounting system, consolidated enterprise reporting system, Big Data based business analytics system, enterprise information portal, enterprise email, and some others. In this case, a deeper level of detail is taken since, for example, the enterprise accounting system actually provides several services (accounting support, human resource management support, material and technical accounting support, etc.) that are involved in various ways in the company's business processes. During the analysis of information services, these services are identified, and their use within the company's business processes and possible violations in the functioning of the services are analyzed; also a preliminary assessment of the significance of information services from the company business point of view is performed.

An example of classification of business processes and IT services is provided in Table 1.5 and Figure 1.29.

Note that "Class 0 Application services" are those services that correspond to the real-time enterprise (RTE) strategy as well as those that, if unavailable, the enterprise will suffer irreparable damage (RTO and RPO are close to "zero").

The analysis of the impact on the business ends with the construction of a model of cause-and-effect relationships between the main and supporting business processes and information resources of the enterprise. This model allows evaluating the information services in terms of:

- criticality for business;
- possible losses for the business depending on the disruption in the operation of the service and the recovery time;
- costs associated with the increased availability of the service.

Table 1.5   Classification of business processes and IT services.

Category	RTO	RPO	Definition	
0	Close to "0"	0–15 minutes	**Category 0: Mission Critical** Requires continuous operation; RPO and RTO are close to "zero." Systems and applications are critical to the company's ability to conduct key business operations and/or data availability must meet strict regulatory requirements. It is characterized by the following requirements and solutions:	
			Users	• The accident has immediate consequences for the company's business – within 10 minutes of downtime • Critical business functions are not possible without IT support • A backup office is required • The means of the work distribution are implemented and ready before the crash • Data loss is not allowed • The business unit must have an alternative method for processing transactions without centralized IT support • The work from home via a VPN is allowed
			IT	**Centralized processing and services** • Continuous data processing centers (CDPC) • Mirroring data between data processing centers • The infrastructure and resources of the CDPC are ready for immediate production work • Automatic/automated switching between CDPC • Parallel data processing on different CDPC • There are no single points of failure for the IT infrastructure • Data loss is not allowed • Synchronous data transfer • Resources are sufficient for a long production load • IT administrators can manage IT infrastructure remotely (from home via VPN) **Network** (data and voice) • Redundancy/availability of an alternative path for data between continuous data processing centers (CDPC) • Permanent network connection to the backup office • There is an available alternative path for voice transmission • Automated switching of transmission routing

Continued

Table 1.5 Continued

Category	RTO	RPO	Definition	
1	15 minutes	Less than 1 hour	Category 1: Business Critical. The following requirements and solutions:	
			Users	• Serious business losses will start 15-30 minutes after aaccident • Users can start working from the back office 2 hours after the accident is declared • Backup office is available and equipped with workstations • Can work without IT support for extremely limited time • Backup office (own or outsourced) • Resources (people, equipment, materials) available to restore business • Limitedlossofdataallowed • Mandatory alternative method of transaction processing without centralized IT support • Allowed to work from home through VPN
			IT	**Centralizedprocessingandservices** • Remote Backup computation centre (own or outsourced) • Production resources are present in the backup computation centre (people, equipment, workstations, network) • Replication of data to the backup computation centre • Data transfer via communication channels to the backup computation centre • Transfer of data changes between the computation centre and the backup computation centre and creation of backups on both computation centres • Backup computation centre is intended for long-term operation • IT administrators can manage IT infrastructure remotely (from home via VPN) **Network (data and voice)** • Redundancy / availability of alternative path for data in the computation centeres • Availability of an alternative path for voice transmission, • Automated switching of transmission routing.

Continued

Table 1.5   Continued

Category	RTO	RPO	Definition
			Category 2: BusinessOperational. Systems and applications are business sensitive, used by departments in daily operations. It is characterized by the following requirements and solutions:
2	1-2 hours	Less than 4 hours	**Users** • Serious business losses start 2-4 hours after the outage • Critical business processes can continue to function without disaster-affected business units and IT systems, data and processes for a short period of time • Data loss reduced to last full or incremental backup • Backup office (in-house or outsourced) • Alternative method of transaction processing without centralized IT support recommended • Allowed to work from home via VPN
			IT **Centralized processing and services** • Remote Backup computation center (own or outsourced) • Data transfer via communication channels to the RVC • IT administrators can manage the IT infrastructure remotely (from home via VPN) • Backup computation center infrastructure ready to go • IT equipment for HDW is available or provided under a service contract • Unique equipment is in HDW (tested and documented) • Equipment used for development can be used for business recovery. Development and testing operations are suspended.
			Network (dataandvoice) • Redundancy / availability of alternative path for data in the VTC • Availability of an alternative path for voice transmission, • Automated switching of transmission routing.

Continued

Category	RTO	RPO	Definition
3	4 hours	Less than 1 hour	**Category 3: Office Production** Systems and applications are sensitive to business and are used by divisions in their daily work, but they can operate without them for some time without noticeable damage. Applications and data are local to the division. It is characterized by the following requirements and solutions:
			Users • Losses will begin 8 hours after the shutdown • Important business processes may continue to function without the affected departments and IT systems, data, and processes working for some time • Data loss is reduced to the last full or incremental data backup • The backup office is defined and can be prepared for use within the declared RTO • An alternative method of processing transactions without centralized IT support is preferable • The work from home via a VPN is allowed
			IT **Centralized processing and services** • Remote backup space (own or outsourced) • Recovery from traditional tape cartridges. Delivering backups from remote storage by car to the recovery location • IT equipment and network equipment for recovery are provided under a service contract • The unique hardware is located in HDW (tested and documented) • The equipment used for the development could be used for business recovery. Development and testing operations are suspended. • The IT staff is ready to put the IT infrastructure into operation on the backup site
			Network (data and voice) • Redundancy/availability of an alternative path for data in the BCC • Alternative ways for voice transmission can be put into action within the declared RTO

Continued

Category	RTO	RPO	Definition
4	8 hours	Less than 24 hours	**Category 4. Not critical** Systems and applications are not critical, important, or sensitive to business and can be restored as needed without time limits. Requirements for business continuity planning by business units are not specified. Data and business records are reserved on request (not regularly). It is characterized by the following requirements and solutions:
			Users • Losses will begin 24 hours after the shutdown • The business process remains functional without IT support for a long time • If necessary, the recovery procedures are documented • If necessary, the hardware and resources needed for recovery are identified • There is a recovery plan • If necessary, the business unit develops an alternative method for processing transactions without centralized IT support
			IT • **Centralized processing and services** • The backup room has been identified or there is a list of potentially suitable rooms • Recovery time – less than 8 hours • Backup equipment and resources are not available, and will be delivered from other locations (branches, suppliers), installed, and put into operation. **Network** (data and voice) • Network resources will be delivered from other locations (branches and suppliers), installed, and put into operation.

*Data is copied and sent to the remote office for secure storage.

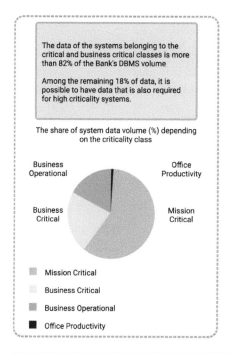
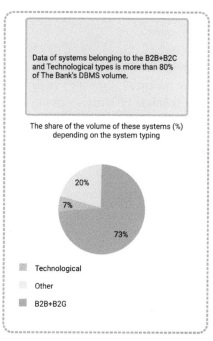

Figure 1.29   Example of classification of business processes and IT services.

In other words, during the BIA, it becomes possible to come to an agreement on the cost of downtime of business processes and its services and determine the appropriate RTO, RPO, MTPoD (MAO), MBCO, and cost-effective business continuity costs [ISO 22301:2019]. Then the BIA results become the initial information for forming the BCM strategy.

Risk assessment (RA):

At the risk assessment stage, the identified business processes (taking into account the data on their significance for the enterprise as a whole obtained at the previous stage) are examined from the point of view of their susceptibility to certain objective external and internal threats. As a rule, the risk assessment scenario methods are mainly used here. Therefore, the well-known statistical methods of risk assessment do not allow to obtain a proper representation of observation samples and identify the required quantitative patterns of business continuity in the context of growing security threats.

The key document of this stage is a threat model for the organization supplemented in the process of analysis with information about the degree of the potential impact of each specific threat on certain processes and on the organization's activities ingeneral.

The main principles of building a threat model are:

- completeness of the analyzed scenarios in order to cover all potentially significant types of threats for the organization;
- sufficient level of scenario detail to enable an adequate assessment of the consequences of threat implementation for each specific business process;
- adequate (if possible-quantitative) assessment of the probability of a particular threat and various scenarios for the subsequent development of the situation in order to prioritize and optimize further processes.

At the analysis stage of the possible impact of a specific threat on the organization's activities, one should be very careful about scenarios that affect several business processes at once because often the resulting impact is not equal to the sum of the consequences for each process separately. For example, one of the reasons for escalating the degree of significance (influence) of a threat is a frequently occurring situation when the same threat, if implemented, can lead to the failure of both the main and backup (including possibly several backup) means of ensuring the business process. Thus, physical damage to the communication cable (breakage, theft, water, or fire damage on the way) can deprive the organization of both interaction with the Internet and, at the same time, telephone communication (which could be defined in BCP as a backup means of communicating with customers).

After risk assessment of certain scenarios of continuity disruption at the level of the organization's management or responsible person, strategic decisions should be made on the choice of the organization's attitude to these specific risks [ISO 22300:2018]:

- accepting the current level of risk as reasonable to the business;
- planning and implementing risk reduction measures, which are the subject of this book;
- transfer of risk to third parties (insurance or other risk management measures);

- risk avoidance (termination of this type of activity in general or in a specific part of it that causes unacceptable risks for the organization (for example, termination of services in certain countries of the world or a certain number of potential customers)).

The results of the risk assessment stage should be documented, and for the risks for which a decision of implementing the measures to reduce the risk level, the main planned directions for achieving this goal are listed.

Choosing a BCP strategy:
There are many options for choosing a suitable solution (Figures 1.30–1.32 and Table 1.6), both for IT services only, and for infrastructure (a geo-distributed datacenter cluster, a metro cluster of two main data centers and a remote backup data center, GridGain, Hadoop, and so on) in general, with an acceptable cost and the required RTO and RPO indicators:

- **Real-time data centers** – continuous availability but minimal preparedness for catastrophic emergencies; contain identical copies of the information infrastructure; usually located relatively close to each other (up to several kilometers).
- **A metrocluster consisting of two main data centers and a remote backup data processing center** – high availability and partial preparedness for catastrophic emergencies; contain close copies of the information infrastructure; usually located up to 60–100 km apart.
- **Geo-distributed data processing center cluster** – acceptable availability and high level of preparedness for emergencies of a catastrophic nature; contain close copies of the information infrastructure; usually located hundreds or thousands of kilometers apart.
- **Backup data processing centers without infrastructure and mobile data processing centers** are auxiliary solutions with the ability to increase capacity (most often with reduced performance).
- **Grid gain redundancy platform is** a solution for reserving high availability platforms based on Apache® Ignite™ that provides real-time data processing by transferring computationcomputations to the RAM of computing facilities.
- **Hadoop redundancy platform** is a solution for reserving Big data storage and processing platforms on clusters of hundreds and thousands of nodes (within the MapReduce computing paradigm, etc.).
- **Hot standby systems** are fully configured infrastructure elements that contain an exact copy of the software and hardware but without data; they are ready to completely replace a failed unit.

- **"Warm" reserve systems** – identical to a hot reserve without hardware. Images of the configured software are archived.
- **Cold reserve systems** – a room with some infrastructure elements but without hardware and software. Requires the acquisition or transportation hardware and a complete deployment and configuration of the software.
  - **Outsourcing or mutual agreements** – use of resources of third parties in the event of incidents or providing such services as one of their activities, or related to the organization by mutual agreements (quite possibly free of charge) on mutual cooperation in BCM. The disadvantages of this option are: the risks of not being able to provide the required level of service in general, and, in particular, when large localization incidents occur, causing multiple cases of resource requirements at the same time; possible problems with the protection of trade secrets and other proprietary information, limited access.

Ensuring business continuity in an IT-dependent company, restoring its business processes with *RTO and RPO* close to "*zero*" can cost millions of dollars. Choosing the best organizational and technical solutions with an acceptable level of costs requires understanding the direct and indirect costs associated with business process downtime. Understanding the acceptable limits of technology costs for each business process and IT service that is critical to the company is very useful for limiting possible recovery alternatives. While it can be expensive to restore RTE quickly, the alternative – to restore it in 2 hours or more – can jeopardize the stability and even the survival of the enterprise. The BIA results will estimate the *total cost of ownership (TCO)* and *return on investment* (ROI) of BCM enterprise programs and take appropriate measures.

The key task that specialists are trying to solve today is to move from defining the general requirements and tasks to the development of quantitative indicators, metrics, and measures of new cyber-stability practices, for example, taking into account the recommendations of *NIST SP 800-34*.

Interrelationship of emergency action plans:
This special publication of the US Institute of standards provides model recommendations. However, the document does not contain any descriptions of quantitative indicators; it is, at best, a percentage. In the absence of clear quantitative estimates, it is impossible to compare or optimize processes and, accordingly, to take adequate measures to move from the "as it is" state to the "as should be" state. Therefore, the experience of

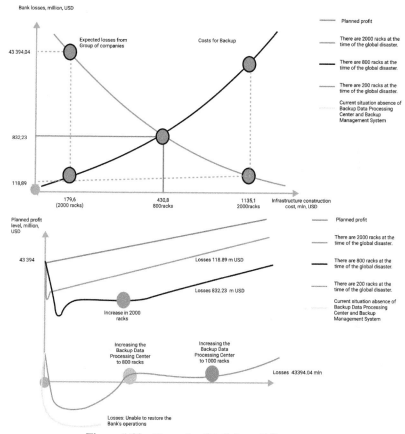

Figure 1.30 Example of defining a BC strategy.

organizations that attempt to build systems of quantitative indicators is interesting. For example, KPI computationcomputation is based on absolute values (time of activation of reserve areas, time of recovery of critical information infrastructure, time of recovery of critical business processes and IT services, etc.) and relative values (percentage and number of permanently ready backup jobs, number of exercises conducted to restore critical business processes and it services, etc.).

Process approach to BCM:
The larger the company, the more important it becomes to inform employees about business continuity issues (Figure 1.33). It is necessary to convey to the company's employees the idea that "ensuring business continuity is the responsibility of all employees." This is achieved by introducing a procedure for familiarization with the requirements of the business continuity

Bank Information Systems Technology Stack	Replication capability	Estimated (possible) replication technology
Oracle Database	Yes	Asynchronous Oracle Data Guard
MS SQL Database	Yes	Asynchronous AlwaysOn
Terradata	Yes	Asynchronous Terradata Replication
Hadoop	Yes	Asynchronous HDFS Replication
PostgreSQL	Yes	Stream asynchronous Replication
SAP HANA	Yes	Asynchronous Replication with SAP HANA
Virtual machines VMWare или Hyper-V	Yes	Asynchronous Replication with VMWare
ШШРБ/GridGain	Yes	Asynchronous GridGain Replication
Split file storages	Yes	Asynchronous Replication with disk array tools (NetApp, EMC, HP, HDS)

 The recommended option of data reservation is the organization of replication of all Bank's platforms based on the first stage of RDC

Parameters	Replication of MC and BC Legacy systems	Hadoop replication	GridGain replication
RTO		< 8 hours	
RPO		< 24 hours	
Racks (8 kW)	29	21	40-51
TCO For 5 years (mln.$)	69,2	5,5	15,2 - 19,4
Total TCO For 5 years (mln.$)		89,9 - 94,1	

Figure 1.31   Example of data representation for selecting a BC strategy.

№	Name/designation	General requirements and technical comments
1	Data center BDPC communication channels	• To ensure high reliability and required RPO and RTO levels, the minimum number of data center- BDPC channels is two • The communication channels should follow geographically different routes
2	Communication channel BDPC – TB/GOSB	• In accordance with the current list of data center channels of the Central Apparatus (with preservation of type, technology and bandwidth)
3	Communication channels BDPC - Internet	• It is necessary to take measures to ensure guaranteed time of switching the addresses of the Bank's autonomous system to the BDPC. The time of the autonomous system address switching shall be determined at the stage of technical design of the BDPC IT infrastructure. The configuration of the network segment providing access to the Internet shall be like the configuration used in the data center of the Central Apparatus
4	Reserve segment of satellite communications in BDPC	• VSP channels only with satellite channels - each of them has 2 channels, one leased; • In order to organize the redundancy it is necessary to construct a new central satellite communications ground station (CSCS, hub) in the BDPC.
5	Channels to the network of self-service devices	• Through the networks of operators • BDPC channels - to the networks of operators according to the current connection scheme at the moment of BDPC implementation.
6	Channels from BDPC to third-party organizations (SWIFT, CB and etc.)	• It is necessary to develop redundant connections with each contractor - most external organizations (like SWIFT and CB) have their own solutions to organize redundant connections to ensure disaster recovery.
7	BDPC channels of communication with subsidiary banks	• It is necessary to organize channels similar in their properties to the data center of the Central Office
8	BDPC channels of communication with contact centers and telephone networks	• Besides channels from Unified Distributed Contact Center sites to BDPC, there is also a need for channels from BDPC to public telephone networks.

Data Center - BDPC Channel Bandwidth Requirements

Year/Data type (platform)	2017		2019	
	Data volume SM, Pb	Changes, TB in a day	Data volume SM, Pb	Changes, TB in a day
Legacy systems (current architecture)	2,5	200	1,5	120
Business Development Support Platform / GridGain	-	-		
• «snapshots backup & WAL-replication	-	-	5,1	400
• Stand-in	-	-	4	100
Data cloud FD / Hadoop	-	-	8,0	23 - 60
Total			Up to 14,6	Up to 580

> Data Center-BDPC backbone channel capacity at different stages of the project should be from 30 to 100 Gbps.

Figure 1.32   Example of calculating the required bandwidth of communication channels.

policy, plans, and procedures and signing a corresponding document stating that employees are familiar with and understand all the requirements for business continuity, and they are committed to complying with them. Business continuity procedures enter requirements for maintaining the necessary level of business continuity in the list of each employee responsibilities. In the course of performing their work duties, employees should be periodically familiarized and trained on business continuity issues.

Table 1.6   Example of comparison and selection of ways to reserve key IS.

	Platform	RTO	RTO with time of developing the recovery infrastructure	RPO	Total cost of ownership (min $)	Time before first backup copie	Time before testing the copie	Migration necessity for target audience	Recovery risks
Backup — D2D	Legacy	96 hours	1,5 years	48 hours	9,71	14–20 months	26–32 months	Yes	Yes
	Hadoop	96 hours	1,5 years	48 hours	97,59	14–20 months	26–32 months	Yes	Yes
	GridGain	96 hours	1–1,5 years	48 hours	20,19	14–20 months	26–32 months	Yes	Yes
53 (own inf)	Legacy	96 hours	1,5 years	24 hours	7,69	14–20 months	26–32 months	Yes	Yes
	Hadoop	96 hours	1–1,5 years	24 hours	9,64	14–20 monhts	26–32 months	Yes	Yes
	GridGain	96 hours	1,5 years	24 hours	13,3	14–20 months	26–32 months	Yes	Yes
53 (provider)	Legacy	96 hours	1,5 years	24 hours	8,12	10 months	26–32 months	Yes	Yes
	Hadoop	96 hours	1,5 years	24 hours	14,82	10 months	26–32 months	Yes	Yes
	GridGain	96 hours	1–1,5 years	48 hours	16,01	10 months	26–32 months	Yes	Yes
Replication	Legacy	<8 hours	Ready infrastructure	<24 hours	65,54	14–20 months	14–20 months	No	No
	Hadoop	<8 hours	Ready infrastructure	<24 hours	9,47	14–20 months	14–20 months	No	No
	GridGain	<8 hours	Ready infrastructure	<24 hours	26,90	14–20 months	14–20 months	No	No

The use of asynchronous replication technology will ensure the preservation of data and applications. This approach minimizes the risk of non-switching due to the possibility of replica testing, as well as will ensure a step-by-step evolutionary development of the RDC to a full-scale and fully functional state.

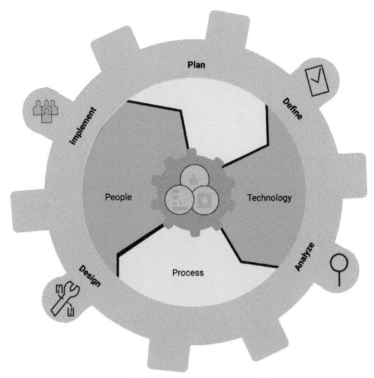

Figure 1.33   Process approach to BCM.

All measures to improve the business continuity management system (BCMS) of the organization constitute a continuous cyclical process (Figure 1.33).

1.3   Example of Task Statement

Let us consider a possible task statement for creating and implementing an enterprise business continuity program (*enterprise continuity program*).

1.3.1   The Purpose and Objectives of Work

The purpose of this work is to develop and implement a set of policies, processes, procedures, and organizational and technical measures to ensure the proper continuity of business processes and IT services of a large multinational company (more than **40** *Hadoop*-class platforms and **20** *GridGain*-class platforms) and reduce possible damage in the event of failures and disasters in the medium-term (up to three years) and long-term (for five years or more).

In order to achieve this goal, there is a need to solve the following tasks:

- Based on the system analysis of the use of information technologies in the company's activities, determine the list and characteristics of factors that affect the continuity of the company's IT services as well as the values of acceptable deviations of the main parameters that ensure the required level of efficiency of IT services.
- Evaluate the current state of continuity indicators for existing IT services as well as the target values of these indicators for the company's future IT services. Determine the list of necessary policies, regulations, processes, procedures, and activities to ensure the required values of the company's IT service continuity indicators for the medium-term and long-term perspective.
- Develop a strategic continuity plan for IT services of the company and reduce potential damage in case of failures and disasters with an assessment of the costs of financial, human, and information resources and which considers a phased improvement plan for the company's infrastructure up to 2024.
- Develop draft regulatory and methodological documents that establish new procedures for ensuring the continuity of the company's IT services.
- Develop awareness programs and training methods for the company's employees on how to ensure continuity of IT services. Conduct training for those responsible for implementing the BCM enterprise program.

Expected effect:
As a result of the work, it is expected to provide:

- improvement in the efficiency of using IT services in the company's operations by improving the continuity and stability of their operation and reducing possible damage in the event of failures and disasters;
- optimal allocation of the company's available resources, according to the "efficiency-cost" criterion when using special tools and methods to ensure the continuity of the company's IT services;
- improvement in the validity of the strategy, plan, and program for the development of information technologies, information systems of the company, and their support tools;

- improvement in the efficiency of using budget resources allocated to IT departments;
- creation of a system for training company's employees in the procedures for ensuring continuity of IT services.

Development requirements:
General requirements:

- Procedures should be developed on the basis of a comprehensive approach to ensuring the continuity of the company's IT services, considering the full range of special tools and methods to ensure the continuity and stability of the functioning of IT services, implemented as part of a phased plan to improve the enterprise information system for the long term.
- Procedures should ensure optimal distribution of tasks to ensure the continuity of the company's IT services between the company's head office and remote offices in accordance with their assigned functions and avoid duplication.
- Procedures for ensuring the continuity of the company's IT services should be focused on improving the company's operations and increasing the efficiency of using modern information technologies.

Requirements for the methodology for determining and evaluating factors that affect the continuity of the company's IT services:

- The methodology should ensure the formation of a list of factors that affect the continuity of the company's IT services and their characteristics (reliability, stability, and survivability) based on modeling the use of information systems, information technologies, and their support tools in the company's business processes.
- Factors affecting the continuity of the company's IT services should be structured according to:
  - ○ main directions, ensuring the continuity of IT services and related activities;
  - ○ management levels of IT services and facilities and organizational structure for company divisions.
- The methodology should provide an opportunity to analyze the continuity of IT services.
- The methodology should include algorithms for calculating the values of the main parameters and acceptable deviations, which ensure the required level of efficiency of the company's IT services.

- The performer must have developed:
  - ○ information support that includes a description of normative reference information, the database model, and information sources that contain the necessary data for evaluating factors;
  - ○ mathematical software that includes a description of algorithms and special methods used for calculating parameters;
  - ○ organizational support, including regulations that must be used to calculate parameters and perform actions to analyze the values obtained.
- As part of the development of the methodology, the contractor should also prepare proposals for using software tools to automate the collection, accounting, and processing of necessary data to ensure the continuity of the company's IT services.

Requirements for evaluating the company's IT service continuity indicators:
  - ○ In accordance with the developed methodology for determining and evaluating factors that affect the continuity of the company's IT services, the following should be calculated:
  - ○ values of continuity indicators for the company's existing IT services;
  - ○ target values of the company's prospective IT services continuity indicators for the medium-term (up to 3 years) perspective;
  - ○ target values of the company's prospective IT services continuity indicators for the long-term (5 years or more) perspective.
- Lists of measures that ensure the achievement of the company's IT service continuity indicators in the medium and long term should be provided and reasoned.

Requirements for the draft strategic plan to ensure the continuity of the company's IT services and reduce possible damage in the event of failures and disasters:

- The company's IT services continuity plan should contain a system of interrelated measures indicating the timing and stages of their implementation, performers, an assessment of the necessary amounts of funding, and implementation mechanisms that ensure the continuity of the company's IT services and reduce possible damage in the event of failures and disasters.
- The plan should include the following sections:
  - ○ preamble;
  - ○ goals, tasks, deadlines, and stages of the plan implementation;

      ○ content and conceptual approach to achieving goals and problem solution;

      ○ system of procedures and measures (including infrastructure development, software improvement, legal regulation, organizational measures, personnel support, etc.);

      ○ regulatory support;

      ○ resource provision;

      ○ plan implementation mechanism;

      ○ organization of management and control of the plan implementation;

      ○ evaluating the effectiveness of the plan and achieving the required values of the company's IT service continuity indicators.

The contractor can supplement the plan structure described above.

- The plan's activities should consider:
  - ○ need for their implementation in the conditions of the company's operation in accordance with the current regulations;
  - ○ activities of the phased plan for the implementation of the business continuity management program, the implementation of which is provided for in the project, the company's plans for the implementation of research and development work, as well as target programs;
  - ○ implementation of the measures envisaged in the plan should be carried out in two stages. The first stage involves the development of policies, strategies, plans, regulations, and procedures for ensuring continuity. The second is the implementation of the company's enterprise IT services management program.

Requirements for draft normative methodological documents establishing new procedures for ensuring the continuity of the company's IT services:

As part of the development of a strategic plan to ensure the continuity of the company's IT services and reduce possible damage in the event of failures and disasters, the contractor should analyze the need to develop new ones as well as make changes and additions to existing documents regulating these procedures. The report should include suggestions for making such changes, as well as drafts of the following documents:

- methodology for determining and evaluating factors that affect the continuity of the company's IT services;
- regulations for calculating the values of the company's IT service continuity indicators and actions of officials to analyze the values obtained and issue control actions;
- drafts of methodological documents that establish standard procedures for ensuring the continuity of the company's IT services.

The above list of documents may be expanded or supplemented.

Requirements for training officials in procedures for ensuring the continuity of the company's IT services:
It is necessary to develop programs and courses to train officials in the order and rules of work using new procedures in the field of ensuring the continuity of the company's IT services on the basis of regulatory documents approved by the company's management, developed within the framework of this assignment.

In order to conduct training, the proper software that will automate the processes of training employees and check the knowledge obtained in the relevant subject area should be applied. The software should allow to study typical operations step-by-step, performed in accordance with procedures in the field of ensuring continuity of the company's IT services, as well as to provide the ability to generate and conduct tests on a given set of topics and questions[32].

Training should be conducted centrally on the basis of the performer's training centers (groups of 10–12 people for each of the courses; the total number of students is 80 –100 people).

Training should be conducted to the fullest extent of the functionality of the developed procedures. The knowledge transferred to the trainees for each of the training courses should be sufficient to fully assimilate the material in order to organize subsequent training of the company officials in the field. At the same time, the average duration of one course (including the time of mandatory credit for the course listened to) should be approximately 30–40 academic hours.

The terms of the contract should provide the possibility of adjusting the amount of training based on real needs.

Requirements for the work scope:
The work scope under this assignment is determined by the work plan for the implementation of the "improving the enterprise information system" project and includes the following works:

- development of the draft strategic plan to ensure the continuity of the company's IT services and reduce possible damage in the event of failures and disasters;
- development of the program and training of company officials in the procedures for ensuring the continuity of the company's IT services.

---

[32]The content of the developed training courses must be presented in a form suitable for use in conjunction with distance learning tools developed outside the scope of this contract.

While preparing an offer, the contractor can add and detail the scope of work and the sequence of their execution, considering their experience and business practice.

The above list of works may be changed and/or supplemented during the implementation of this contract due to the revision of the functional tasks of the project due to changes in local legislation caused by the adoption and implementation of relevant acts, guidelines, resolutions, and orders regarding the company's activities and other legal acts.

Requirements for reporting materials:
During this task, the performer prepares two types of reports: process and content reports.

Process reports are the one on the implementation of the work specified in the terms of reference and include:

- initial report that is provided within 30 days from the start date of the task and contains a detailed work program and a list of reporting materials agreed with the customer;
- monthly reports, the last one of which must contain an appendix with a list of all materials provided by the performer and defined in the initial report.

As a result of this work, the following reporting materials should be prepared:

- methods for determining and evaluating factors that affect the continuity of the company's IT services with the results of modeling; a list of factors that affect the continuity of the company's IT services, their characteristics, and acceptable deviations of parameters that ensure the required level of efficiency of information systems;
- assessment results of the current state of the company's IT service continuity indicators as well as the target values of these indicators for the company's future IT services; a list of necessary procedures and measures to ensure the required values of the company's information system continuity indicators for the medium and long term;
- draft strategic plan for ensuring the continuity of the company's IT services and reducing possible damage in the event of failures and catastrophes, with an assessment of the necessary costs of financial, human, and information resources and linking with the activities of the step-by-step plan for improving the enterprise information system;
- draft regulatory and methodological documents that establish new procedures for ensuring the continuity of the company's IT services;

- program and training of company officials in the procedures for ensuring the continuity of the company's IT services.

Reporting materials must be prepared on paper and electronic media and meet the requirements for registration of scientific and technical products of the company.

Scientific and technical products (a set of design, design, and operational documentation, documentation on machine media, hardware, and software and technical equipment) created and used within the framework of this assignment are transferred to the property of the customer. The contractor is obliged to ensure the patent purity of scientific and technical products.

1.3.2 Work Duration

It is expected that the duration of work on the task will be up to 12 months, without considering the terms of approval by the customer of the developed methods and draft normative methodological documents.

Implementation area:
The scope of the contract is all divisions and remote offices of the company that have IT departments

Additional requirements:
The functional customer under this contract will be the company's information technology department with the involvement, if necessary, of specialists from interested divisions of the company and remote divisions in the areas of activity.

In order to integrate the work carried out, it is necessary to ensure close interaction of performers under related contracts. Coordination groups, headed by representatives of the customer, will be created, to implement the above-mentioned part. The consultant under this contract is required to participate in the work of the relevant coordination group, including:

- participation in the planning of joint/interrelated activities;
- provide interim materials in a timely manner and inform about the progress of work;
- when performing the work provided in this assignment, consider the results of other performers' work;
- inform the customer in case of incompatibility of solutions proposed by other performers participating in the work of the coordination group.

The list of performers will be determined as a result of competition

Requirements for the performer:
The following requirements are applied to the performer:

- be a company working in the field of consulting on improving the efficiency of information systems and IT services, including improving the continuity of complex, geographically distributed information systems, including for government customers, for at least 5 years;
- have experience in optimizing the use of methods and tools to ensure a given continuity of complex, geographically distributed information systems over the past three years;
- have at least four major successful contracts in the area under consideration for major regional structures over the past three years;
- have a key group of full-time specialists who are able to perform both short-term and long-term work in accordance with the requirements of this assignment;
- have sufficient resources to manage large projects and a well-established project management mechanism;
- have experience of working on projects implemented with the involvement of international financial organizations, preparing substantive and process reports;
- have good skills in organizing and conducting employee training, involving the customer's management team in the skills transfer program;
- have the appropriate licenses of authorized bodies and regulators in the field of business continuity and it services, as well as to conduct training;
- have an understanding of the specifics of the company's activities, taking into account the requirements of national and international legislation.

1.4   Analysis of BCM Technologies

Currently, almost all manufacturers of both hardware and software implement various features in their solutions aimed at ensuring fault tolerance and disaster tolerance. Some of the most popular approaches have received the status of standards, which are considered "a matter of honor" by leading companies in their decisions. The competition is still ongoing in other tasks and technologies, and each vendor offers its own solutions with specific advantages and disadvantages [88, 89, 90, 91, 92, 93, 94, 95, 96, 97, 98, 99, 100].

1.4.1   General Approaches and Directions

Before implementing technical measures to ensure business continuity, it is necessary to properly classify the technical solutions under consideration and correlate them with the tasks set.

Classification by assignment level:
The classification criteria for the assignment level can be attributed:

- number and composition of system users;
- degree of information aggregation;
- the method of data storage and processing;
- features of the tasks to be solved;
- the level of difficulty, etc.

We recommend allocating the following classes or levels of IT infrastructure:

1. Enterprise group data center (level I Data processing center) – provides centralized data storage and processing across the organization. The resources of this data processing center are used to consolidate information from lower levels, provide information interaction, and provide IT services.
2. Enterprise data center (level II Data processing center) – provides centralized storage and processing of data at the scale of one fairly large enterprise.
3. Division data center (level III Data processing center) – provides centralized data storage and processing within a small enterprise or division.

Classification by continuity level:
The following classification of IT systems is possible:

- **Mission Critical** systems operating in "combat" mode (*RTO* and *RPO* are close to zero). These systems include:
  - business and environmental critical systems and software applications;
  - network management centers (monitoring, security, administration, etc.);
  - technological applications that can run in real time.

Failure of these systems entails irreparable losses for business and threatens the life and health of the company's employees. The specialized server platforms and infrastructure layers must be used with multiple redundancies of components, including the use of redundant remote data centers for mentioned systems.

- **Business Critical** systems that are critical for business, with a 24×7× 365 operating mode. Failure of these systems leads to serious losses for the business. The recommended disaster recovery time for such systems is no more than 2 hours. The cluster solutions and infrastructure layers with partial redundancy of the applied components should be used for the mentioned system.
- **Business Operational** – normal business applications – systems with 8 × 5 operation mode. The recommended disaster recovery time for such systems is no more than 8 hours. We recommend using backup data storage and power supply for such systems.
- **Office Production** – applications that are not critical for running a business. Failure of these systems does not affect the dynamics of the company's key performance indicators (KPIs). The recommended disaster recovery time for such systems is not over 24 hours.

Note that as in the case of cybersecurity, the overall business continuity indicators depend on the level of continuity and fault tolerance of the weakest link.

Ways to ensure continuity:
The main ways to ensure business continuity in the field of information technology include:

- consolidation of IT resources;
- virtualization of IT resources;
- fault tolerance and disaster tolerance technologies.

Consolidation of resources:
The main types of the consolidation of IT resources are:

- *Centralization* – consolidation of geographically distributed servers in one or more data processing centers.
- *Data consolidation* – consolidation of databases and/or storage devices to achieve higher availability and data management.
- *Physical consolidation* – combining servers running the same operating system and with similar applications on more powerful systems.
- Application consolidation *and data storage* – hosting various applications on "large" servers with shared partitions or mainframe.

One can use the so-called "consolidation rating," defined as the number of entry level, low end, and middle range class servers that have been replaced or can be replaced with a single higher-class system to evaluate current or planned resource consolidation. The recommended values for this indicator should be in the range of 15–20 units.

Resource virtualization:

Possible options for virtualization of IT resources depend on the goals and platforms for which it is used (Figure 1.34). The main virtualization tools for Windows platforms are virtual machines or operating environment emulators, supplemented by clustering technologies and virtual machine migration technologies: *P2V (Physical-to-Virtual), V2P (Virtual-to-Physical), and V2V (migration between virtualization platforms).* However, virtualization technologies for enterprise class servers and high availability operating systems differ from one server manufacturer to another and are based on dividing computing resources into logical partitions.

Hardware and software partition technology (*partitioning*) allows virtualizing the hardware resources and make them available to a variety of independent operating environments. Originally developed for mainframe, this technology allows splitting a single server into several fully independent hardware or software virtual servers or logical partitions. The technology is supported by leading IT equipment manufacturers with minor differences.

Implementation of virtualization mechanisms allows IT infrastructure to acquire a level of abstraction of the logical model from physical components.

Failure of a specific physical object does not lead to serious and long-term system downtime – the virtual object is transferred to a physical host

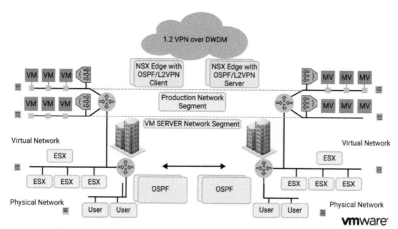

Figure 1.34 Example of a solution for resource virtualization.

that currently has resource reserves and continues to function with minimal downtime.

In addition, the implementation of the virtualization mechanism provides a much higher level of measurability of processes; first of all, their resource requirements and, as a result, the ability to evaluate and plan the resource reserves of the enterprise as a whole.

Partitioning technology together with *disaster recovery plan* for two separate data centers (main and backup) creates the basis of a disaster-resistant IT infrastructure.

Inset 1.2

Possible technical solutions for continuous availability:
- load distribution between sites using Cisco LISP, F5 Global Load Distribution, NetScaler, and other technologies;
- distribution of traffic on a platform farm with GTM or LISP for Cisco stores instead of a traditional load balancer;
- using VMware®computing farms vSphere® and vSphere Metro Storage Cluster (vMSC);
- using virtual LANs with second-level solutions, such as Cisco OTV or IETF Standard VPLS;
- clustering databases, for example, using Oracle RAC;
- application of elastic clustered NAS head systems. For example, Quantum StorNext for CIFS or NFS-based access via a block storage system running VPLEX™;
- distribution of RAID mirrors between sites using VPLEX for block access.

Possible technical solutions for data storage:
- using converged solutions from Cisco, NetApp, IBM, Pure, etc.;
- application of hyperconverged systems for HyperFlex virtualization platforms, etc.;
- use of integrated systems for Big Data and predictive analytics;
- use of integrated backup and copy storage solutions (Veeam, Commvault, Coherence, etc.);
- use of integrated platforms for storing archives and secondary data (Coherence, Scality, SwiftStack, etc.);
- using the SDS platform (Ceph, Gluster, IBM, etc.);
- application of a data storage network (IP, FC, FCoE, etc.).

1.4.2   Infrastructure Decisions

Requirements for data center infrastructure:
General requirements for data processing, storage, and backup systems for
all types of data centers (Figures 1.35 and 1.36 and Table 1.7):

- Hardware performance must consist of the main subsystem's performance. It is necessary to monitor the load of the main subsystems,
  identify bottlenecks, and increase performance, as necessary, by
  installing additional modules. In operational mode, the server must
  have a maximum load of 70% on the main resources to withstand
  peak load if necessary.
- Virtualization.
- Scalability.

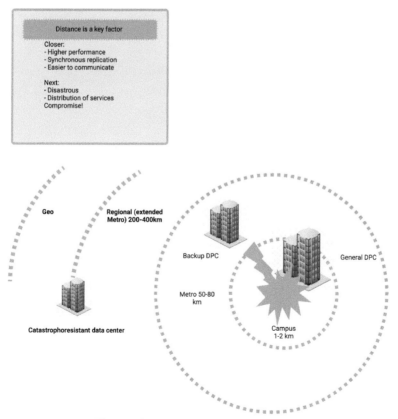

Figure 1.35   Data processing center design.

Table 1.7   Characteristics of data processing center solutions.

Data center type	Distance between platforms	Typical scenario usage	Disaster-resistance	LAN connection	SAN connection	Optimization paths	Nuances
"Campus"	Up to several km	Multiple data centers in a complex of buildings or on the territory of an enterprise	Minimum	Optical fiber 10GBASE-LR/ 40GBASE-LR/ 100GBASE-LR	LW Optical fiber	is not required	can be considered as part of one modular data center
"Metro"	Several dozen km (up to 60–100 km)	"Backup data center" within the city or "subway destination" – region	Partial	Optical fiber 10GBASE-ER/ ZR/ DWDM, if necessary, IP+OTV OR MPLS/VPLS	CWDM/ DWDM fiber, if necessary FCIP if possible	If possible	If possible, use "dark fiber" for LAN and SAN, synchronous replication, metrocluster technologies (Vmotion).
"Region"	Up to several hundred km (300–400)	Backup data center in another city within the region	Significant	IPTV or MPLS/ VPLS, DWEM if available	FCIP, DWDM if available	Desirable	It can be used for a number of metro cluster technologies (Vmotion). Asynchronous replication or synchronous replication with restrictions and additional tools
"Geo"	Many hundreds and thousands of km	Data center in the event of a disaster (DR) in another region of the country	High	IPTV or MPLS/ VPLS, if required by the cluster technology	FCIP if required by the cluster technology	Necessary	Asynchronous replication, "log shipping", or other disaster management tools. Long recovery time (hours or more). Direct communication between data centers only if required by Geocluster

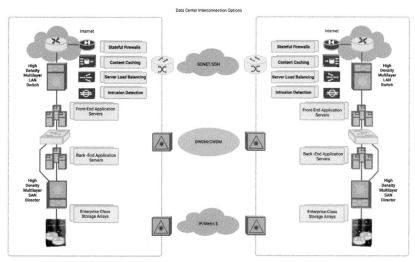

Figure 1.36 Typical scheme of a fault-tolerant data center.

- Readiness. The degree of the equipment readiness must be ensured by:
  ○ reducing single points of failure;
  ○ technologies for combining multiple servers into a cluster;
  ○ use of high-availability systems from leading manufacturers.

Servers:
The choice of server hardware should depend on the tasks (applications) that they will solve. Given that the spread of tasks is huge and, as the number of users and data volumes increases, the requirements for computing resources increase dramatically. We recommend to:

- select servers that allow gradual scaling resources and increasing performance.
- use virtualization technology that will divide the resources of a high-performance server between applications that do not require very large resources for their implementation at the hardware and software levels. This approach, which combines the installation of scalable servers and virtualization technology, will reduce TCO and increase the transparency and manageability of the entire computing infrastructure by dynamically reallocating resources.

Servers must provide:

- high data processing speed with reduced maintenance costs;
- easy management to quickly change and reallocate resources based on needs;

- high reliability and continuity of processing and access to information;
- integrate them into existing infrastructure and work together with existing data processing systems.

High-performance server platforms must have:

- backup fans and hot-swappable power supplies;
- hot-plug I/O disks and adapters;
- services for dynamic cleaning and reallocation of memory pages;
- dynamic processor reallocation and recovery services;
- a real-time event notification service;
- fault detection system with dedicated service processor and bus;
- remote management console.

All server-side solutions should be given great attention to the prevention of possible failures. For data processing centers of level I and II, appropriate functions should be implemented, which are used to continuously monitor the status of all server components and analyze trends in the monitored indicators. If a potential problem is detected, such as possible CPU overheating, special dynamic resource reallocation functions should ensure that processes are transferred from a potentially failed component to a healthy one without interrupting application execution. In this case, the system administrator and/or technical support service should receive a notification and a detailed report about the event.

Software implementing server virtualization technology should make it possible to quickly and easily divide computing resources depending on application requirements as well as reduce the total number of servers, allowing multiple virtual servers to be hosted on a single physical server, efficiently using its computing resources and memory.

Software that implements server virtualization technology must implement the following functionality:

- decomposition:
  - computer resources should be considered as a single homogeneous pool distributed among VMS;
  - multiple applications and operating systems must coexist on the same physical computer system;
- insulation:
  - VMS should be completely isolated from each other; a crash of one of them should not affect the others in any way;
  - data should not be transferred between VMS and applications, except when using shared network connections with a standard configuration.

When implementing fault-tolerant solutions, consider monitoring hardware, partial hardware failure, and ensure recovery from failures at the levels:

- central processor;
- RAM;
- motherboard/system bus;
- I/O ports;
- network hardware interfaces;
- power supply;
- internal and external cables.

A possible comprehensive approach to data storage and building highly reliable, fault-tolerant, and cyber-resilient computing systems is detailed in Figures 1.37 and 1.38.

Networks and data storage systems, electronic archives:
Priorities in the development of data storage systems should be:

- scalability of storage systems;
- concentration of data storage systems in a single location, and the number of geographically remote locations should be limited;
- expanding disaster recovery capabilities;
- reducing recovery time;
- reduce backup Windows (the time intervals allowed for preparing a backup) for critical applications.

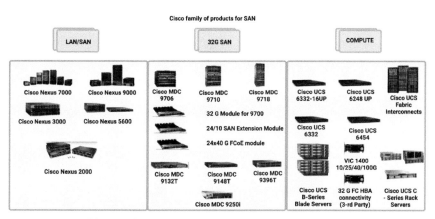

Figure 1.37 Comprehensive approach to data storage.

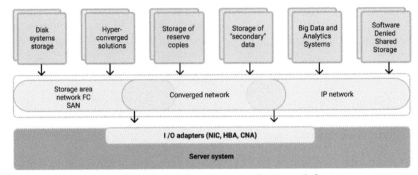

Figure 1.38   Possible solutions for network storage infrastructure.

It is recommended to allocate the following levels of data storage systems:

- operational level;
- level of long-term data storage;
- the level of e-archive;
- the level of data backup.

Operational level-data from this level is often used by users. Accordingly, the equipment must be fast enough and have a high degree of availability. It is recommended to use fast SATA drives.

The long-term data storage layer is a permanent data storage location that must have reliability in addition to performance.

An electronic archive is a data store that provides physical security of data regardless of user actions. Data stored in an electronic archive cannot be erased or changed. When changing data, the electronic archive must store both the original copy and all its modifications. An electronic archive must be created on non-rewritable optical media. This device must have a fiber channel interface and be connected to a SAN network.

We recommend implementing information lifecycle management (ILM) software, which, according to the configured rules, should automatically move data between different storage levels, depending on their demand.

In addition, the content management software must transfer data to a single storage system according to pre-configured rules, depending on the business-critical nature of the information, directly from the users' PCs.

The data storage system must have:

- unified data replication tools that transfer data between storage layers, aligning the value of data and its lifecycle stage with the availability, performance, security, and cost of the storage layer;

- scalable virtualization that allows managing multilevel storage resources as a single pool that must be shared between users and maintained as a single unit.

The storage system must be able to manage logical partitions of external memory. Logical partitions must allocate resources from one physical storage device to multiple virtual devices, each of which must be independently configured for individual applications and/or user groups. This strategy works effectively when storing large amounts of different types of data; so logical partitions should become part of a multilevel data storage infrastructure.

The classic data storage technologies are:

- the concept of a network attached storage (NAS) device that connects directly to a local or global network, focused on file services;
- a data storage network (SAN) concept focused on processing blocks of data stored in databases;
- Internet small computer system interface (iSCSI), which will create data storage networks without using expensive fiber-optic equipment, etc.

In a level I Data processing center, it is recommended to implement SAN technology as the most promising, developing, and satisfying of all the necessary requirements. SAN technology is also recommended for level II Data processing centers, but NAS devices can be deployed to organize file services. At the same time, a single technology must be strategically selected or a well-thought-out and justified approach to simultaneous use of both SAN and NAS without mutual conflicts, which provides virtualization for server systems and overall management of a single storage pool.

In a level III Data processing center, we recommend using wide-area file services (WAFS) technology – file services for global networks, when organizing remote access to a unified data storage system for collaborating in work with files. In this case, WAFS technology must be supported in a level II Data processing center or in SAN and/or NAS devices, or by network access equipment. It is not recommended to use iSCSI technology and devices because it is recommended to use special iSCSI-FC bridges to combine iSCSI and SAN resources, and it makes system management and administration more complex. In addition, using iSCSI technology creates an additional load on IP networks.

The goal of implementing and using SAN technology should be to ensure real consolidation of storage resources and their sharing since the storage capacity should be connected to many servers, including remote ones, and the machines that process data should be out of the resource management and storage tasks.

When implementing SAN technology, the following must be ensured:

- SAN topology independence from storage systems and servers;
- convenient centralized management;
- convenient data backup without overloading the local network and servers;
- fast response;
- high scalability;
- high flexibility;
- high availability.

When selecting and implementing specific hardware and software for SAN implementation, the following requirements must be met: the storage system must support logical abstraction (virtualization) levels between physical ports on a given disk array, data blocks on specific disk groups, and logical volumes or files available for servers or applications.

In particular, the following virtualization services should be implemented:

1. Virtualization of the SAN connection. Multiple servers should be able to access the disk array via the SAN, and the distribution of physical ports of the array between them should not be a management problem and should not become an obstacle to the full use of the storage system's capabilities. The storage system must allow creating multiple virtual ports on a single physical fiber channel port and manage these ports.
2. Virtualization of logical disks and volumes. Any modification of the application (for example, adding new servers, storage devices, or functions) requires a complex set of actions to change settings, both on servers and on disk arrays. These actions should not cause errors or downtime and should not increase the time required to deploy and modify the application.

The storage system must provide reliable logical disk management (LUN) and volume management services to extend information management and storage services to modular storage systems that support different types of disks (Figures 1.39 and 1.40).

One must have the appropriate hardware with sufficient performance to significantly increase the scalability and flexibility of virtualization solutions without compromising data availability or increasing the cost of managing the data storage system.

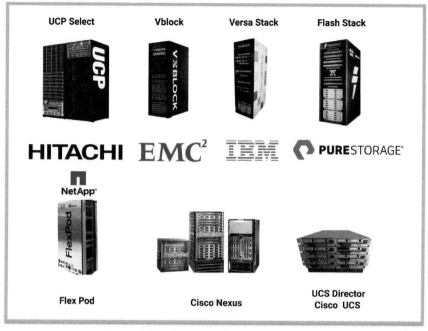

Figure 1.39   Integrated data storage solutions.

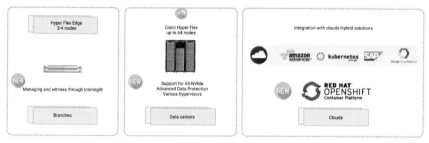

Figure 1.40   Hyper-converged storage solutions data.

Out of all possible SAN architectures, we recommend using switches with the core-edge method of combining switches, which is optimal in terms of performance, scalability, and reliability.

When building a SAN in a level I Data processing center on switches (with the exception of the director class), it is recommended to create two independent fabrics (this design is called dual fabric). Dual fabric avoids a single point of failure in the SAN, providing a very high level of reliability and fault tolerance. In addition, configuration changes or routine maintenance (for example, installing a new firmware) on one of the fabrics does not affect the operation of the other. Using dual fabric in conjunction with software that supports alternative access paths and load balancing to connect servers and storage devices (paths must be distributed between different fabrics) will create a reliable SAN. It is also necessary to provide dynamic multipath software on the servers to ensure continuous operation of applications with two factories.

It is recommended to choose equipment that supports fiber channel with a bandwidth of at least 4 GB/s and supports data transfer rates of at least 2/1 GB/s. One must use trunking – combining multiple data channels into a single channel to provide higher data transfer rates over trunk channels.

Data backup:
Backup of IP data for all data centers should be performed in accordance with the backup methodology, which should contain a description of the backup processes of various IS and the rules for their implementation.

The following approach for data backup is recommended:

- local data backup for quick recovery;
- remote data backup to ensure disaster tolerance.

Local data backup must be performed to mobile media from the backup server. The backup server is also an intermediate hardware that should reduce the backup window by pre-copying data to the SATA fast disk system, which should be located on the SAN network.

The automatic block of backup data on a different physical device than the original (data mirroring) in the SAN should be organized for level I Data processing center (for level II Data processing center is recommended).

A backup procedure must be adopted and approved, which should regulate the backup periods and Windows for full and incremental backups. The backup methods should be selected depending on the following parameters:

- *RTO* – target recovery time, i.e., the time required for data recovery.
- *RPO* – target recovery point, i.e., the point to which data needs to be restored or, in other words, it is the actual allowable amount of lost data.

Note that for some applications, these parameters may differ by an order of magnitude. For example, a Web server may have lost data (RPO) over the past few days, but usually, the break in their work (RTO) should not exceed a few minutes. Therefore, we recommend paying close attention to the RTO parameter.

The most common backup technologies today are:

- General (consolidated) backup, usually performed on tapes or high-capacity disks, the least expensive solution, and the highest RTO and RPO values.
- Periodic replication of changes is usually performed over communication channels to a remote site (another city) without guarantees to the response time. The RTO value is lower than that of a consolidated backup; however, the RPO is non-zero and is within the replication execution period.
- Asynchronous replication supports RPO within a few seconds–minutes, while not reducing system performance in the event of a significant delay on the transmission lines between the main and backup data copies.
- Synchronous replication is performed over high-speed communication lines with minimal delays (since this value directly affects the performance of the working system as a whole); so it has a limit on the territorial distance of the backup from the main one, and the RPO is 0.

When performing scheduled work on the main or backup copies of data in synchronous replication mode, there is a need to temporarily switch the system to asynchronous mode. Restoring synchronous mode after such a break causes an additional load on the storage subsystem in order to synchronize both copies. This, in turn, imposes additional requirements both on the schedule of planned work and on the stock of equipment performance and bandwidth of the communication channels used.

It is recommended to create a single backup plan to implement backup of data from various IS systems (Figures 1.41 and 1.42), which reflects the types of backups being performed, their frequency, the amount of data being backed up and the duration of the backup, as well as the chronological sequence of backup processes. Also, the backup of system data should be provided to restore the OS of servers, and, if necessary, client workstations.

Ensuring disaster tolerance:
Disaster tolerance must be ensured by copying data to a geographically remote data center. The degree of remoteness of the backup data center should be determined based on the threats that you plan to protect your data from. This list of threats should be formulated in the appropriate "business continuity plan."

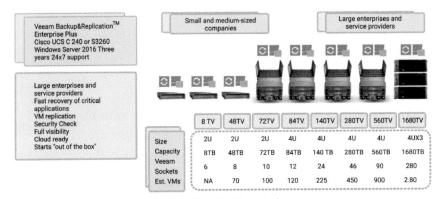

Figure 1.41   Backup solutions.

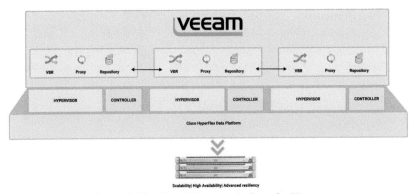

Figure 1.42   HyperFlex as a platform for Veeam.

We recommend copying (replicating) all business-critical data or its backups (depending on the value of the RPO data recovery point parameter) from a level II Data processing center to a geographically remote level I Data processing center at a set frequency to ensure disaster tolerance. It is recommended that a level I Data processing center has a backup center, to which data from the level I Data processing center should be replicated.

Ensuring disaster tolerance should be carried out in accordance with the IP recovery methodology, which should contain a description of the recovery processes of various IP and the rules for their implementation.

All business-critical and technologically important data from the level III Data processing center must be copied synchronously to the level II Data processing center.

Regardless of the accepted model of remote replication of data, the procedure for the storage and handling of backup data, no matter where they were, should be established. The established procedure should guarantee the safety of backup copies of data as well as their high availability when data recovery is necessary.

According to the principle of storage and use, there are two types of data:

- unstructured data (files);
- structured data that is stored in the DBMS.

The use of specialized software-backup agents is necessary to replicate unstructured data. DBMS tools or software that is compatible with this DBMS are required to replicate structured data.

We recommend choosing the type of replication (full or incremental) depending on the volume of data, the width of the communication channel (replication time), and the criticality of the data recovery time in case of an accident. The recommended order can be as follows: daily incremental data copying and weekly full data copying. This procedure should be recorded in the documents on business continuity and recovery after accidents.

Electronic storage provides daily backup of data to disk arrays and tape drives. Using electronic storage reduces RTO but does not affect RPO.

The electronic log provides for transmitting the latest changes in database records and/or file systems to a dedicated server in the backup computing center. However, the entire database and file system are not updated. These changes are transmitted via dedicated communication network channels, first, to the server in the backup computing center and then to magnetic tape. Electronic logs will reduce RPO since recovery occurs in a short period of time after an incident. Logs also reduce RTO by reducing the time it takes to deliver logs to the backup computing center.

Parallel copying (shadow memory) creates an exact copy of databases or file systems based on information about current data changes. The method involves moving data that is not critical for access time to slower media-magnetic tapes. The RTO is usually between one and eight hours, and the RPO depends on the latest data changes.

Mirroring is performed at the disk, virtual device, logical units (LUN), or storage level. There is a distinction between synchronous and asynchronous mirroring. Synchronous mirroring instantly creates an exact copy of the main disk data on the backup disk. A copy of the data is created with a small (often regulated) time delay for asynchronous mirroring. If the main data storage system fails, all data requests are automatically forwarded to the backup system. As a result, the RTO can take from 20 minutes to

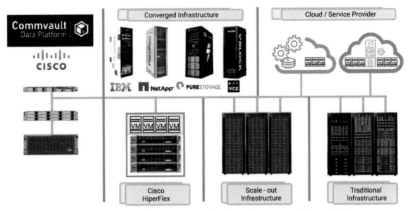

Figure 1.43   Option for a comprehensive data protection solution.

several hours, while the RPO is reduced only for the time of outstanding work. The overall performance of storage systems is determined by the speed of the least productive device. The obvious advantages of the method include a high availability rate, and the disadvantages are the high cost of the solution as well as the inability to neutralize errors and failures of the main system.

Load balancing between multiple physical devices is a way to manage a load of devices to ensure the desired performance and application continuity.

Hot standby systems provide efficient redundancy of the physical infrastructure of information services (Figures 1.43 and 1.44) and ensure the required performance and continuity of services, and, if necessary, data recovery, usually from a few minutes to an hour.

Multiservice network:

General requirements:

The general requirements for the multiservice network of the enterprise include high performance indicators:

- safety;
- availabilities;
- reliabilities;
- manageabilities;
- survivabilities.

In emergency situations, network resources should be spent to the extent necessary to ensure the smooth operation of the main process equipment. These rules must be implemented in the equipment configuration

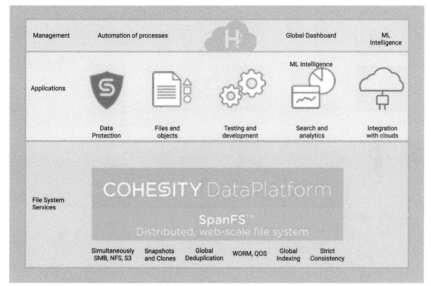

Figure 1.44 Version of the platform for storing "secondary" data.

and to enter into force automatically upon the occurrence of a failure or malfunction.

Three-level switching model:

A multiservice network for a data processing center is recommended for design based on a typical three-level switching model:

- access layer;
- distribution layer;
- core layer.

The classic three-level network model is shown in Figure 1.45.

The general rules for designing a three-level structure must be followed:

- problems with equipment and communication channels at the lower levels should not affect the performance and quality of service at the upper levels;
- transit backup routes of this level must not pass through the lower levels;
- traffic classification and prioritization must occur at the access level;
- distribution level should only aggregate traffic;
- network core should only perform fast packet switching;
- convergence time of routing tables and their volume should be optimized for each level by optimizing the redundancy scheme;

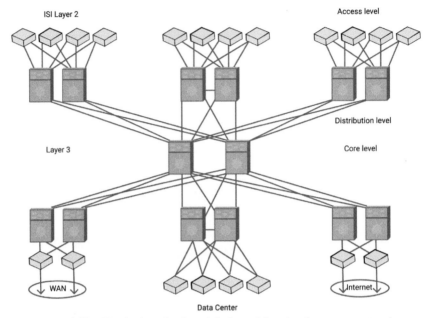

ISI Layer 2

Access level

Distribution level

Layer 3

Core level

WAN

Internet

Data Center

Figure 1.45   Classic three-level model of a multiservice data center network.

- remote users and external communication channels should not be connected directly to the network core – access switches should be used to prevent avalanche-like rearrangements of the routing tables of the entire network.

We recommend ensuring the network reliability by the following rules:

- failure of any network device must not result in failure of more than 48 ports;
- trunk level equipment must have redundancy;
- network must be designed to meet information security requirements.

Scalability support for the data processing center network must be provided by correctly implementing the three-level switching model and switch scalability (grouping (stacks)). In this case, each switch in the stack can operate in two modes – as the main switch of the stack and as a packet switching processor. The system must be fault-tolerant according to the scheme $1:N$ – if one of the stack switches fails, regardless of the function it performs, the others will continue to perform their functions without stopping the entire network.

Dynamic switching and routing:
The OSI model for level I Data processing centers is necessary to ensure high network availability at the network level, and for level II and III Data processing centers, it is recommended to use:

- dynamic internal routing protocol OSPF as having good scalability, fast convergence, considering the quality of communication channels and occupying the minimum channel bandwidth.

The OSI model for level I Data processing centers is necessary to ensure high network availability at the channel level, and for level II and III Data processing centers, it is recommended to use:

- spanning tree protocol (STP) with support for VPNs (i.e., it must be possible to use a separate spanning tree algorithm in each virtual network to manage traffic paths up to a separate subnet);
- managing the port weight parameter in the spanning tree algorithm for a virtual network on trunk ports to load both channels of a dual homing connection (a method for connecting devices that provide a primary and backup connection);
- support for multiple spanning tree groups on the same network;
- optimization of the spanning tree algorithm for switching to backup channels in less than 5 seconds;
- no delay in the spanning tree algorithm when enabling the user port.

We recommend the following features in order to support applications based on IP multicast technology from level I and II Data processing center networks:

- at the access/distribution level – transmitting IP multicast packets at the channel level at the speed of the physical channel, dynamic registration via IGMP, and PIM protocols;
- at the backbone level-IP multicast packet transmission at the channel and network levels at the speed of the physical channel and scalable IP multicast traffic routing protocols.

Equipment requirements:
Access-level hardware must be able to provide traffic classification by application type, source and destination physical and network addresses, and switch ports to ensure priority. Classified traffic should receive a label indicating the priority level assigned to packets, thus enabling network devices to serve this traffic appropriately. Packages must be reclassified

based on the quality of the service policy set by the administrator. For example, a user assigns a high priority to their traffic and sends it to the network. This priority can then be lowered in accordance with network policy, rather than based on user requirements. This mechanism should be key in ensuring the quality of service throughout the network.

The main level equipment must have the following functionality:

- Congestion prevention and management, i.e., it should be possible to manage network behavior during congestion by dropping certain packets based on classification or policy at times of network congestion and multiple queues on interfaces. The administrator must set thresholds for different priority levels.
- Scheduling, i.e., it should be possible to prioritize packet transfers based on classification or quality of service policy using multiple queues.
- Redundancy of the main nodes, which can include power supply, fan unit, processor module, etc.

In level I and II Data processing centers, in addition to providing redundancy for the main nodes of backbone-level equipment, it is recommended to provide the same redundancy for distribution-level equipment (Figures 1.46–1.48).

All active network equipment of level I and II Data processing centers must have tools for monitoring the quality of service and security policy, network, and service planning must be secured:

- ability to collect RMON statistics up to the network port for network performance analysis;
- ability to redirect traffic from individual ports, groups of ports, and virtual ports to the protocol analyzer for detailed analysis;
- advanced event monitoring in real time for expansion of diagnostic capabilities, in addition to external analyzers;
- ability to collect and save information about significant network events, including device configuration changes, topology changes, and software and hardware errors;
- ability to access the device management interface and reports via a standard WEB browser in emergency situations;
- automatic configuration of fast/gigabit Ethernet ports, virtual networks, and VLAN trunks;
- automatic network topology recognition via topology recognition agents to reduce network connectivity recovery time in case of damage.

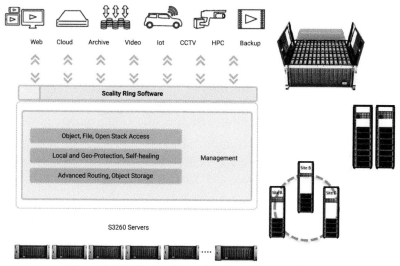

Figure 1.46   Option for storing large amounts of data.

Requirements for external communication channels:
Communication channels for point-to-point or point-to-multipoint connections between level I and II Data processing centers should only be organized using technologies that provide high fault tolerance and flexible quality of service management (for example, MPLS). At the same time, a contract that provides QoS for regular and emergency operation modes of both the organization and the operator must be signed. The requirements must be specified in the corresponding SLA.

Connection to two independent operators for level I Data processing center is required for backup communications, and for level II Data processing center is desirable. When connecting the data processing centers of levels I and II, the operator must provide round-the-clock technical support, which, at any time of the day, not only accepts requests but also eliminates incidents.

Premises and engineering systems:
General requirements for premises:
Server rooms in all data centers must meet the following general requirements:

 • it is forbidden to place server rooms in basements and other rooms equipped with a large number of engineering structures that pose a potential danger to the equipment;

- it is forbidden to place server rooms under the premises of the dining room, toilets, and other premises associated with water consumption;
- it is forbidden to place server rooms on the top floor of the building to avoid water leaks from the roof;

The server room must be a restricted space intended for hosting server hardware.

The server room design must meet the following requirements:

- maintain the required business continuity;
- maintain the required weight of server room hardware;
- protect valuable hardware and data.

Only authorized employees of IT departments and service organizations should have physical access to the server room. Automated access control systems must be used to restrict physical access to the server room.

Depending on the level of the data processing center, the server room must be equipped with:

- uninterruptible power supply;
- conditioning system;
- diesel generator;
- air cleanliness and humidity control system;
- server and telecommunication cabinets, racks;
- control of the internal environment;
- early smoke detection system;
- access sensors;
- sensors for the physical state of equipment;
- temperature/humidity sensors;
- surveillance system.

Power supply

General requirements for the power supply system of data processing centers at all levels:

1. The power supply to the server room must be supplied from the main board of the building, regardless of the location of the data center room. Also, a ground cable runs from the main building ground bus to the server room ground loop. All wires must have the appropriate cross-section according to the technical design and color coding, according to the regulatory documents.
2. Power to PCs, peripherals, office equipment, servers, storage systems, and active network equipment must be separated from power to industrial installations. Power must be provided from separate floor-level

Validated by ISV
partners, supported
jointly

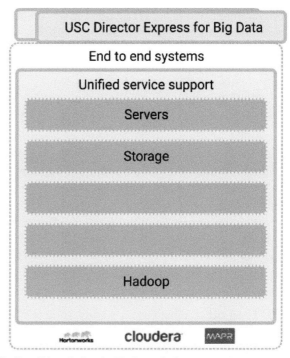

Figure 1.47 Possible solution for Hadoop platforms.

automatic machines and those, in turn, must be connected separately
to the main board of the building through a separate system of automatic machines.

3. When powering PCs, peripherals, and office equipment, it is recommended to install a protective cutout device (PCD) in accordance with
   the current regulations after automatic machines in floorboards.

The power supply system for a level I Data processing center should be
organized from two geographically separated transformer substations.
Cable lines must run on independent routes. It is recommended to use
automatic load transfer (ALT), which selects and switches between the
main and reserve lines.

For a level I Data processing center, it is necessary, and for a level II
Data processing center, it is recommended to use a diesel-driven generator
(DDG). In the power supply scheme, they must be located parallel to the
input of power cables to the building. For the correct operation of the DDG
and two independent inputs, a device for automatically switching on the
backup power must be provided. In case of complete loss of power supply,
or its non-compliance with the required parameters (voltage, frequency,

"cleanliness," etc.), the DDG must be automatically started, and the load is transferred to it. DDG must have a fuel reserve calculated for at least 8 hours of continuous operation and the ability to replenish fuel without stopping the generator. DDG should be able to operate continuously for up to 3 months, provided that the fuel supply is well established.

For I and II level Data processing centers, centralized uninterruptible power supplies of the online type must be installed after the power cables are inserted into the building or after the DDG if available. This is a mandatory requirement since the most dangerous are short power surges lasting 2–3 seconds. In addition, there are some negative factors, causing significant damage, as the excess voltage, frequency, harmonics, ground fault, interfacial potential, etc., is allowed to install the UPS batteries if it is not economically feasible, and provided the UPS (individual or group) with rechargeable batteries which will provide the desired hold time system health.

Power supply requirements for data processing center cabinets at all levels as follows:

1. Each office must have a supply voltage of 220±10% AC power from two independent sources via individual circuit breakers.
2. Connect equipment that has two power supplies to two independent sources. Connect equipment that has a single power supply to one of the power sources, evenly distributing the load in accordance with the power consumption specified in the equipment passport.
3. The power consumption must be calculated in the technical design. If empty cabinets are installed and there is not yet a project for placing equipment in them, it is recommended to estimate the average power consumption of 4 kW per cabinet. If it is planned to install the blade servers or other equipment with increased power consumption in cabinets, the power consumption must be adjusted according to the manufacturer's documentation.

Air conditioning and cooling system:
Redundancy of the cooling system of a level I Data center is mandatory, and for a level II Data center, it is recommended to implement the scheme with $N+1$ (with one spare air conditioner). All air conditioners must be connected to a single control system. The software should allow rotation of the spare air conditioner, which allows more efficient use of the resource of the cooling system as a whole.

For a level I Data processing center, it is necessary, and for a level II Data processing center, it is recommended to organize an influx of fresh air from the street since the air that constantly circulates through computer office and air conditioners "burns out" and requires updating. The inflow is recommended to be carried out through a special installation that heats and dehumidifies the street air. In addition, it must create an additional pressure inside the room, which prevents the penetration of dust.

It is recommended to use steam generators to humidify the air in data processing centers of levels I and II. Dry air is not effective for cooling by the cooling system due to the physical principles of air conditioning. By lowering the humidity, electrostatic potential increases, which may be the reason for the withdrawal of equipment failure.

Early fire detection and extinguishing systems:
Data processing centers at all levels must be equipped with an automatic fire extinguishing system. The fire extinguishing system must not cause damage to the equipment.

The gas fire extinguishing system must work in the rudimentary phase of fire development, i.e., when there is a smoldering of heating elements or initial ignition and in less than one minute to extinguish the fire centers.

The fire warning and extinguishing system should inform of the potential ignition source much earlier than a need to activate the extinguishing system. This should be achieved by installing a large number of highly

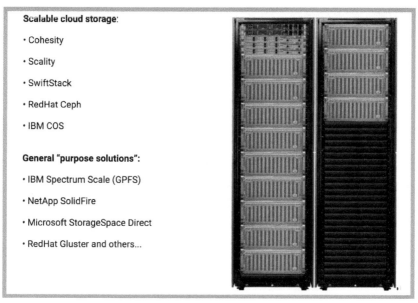

Figure 1.48   Possible platform for software-defined data warehouses.

sensitive smoke, optical, chemical, spectral, and other fire detectors linked to a single intelligent fire warning and extinguishing system, as well as a set of organizational measures. It should include a constant visual inspection of equipment, compliance with fire regulations and rules, as well as the rules of operation of electrical installations.

It is recommended to use fire extinguishing mixtures based on refrigerants or inert gases since they cause the least damage to the equipment.

It is recommended to provide a system for removing gas from the room after the fire extinguishing system is activated.

When the gas fire extinguishing system is activated, all systems that pump air into the data processing center room must be disabled.

1.4.3 Software

Operating system level:

Modern operating systems will organize fault-tolerant systems for most types of IT services required in daily practice. Two main areas of continuity in this case are:

- formation of a cluster with redundant computing capacity (fail-over) using OS tools, designations – "AP," "AAP," and so on, depending on the cluster composition;
- forming a cluster with load balancing by means of the OS, designations – "AA," "AAA," and so on, depending on the cluster composition.

Both methods imply achieving the goal of ensuring continuity at a level higher than hardware by reserving entire computing systems that are as independent of each other as possible and, therefore, have a minimal chance of simultaneous failure.

Features of clusters with redundant computing capacity:
The key issues of quality assurance of continuity and fault tolerance by means of fail-over clusters are:

- the algorithm for detecting the failure of the main equipment (it can be implemented by both built-in and external means; it most often used periodic verification of correct functioning-«heartbeat», and the verification interval can be set from a few minutes to a fraction of a second);
- the algorithm for commissioning backup equipment (includes activation of application software, possible switching of IDs, including network IDs, etc.);

- method of distributed storage of the current state (for stateful services) between the main and backup equipment (including issues of protecting information about the current state from partial damage and/or distortion at the time of hardware failure).

Limitations of the fail-over cluster technology are as follows:

- there may be single points of failure (usually associated with storing a single current state);
- there may be restrictions on using the method for stateful services with a large amount of state information;
- increasing the cost of the system, which is not associated with an increase in its performance.

Features of load-balancing clusters:

The key issues of quality assurance of continuity and fault tolerance by means of load-balancing clusters are:

- Algorithms for selecting the server to use in the normal operation of the cluster (the choice can be made depending on the technology used, either by an application service/software using this IT service or by automatic dispatching technology), the most common algorithms:
  - random selection;
  - alternate (round-robin) selection;
  - weighted sequential selection (used in cases of unequal computing power of servers in the cluster to equalize the relative load on each server);
  - choosing the fastest response;
  - selection of the minimum load on the server;
  - selection with caching (if the server through which the previous successful connection was made is active, the next one will be made through it).
- The detection algorithms applied by the service/bY the fact that one of the servers fails before starting the connection and during operation.

DBMS level:

All modern database management systems (DBMS) implement, in order to provide certain guarantees to the applied software, an ideology that is named after the first letters of the properties it implements in relation to processed and stored information – ACID. These properties include:

- *Atomicity* of operations – a guarantee that a user-defined indivisible block of operations (transaction) will either be executed in full or rejected if it fails in full.
- *Consistency* – consistency or data integrity guarantees that the database is in a state before the transaction is executed and after its completion, when none of the restrictions on the syntax and semantics of the data stored in it is violated by the user at the design stage.
- *Isolation* of transactions – a guarantee that any operation performed on data that is external to the current transaction does not see intermediate states generated by it during execution. External operations can access either the state of data before the start of the transaction or at its end.
- *Durability* – data security – a guarantee that after notifying the user of the successful completion of the transaction, the data will be stored in the final state until the next authorized modification.

The ACID ideology significantly simplifies the formalization of data continuity and security requirements at the DBMS level, essentially introducing certain quality of service (SLA) guarantees similar to the OSI open system interaction reference model but with respect to data integrity and security. In order to ensure these properties, DB management systems use logging of operations performed as well as data states before and after transaction execution. In addition, the DBMS actively uses a variety of replication schemes for modifiable data.

There are two approaches to implementing the cluster solutions at the DBMS level:

- shared data storage (multiple cluster nodes access the shared storage to read and write data and replicate changes to the cache supported for faster access to information);
- replicated data storage (multiple cluster nodes independently store copies of data) – this schema raises questions about replication of user-made changes to data, for example, by using:
  - centralized replication (replication) layer for data modification operations across all cluster nodes;
  - definitions of only one cluster node from the currently active ones (for example, according to the voting scheme) that allow changes to be made to stored data with subsequent replication – data is read through any cluster nodes without restrictions.

Application software level:

The level of application software is the highest-level point of possible measures to ensure continuity and disaster tolerance of the IT service. At the design stage of a software product or IT service, to achieve this goal, special attention should be paid to the Project Manager:

- automatic detection of failure of active equipment, communication channels, and subordinate IT services in order to switch to backup equipment or emergency service operation schemes (including alternative means of achieving obligations to support the required business service functionality);
- for stateful services, which, in practice, make up the vast majority of applied IT services – a single storage of information about the current state of user sessions or, better, replication of states to all nodes in the cluster to switch users to another node without data loss in automatic mode;
- transactional semantics of all critical executable operations on data;
- manage resource inventory and scalability of the selected software implementation.
- During the implementation, commissioning, and day-to-day use phase, attention should be paid to:
- the quality of the software code, especially with regard to changes to software versions that are already in use;
- testing the implementation before commissioning and when making changes on a representative sample of tests that are as close as possible to the operating conditions;
- periodic scheduled testing of the functionality of the built-in mechanisms for ensuring continuity and fault tolerance for various scenarios of violation of the normal mode of operation of equipment, communication channels, subordinate IT services, etc.

Management and monitoring system:
Structure of the management and monitoring system:
The management and monitoring system should consist of the following main subsystems:

- subsystem for monitoring and managing server complexes, operating systems, and applications;
- subsystem for monitoring and administration of backup equipment and processes;

- subsystem for monitoring and managing the enterprise data network and peripheral equipment;
- PC monitoring and administration subsystem.

Functionality of the management and monitoring system:
The management and monitoring system should provide the following functions:

- remote access to the management server via active consoles;
- support for parallel operation of multiple operators (with their own permissions and area of responsibility) with the management server;
- protection of access to the management server by unauthorized persons using any login options;
- differentiation into areas of competence for solving emerging problems;
- different level of graphical representation of information for different operating personnel, depending on their role in the operational process;
- remote monitoring of management objects;
- monitoring of the controlled objects with the help of agents;
- selecting monitoring parameters and setting agent response thresholds to assess the current state of systems;
- centralized registration of events that occur in controlled objects, operating systems, DBMS, applications, and information services;
- expanding the list of registered events and adapting to the applications and existing technologies used;
- centralized processing of all registered events;
- notifying system operators about the operation of information resources by issuing an information message to the operator console;
- analysis of the performance of management objects;
- automatic processing and graphical representation of operational information on the state of information services;
- collection, storage, and analysis of parameters for the functioning of management objects.

Managing and monitoring a multiservice network:
Monitoring of IP networks, configuration management, failures, performance, and methods and tools for IP network inventory must be strictly regulated and automated.

The corresponding database for data processing centers of levels I and II must contain complete and reliable information about all network elements in a hierarchical form, taking into account the geographical hierarchy on the one hand and the functional hierarchy (applications, IP networks, base networks, etc.) on the other.

Functionally, a database can be represented as two interacting databases: an inventory database and a network event management system.

The inventory database should combine information from various sources and provide it in a convenient form. Recommended set of information to include in the inventory database:

- network topology and the installed equipment in managed networks;
- information about the communication channels used (in the case of leased channels-information about the organization that provided the channel and methods of communication);
- hardware specification: current software version, factory and inventory numbers, current and previous configuration files, software version, contact information of the service organization, installation location, and responsible persons;
- for network hardware, the interfaces are described in a table with IP addresses, connected networks, or application servers for LAN interfaces, or used communication channels (physical and virtual), routing protocols, and connected remote hardware for WAN interfaces.

Automated event processing system:

An automated event processing system for level I and II Data processing centers should provide:

- collecting and storing information about failures, malfunctions, exceeding critical thresholds, etc., for active equipment and communication channels;
- notifying service personnel of any problems and transmitting this information in a hierarchical structure (administrative, topological, system, etc.) in accordance with the established administrative rules;
- displaying the event processing history (by whom, when, and what actions were taken);
- saving the event history for each control object.

In addition, the event processing system must provide analytical information in accordance with the specified administrative requirements (for example, a sample for objects that did not have timely maintenance, a sample for events that were not closed for a month, and so on).

1.5   Business Continuity and Cyber Resilience

1.5.1   Basic Concepts and Definitions of Cyber Resilience

The *cyber resilience* characteristic is a fundamental feature of any cyber system created on the *Industry 4.0* breakthrough technologies (and *Society 5.0 – SuperSmart Society*). The characteristic can intuitively be defined as a certain constancy, permanence of a certain structure (*static resilience*), and behavior (*dynamic resilience*) of the named systems. As applied to technical systems, the resilience definition was given by an *outstanding Russian Mathematician and Academician of the St. Petersburg Academy of Sciences, A. M. Lyapunov* (1857–1918): *"Resilience is a system ability to function in conditions close to equilibrium, under constant external and internal disturbing influences"* [101, 102, 103, 104, 105, 106, 107].

In the monograph, it is proposed to clarify the above definition since the cyber resilience of *Industry 4.0 systems* does not always mean the ability to maintain an equilibrium state. Initially, the resilience feature was interpreted in this way since it was noticed as a real phenomenon when studying homeostasis (*returning to an equilibrium state when unbalancing*) of biological systems. The system analysis apparatus use implies a certain adaptation of the term *"resilience"* to the characteristic features of the studied cyber systems under information and technical influences, one of which is the operation purpose existence. Therefore, the following **resilience definition** is proposed: *"Cyber Resilience is an ability of the cyber-system functioning, according to a certain algorithm, in order to achieve the operational purpose under the intruder information and technical influences."*

Indeed, according to *Fleishman B. S.* [108], it is necessary to distinguish the active and passive resilience forms. The *active resilience* form (*reliability, response and recovery, survivability*, etc.) is inherent in complex systems, whose behavior is based on the *decision act*. Here the decisive act is defined as the alternative choice, the system desire to achieve its preferred state that is *purposeful behavior*, and this state is its goal. The *passive form* (*strength, balance, homeostasis*, etc.) is inherent in the *simple systems* that are not capable of the *decision act* [109].

Additionally, in contrast to the classical equilibrium approach, the central element here is the *concept of structural and functional resilience*. The fact is that the normal cyber system functioning is usually *far from an equilibrium* [110]. At the same time, the intruder external and internal information and technical influences constantly change the equilibrium state itself. Accordingly, the proximity measure that allows deciding whether the cyber system behavior changes significantly under the disturbances, here, is the *performed function set*.

After the work of *Academician Glushkov V. M. (1923–1982), the researches of V. Lipaev (1928–2015), Dodonova A. G., Lande D. V., Kuznetsova M. G., Gorbachik E. S., Ignatieva M. B, Katermina T. S.*, and a number of other scientists [111, 112] were devoted to the resilience theory development. However, the resilience theory in these works was developed only in regard to the structure vulnerability of the computing system without considering explicitly the system behavior vulnerability under *a priori* uncertainty of the intruder information and technical influences. As a result, in most cases, such a system is an example of a predetermined change and relationships and connection preservation. This preservation is intended to maintain the system integrity for a certain time period under normal operating conditions [113, 114, 115, 116]. This predetermination has a dual character: on the one hand, the system provides the best response to the normal operating disturbance conditions, and on the other hand, the system is not able to withstand another, *a priori* unknown information and technical intruder influences, changing its structure and behavior (Figures 1.49 and 1.50) [117, 118, 119, 120, 121, 122].

Cyber resilience challenges:
The main cyber challenges of modern *Industry 4.0* cyber systems under the unprecedented cyber threat growth include:

- insufficient cyber resilience of the mentioned system;
- increased complexity of the *Industry 4.0* cyber system structure and behavior;
- difficulty of identifying quantitative patterns that allow investigating the cyber system resilience under the heterogeneous mass cyber-attacks.

We will give a detailed comment on these problems.

The first (and most significant) problem is the lack of the *Industry 4.0 cyber systems* resilience, which is often lower than required. In many cases, the hardware and software components of the mentioned system are not able to fully perform their functions for a variety of reasons (Figures 1.51–1.53). There are the following reasons:

- inconsistency of the actual system behavior parameters in the software and hardware specifications;
- current level reassessment of the programming technology development and computer technology;

Cyber Resilience Serves a Number of IT and Risk Management Disciplines

Cyber Resilience Combines Multiple IT Disciplines

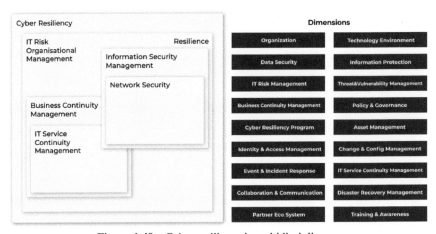

Figure 1.49   Cyber resilience is multidisciplinary.

- destructive information and technical impact of external and internal factors on the system, especially under mass intruder attacks;
- capability reassessment of the modern cyber systems information protection methods and technologies, infrastructure resiliency, and software reliability.

Ignorance or neglect of these reasons leads to a decrease in the effectiveness of the *Industry 4.0 cyber systems* functioning. Moreover, this problem significantly aggravates under the group and mass cyber-attacks [123,124, 125, 126, 127, 128, 129, 130, 131, 132].

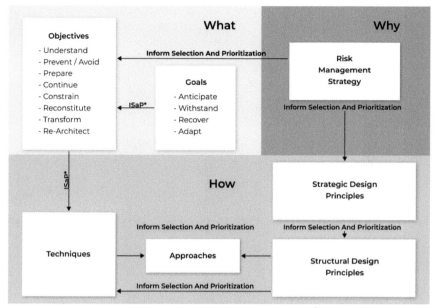

Figure 1.50   The main cyber resilience discipline goals and objectives.

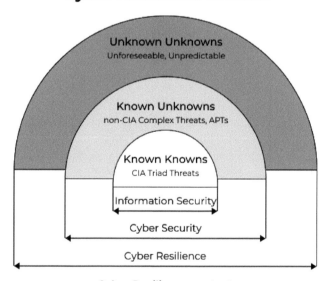

Figure 1.51   The focus on the previously unknown cyber-attacks detection and neutralization.

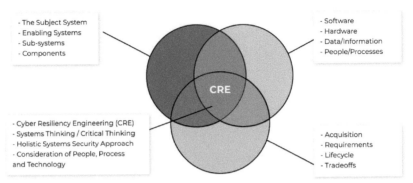

Figure 1.52   The enterprise cyber resilience management program components.

Figure 1.53   The main risks of the Industry 4.0 enterprise business interruption.

The second is the growing structural and behavioral complexity of the *Industry 4.0 cyber systems* (Figures 1.54–1.58).

The system structure features include the following. As a rule, the modern cyber systems are heterogeneous distributed computer networks and systems, consisting of many different architecture components. According to the author, the composition of the mentioned systems includes more than the following:

• 28 BI types based on Big Data and stream data processing;
• 15 ERP types;
• 16 systems types electronic document management;
• 28 varieties of operating system families;
• 1040 translators and interpreters;
• 2500 network protocols;

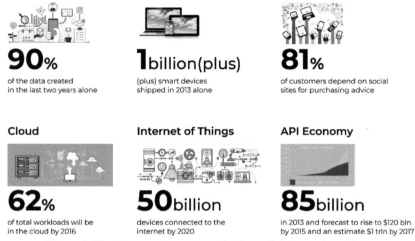

Figure 1.54 Prospects for state and business digital transformation.

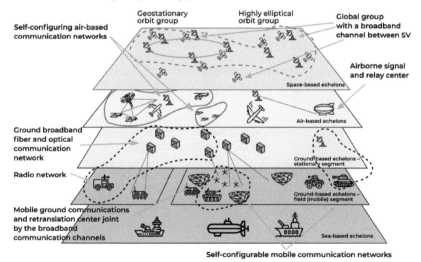

Figure 1.55 The research object characteristics.

- 20 network equipment types;
- 28 information security tool types (*SOC, SIEM, IDS/IPV, DPI, ME, SDN/FPV, VPN, PKI, antivirus software, security policy controls, specialized penetration testing software, unauthorized access security tools, cryptographic information security tools,* etc.)

Digital railways

Digital railways solution:
- Internet of Things, IoT
- Big Data/Cloud
- Digital Train
- Digital Depot
- GSM-R/LTE/Wi-fi
- SDH/WDM/xPON/IP

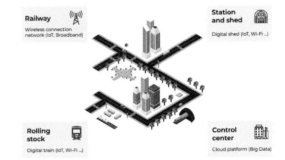

- The main tasks: process automation and optimization, increase of railway reliability and security
- Key technologies: Internet of Things, Big Data, modern systems of connections / storage and data processing

Figure 1.56   The complexity of the research object structure.

Technological platforms

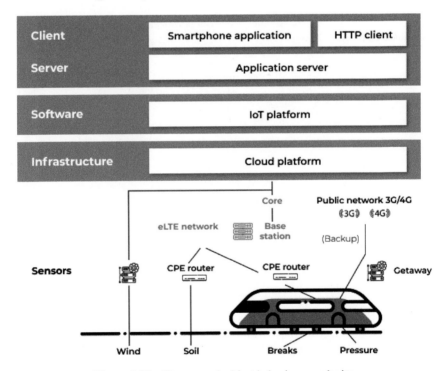

Figure 1.57   The research object behavior complexity.

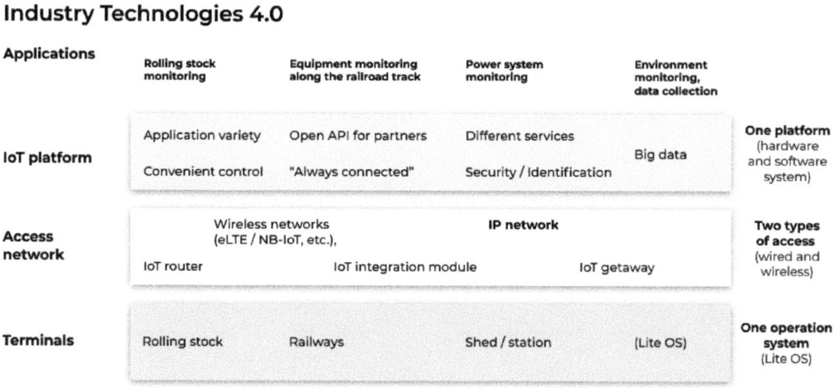

Industry Technologies 4.0

Applications	Rolling stock monitoring	Equipment monitoring along the railroad track	Power system monitoring	Environment monitoring, data collection	
IoT platform	Application variety	Open API for partners	Different services	Big data	**One platform** (hardware and software system)
	Convenient control	"Always connected"	Security / Identification		
Access network		Wireless networks (eLTE / NB-IoT, etc.),	IP network		**Two types of access** (wired and wireless)
	IoT router	IoT integration module		IoT getaway	
Terminals	Rolling stock	Railways	Shed / station	(Lite OS)	**One operation system** (Lite OS)

Figure 1.58 The diversity of the research object representation levels.

The cyber system functioning features include the following:

- slightest idle time in the system can cause a complex technological process shutdown – significant disaster recovery costs;
- system failure consequences can be catastrophic;
- proprietary technological protocols use of equipment manufacturers that hold difficult-to-detect vulnerabilities;
- false positives, leading to interruptions in the normal functioning maintenance of the technological processes, are unacceptable;
- use of the buffer, demilitarized zones to organize the interaction of *MES, ERP, BI,* and other systems with the enterprise system;
- need to provide remote system access and management by contractors, etc.

The listed cyber system features cause the expansion of the threat spectrum to cyber security and determine the high system vulnerability [133, 134, 135, 136, 137, 138, 139, 140].

The third problem is the difficulty of identifying quantitative patterns (Table 1.8) that allow investigating the *Industry 4.0* system cyber resilience under group and mass cyber-attacks [141, 142, 143, 144, 145]. The fact is that the external and internal environment factors significantly affect the above system functioning processes. These factors within the considered structure framework are either fundamentally impossible to control or manage with an unacceptable delay. Moreover, the external and internal environments have the property of incomplete definiteness of their possible states in the future periods, i.e., factors affecting the cyber system behavior are subject to such changes in time that can fundamentally

Table 1.8 Complexity factors in ensuring cyber resilience.

No.	Complexity factors	Generated difficulties
1	Complex structure and behavior of the *automated systems of critically important in objects (AS CIO)*	Solved problem awkwardness and multidimensionality
2	AS CIO behavior randomness	System behavior description uncertainty and complexity in the task formulation
3	AS CIO activity	Limiting law definition complexity of the potential system efficiency
4	Mutual impact of the AS CIO data structures	Cannot be considered by the known type models
5	Failure and denial influence on the AS CIO hardware behavior	System behavior parameter uncertainty and complexity in the task formulation
6	Deviations from the standard AS CIO operation conditions	Cannot be considered by the known type models
7	Intruder information and technical impacts on AS CIO	System behavior parameter uncertainty and complexity in the task formulation

change the algorithms of its functioning or make the set goals unattainable. The changes, which the external and internal environment factors undergo, occur both naturally and randomly; therefore, in the general case, they cannot be predicted exactly, as a result of which there is some uncertainty in their values. Cyber systems that face a specific purpose have a certain "safety margin," such features that allow them to achieve their goals with certain deviations of the influencing external and internal environment factors [146, 147, 148, 149, 150, 151, 152].

Until recently, mainly two main approaches were applied to identify the technical system functioning patterns: an *experimental method* (for example, *mathematical statistics methods and experiment planning methods*) and *analytical* one (for example, *analytical software verification methods*). In contrast to the *experimental methods*, which allows studying the individual cyber system behavior, the *analytical verification methods* allow considering the most general features of the system behavior that are specific to the functioning processes class in general. However, the approaches have significant drawbacks. The disadvantage of experimental methods is the inability to extend the results obtained in the experiment to a different system behavior that is unlike the one studied. An analytical verification method drawback is the difficulty of transitioning from a system functioning process class characterized by the derivation of the universally significant attributes to a single process that is specified by additionally relevant functioning conditions (in particular, *specific parameter values of the cyber system behavior in group mass cyber-attacks*).

Consequently, each of the approaches separately is not sufficient for an effective resilience analysis of the cyber system functioning under group and mass cyber-attacks. It seems that only using the strengths of both approaches, combining them into a single one, it is possible to get the necessary mathematical apparatus to identify the required quantitative patterns.

The problem solution idea:
The design and development practice of Industry 4.0 cyber system indicates the following. The modern confrontation conditions in cyberspace assign these systems features that exclude the possibility of designing cyber-resilient systems in traditional ways [153, 154, 155, 156, 157]. The complexity factors arising at the same time and the generated difficulties are given in Table 1.8.

Here the factors 1, 4, and 7 are determinant. They exclude the possibility to be limited by the generally valid features of *Industry 4.0 cyber systems* in group and mass cyber-attacks. However, traditional cyber security and resilience methods are based on the following approaches:

- simplification the behavior of cyber systems before deriving generally valid algorithmic features;
- generalization of the empirically established specific behavior laws of the named systems.

The use of these approaches not only causes a significant error in the results but also has fundamental flaws. The analytical modeling's lack of the cyber system behavior, under group and mass cyber-attacks, is the difficulty of the transitioning from the system behavior class, characterized by the derivation of general algorithmic features, to a single behavior, which is additionally characterized by the operating conditions under growing cyber threats. The empirical simulation's disadvantage of the cyber system behavior is an inability to extend the results of other system behavior that differs from the studied one in the functioning parameters.

Therefore, in practice, traditional cybersecurity and fault tolerance approaches can only be used to develop systems for approximate forecasting of system cyber resilience in group and mass cyber-attacks.

In order to resolve these contradictions, there is a proposed approach, based on the dimension and similarity theory methods [158, 159, 160], which lacks these drawbacks and allows the implementation of the so-called cyber system behavior decomposition principle under group and mass cyber-attacks, according to the structural and functional characteristics. In the dimension and similarity theory, it is proved that the relation set between the parameters that are essential for the considered system behavior is not the natural studied problem property. In fact, the individual factor influences of the cyber system external and internal environment, represented by various quantities, appear not separately but jointly. Therefore, it is proposed to consider not individual quantities but their total (the so-called similarity invariants), which have a definite meaning for the certain cyber system functioning.

Thus, the *dimensions and similarity theory method application* allow formulating the necessary and sufficient conditions for the *two-model isomorphism* of the allowed cyber system behavior under group and mass cyber-attacks, formally described by systems of homogeneous power polynomials (*posynomials*).

As a consequence, the following actions become possible:

- Producing an analytical verification of the cyber system behavior and to check the isomorphism conditions.
- Numerical determination of the certain model representation coefficients of the system behavior to achieve isomorphism conditions.

This, in turn, allows the following actions:

- Controlling the semantic correctness of the cyber system behavior under exposure by comparing the observed similarity invariants with the invariants of the reference, isomorphic behavior representation.
- Detecting (including in real time) the anomalies of system behavior resulting from the destructive software intruder actions.
- Restoring the behavior parameters that significantly affect the system cyber resilience.

It is significant that the proposed approach significantly complements the well-known *MITRE*[33] [161, 162, 163] and *NIST* [164, 165, 166, 167] approaches (Figures 1.59 and 1.60) and allows developing the *cyber resilience metrics and measures*, including *engineering techniques* for *modeling, observing, measuring, and comparing cyber resilience*, based on *similarity invariants*. For example, a new methodology for modeling standards semantically corrects the cyber system behavior, which will consist of the following four stages.

---

[33] www.mitre.org

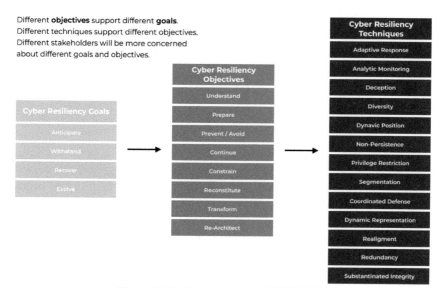

Figure 1.59   Recommended MITRE 2015.

Cyber Resiliency Constructs in System Life Cycle

Figure 1.60   Cyber resilience lifecycle, NIST SP 800-160.

The cyber resilience control methodology:

The *first stage* is the π-*analysis* of the cyber system behavior models. The main stage goal is to separate the semantic system behavior correctness standards, based on *similarity invariants*.

The step procedure includes the following steps:

1. structural and functional standard separation;
2. time standard separation;
3. control relation development necessary to determine the semantic system behavior correctness.

The *second stage* is the algorithm development of the obtaining *semantic cyber system behavior correctness standards*. Its main purpose is to obtain the system behavior probabilistic algorithms of standards or similarity invariants in a matrix and a graphical form.

The step procedure includes the following steps:

1. construction of the standard algorithm in the tree form;
2. algorithm implementations listing;
3. weighting of algorithm implementations (a probabilistic algorithm construction);
4. algorithm tree rationing.

The *third stage* is the *standard synthesis* of the semantic cyber system behavior correctness, adequate to the application goals and objectives. Its main goal is to synthesize algorithmic structures formed by a set of sequentially executed standard algorithms.

This procedure is carried out in the following steps:

1. structural and functional standard synthesis;
2. time standard synthesis;
3. symmetrization and ranking of matrices describing standards.

The *fourth stage* is the *simulation of the stochastically* defined algorithmic structures of the semantic cyber system behavior correctness standards. The step procedure includes the following steps:

1. analysis of the empirical semantic correctness;
2. determining the type of the empirical functional dependence;
3. control ratio development sufficient to determine the semantic system behavior correctness and to ensure the required cyber resilience.

As a result, the dimensions and similarity theory method applicability to decompose *Industry 4.0 cyber systems* behavior algorithms, according to functional characteristics and the *necessary invariants formation of semantically correct systems operation*, was shown. The *self-similarity property* presence of *similarity invariants* allowed forming static and dynamic standards of the semantically correct system behavior and uses them for engineering problems solution of *control, detection, and neutralization of intruder information and technical influences.*

1.5.2 Cyber Transformation Trends

Currently, the technologically advanced world companies are implementing the most extensive technological transformation [168, 169, 170, 171, 172], the key goals and objectives of which are to scale up business and boost profitability and efficiency while increasing flexibility, speed, and customer focus based on the *Industry 4.0* introduction. However, the mentioned digital transformation brought new threats to cybersecurity (Figures 1.61–1.64) [173, 174, 175].

Let us take a brief look at the main technological trends of the above digital transformation on the example of one of the most technologically advanced banking industries. At the same time, we will also show what impact (Figure 1.62) the digital transformation has on the creation, support, and development of appropriate enterprise cyber resilience management programs [176, 177].

Social computing:
Historically, the "*social computing*" was first to form a global trend, which refers to a whole set of special software and hardware solutions, that

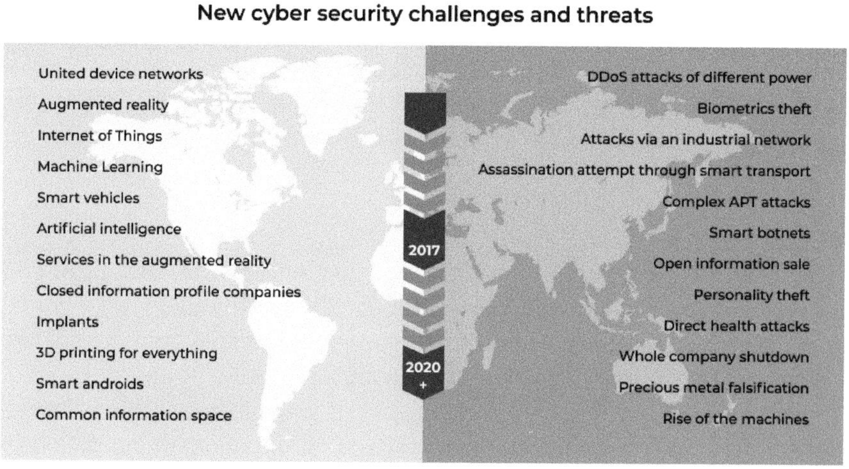

New cyber security challenges and threats

United device networks	DDoS attacks of different power
Augmented reality	Biometrics theft
Internet of Things	Attacks via an industrial network
Machine Learning	Assassination attempt through smart transport
Smart vehicles	Complex APT attacks
Artificial intelligence	Smart botnets
Services in the augmented reality	Open information sale
Closed information profile companies	Personality theft
Implants	Direct health attacks
3D printing for everything	Whole company shutdown
Smart androids	Precious metal falsification
Common information space	Rise of the machines

Figure 1.61 New cybersecurity challenges and threats.

Barriers	Business Ownership
•Lack of investment - 60% •Inability to hire skills - 56% •Lack of Visibility into assets - 46% •Lack of end user training - 31% •Lack of training for IT staff -28% •Silo and Turf issues - 24% •Lack of governance practices - 22% •Lack of Board reporting - 17% •Lack of C Level Buy in - 15%	•CIO - 23% •BU Leader - 22% •CISO - 141% •NO ONE PERSON - 11% •BC Manager - 8% •CRO - 7% •CEO - 7% •CTO - 6%

Figure 1.62   Who is most interested in a cyber resilience?

promote a social behavior online [178]. An example of this is the various social networks and instant messengers. Other forms of social activity are also known, for example, the *Branchout application* for *Facebook*, which is another communication opportunity in the professional sphere and is already competing with *LinkedIn*. In other words, today, there are all possibilities to behave online just like in the real world.

Banks use social media not only to communicate with customers, for example, answering user questions on *Facebook*, but also to integrate with customer service and contact centers, and, of course, can use them to solve the cyber resilience problems. Here the social media is one of the possible and supported communication channels. How will the social networks and instant messengers affect business sustainability as a whole? Banks are actively seeking an answer to this question and conducting a large number of experiments with relevant business models based on "social computing."

Today, the social networks are becoming a source of an important information about potential and current threats to cybersecurity. A "social scoring" is developing, which is only at the very beginning of the journey. Many issues should be resolved here, including those related to sphere regulation [179].

Big Data:
The second global trend is the *Big Data* – the technologies that allow collecting, storing, and analyzing huge amounts of the structured and unstructured information [180]. They already have an industrial quality standard in a number of banks; however, most companies are still on the verge of the *Big Data revolution*. Here, it is important to explore possible solutions for solving cyber resilience problems based on the in-depth big data analytics as well as services based on such analytics. This is important to form

a *"better model"* of the cyber resilience management in the context of an unprecedented increase in cybersecurity threats.

It is clear that in a couple of years in the banking industry, they will stop talking about Big Data as something special, just as the need to use personal computers in the business is no longer discussed today. The *Big data* use in business will become obvious. Companies will learn how to collect a huge amount of data, analyze them, and conduct business with the required resilience, based on this knowledge. At the same time, those companies that do not understand how to use *Big Data* will lose their position in the business. Until now, in a number of companies, there is no proper understanding of the need to collect and process *Big Data* to solve cybersecurity and cyber resilience problems; data volumes can be destroyed every day. It remains a prejudice that their storage is quite expensive. That is not the way it seems to be: the cost of the *Big Data storage* is rapidly falling [181]. Currently, all the necessary technologies for collecting, storing, and processing *Big Data* are known. There are many interesting startups in this area, for example, *Cloudera*. Companies that have not found out, yet, how to work with Big Data should first start thinking in the *Big Data* categories. Often, managers begin to choose which information to store and analyze. They talk about *return on investment (ROI)* in relation to *Big Data*, forgetting that it is impossible to predict *ROI* in advance. Initially, it is required to collect all the available data and analyze them and then build new services that will bring a new profit. For example, *Google* indexes the entire Internet, without selecting only *"interesting"* sites for this procedure. It is also necessary to deal with large data to ensure the cyber resilience [182]. If a transaction has occurred, it should be recorded and saved with all the logs, which can help to understand its place in the information array and relationships with other elements. This will help, first, to make better business decisions and, second, to better provide the required cyber resilience. For example, by collecting the information on cybersecurity incidents, banks can protect their customers much better (Figure 1.63).

Mobility:
The third trend is mobility, a global reach of mobile communications. *"A person can forget a wallet at home, but not his smartphone"* is the principle of the 21st century, when a small device replaced about a dozen devices: a phone, a clock, a player, a radio, cards, a camera, a video camera, and more [183]. Users are increasingly using various types of mobile devices. Today, banks offer a variety of mobile banking applications. Initially, a fairly simple solution had appeared, when the person could enter the mobile bank

Addressing your most challenging IT initiatives

With technology breadth and depth

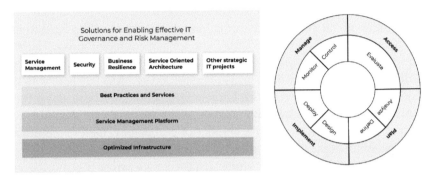

CIO Note: Establishing an enterprise wide architecture initiative is an important project for enabling better IT governance and risk mitigation.

Figure 1.63   Approach to ensure cyber resilience, IBM.

through a browser or through a mobile application. Then the developers learned how to consider various platforms, for example, in *BackBase* solution.

Then this process began to grow exponentially. Today, the mobile payments are a whole ecosystem, two elements of which can seriously change the mobile devices. The first is the card issue. Instead of having the account details and the account number on the chip in a plastic piece, one can have them on the chip in a phone. The second element is receiving payments. There is software that is installed on the phone, and there is "hardware," which acts as a reader. If the phone is equipped with a pre-installed chip, then it remains only to install the appropriate application, and each phone thus turns into a *POS*-terminal.

There are startups that want to link even more functions to the cards and make them even more expensive, but all the hard drive is already in the phone; so software is the best way to go. Thus, the *Canadian company Mobeewave* has developed an application that turns any *NFC*-enabled smartphone into a terminal to pay for purchases using contactless plastic cards like *MasterCard PayPass or VisapayWave*. The *Mobeewave* development relevance is explained by the insufficient mobile payment acceptance level. In the *USA* alone, the ratio of contactless plastic cards and corresponding terminals is *600 to 1*. According to the service, there are more than *8.7 million* payment terminals for contactless plastic cards in the world today, and the number of corresponding cards is more than *4.2 billion*. The *NFC technology* is supported by more than *300 types* of modern smartphones and tablets and about 100 device types already out of production. Thus, the *Mobeewave* development allowed *"blowing up"* the mobile payment market by a simple application. It did not require any additional equipment and did not have any complicated logistics; it was enough to download the application, subscribe to the services, and immediately start accepting payments. At the same time, *NFC* is just another contactless payment technology, another protocol that was required to preserve the existing payment infrastructure to the maximum. Payments have become more convenient and faster because plastic cards with integrated *NFC* chips, unlike previous technologies, did not require pulling the card through the card reader. It also became possible to emulate a card in the phone and use it in the future transactions. However, the mobile phone opens up other possibilities. For example, if the telephone numbers of the buyer and the seller are known, then it becomes possible to transfer money from one account to another, using the phone number as an identifier. Perhaps, this operation will take a little more time than usual, but it does not need any infrastructure. Therefore, the future belongs to the technologies in which cash is not involved at all, which allow transferring electronic money from the sender's account to the card or to the recipient's mobile application. This will be a real "mobile wallet." In addition, the phone allows seeing the balance and account statement, which is impossible in case of a credit card. Experiments are continuing now with coupons and special offers, the purpose of which is to obtain confirmation that the consumer will want to receive special offers from the establishments he visits on his mobile phone. Thus, the new possibilities of using mobile devices are opening up, a search for suitable business use models is being carried out, and appropriate experiments are being conducted. It is expected that *Mobility* will increasingly influence our lives and behavior, including solving the problems of ensuring business sustainability and the information infrastructure cyber resilience.

Artificial intelligence:
The fourth trend is the active AI implementation in all places [184]. This is especially noticeable in the analysis of customer data accumulated and obtained by banks online. AI makes decisions about issuing loans, gives advises to clients on what is more profitable for them to purchase, offers ways to cut costs and manage finances, communicates with clients, and solves their issues. It is clear that in the current state, the *bots* in various messengers are far from being perfect, but the efforts and funds invested in this direction by all players (from niche to market makers) will be

fruitful in the next 3–4 years, especially since the algorithmic component is improving at a rapid pace [185].

The intelligence development in automated systems supports another clearly emerging trend that is the transition to the natural languages of customer communication with financial organizations. Banks and payment systems are actively experimenting with new mobile application forms and new interaction formats with consumers.

Many players are betting on messengers as the new generation mobile applications, as well as chatbots and marketplace. In many ways, this is a response to the challenge of the large technology companies (*Google, Facebook, Apple, Yandex,* etc.), which introduce financial services into all their products. The next step in this direction is the development of full-fledged voice-controlled systems that allow performing many operations remotely. *Google* and *Apple* are still improving their voice control systems for smartphones, but *Amazon* went further by offering *Echo*, an assistant system for everyday use at home and at work, which allows managing home, bank accounts, make transfers, order food, etc. using only voice interface.

As the instantaneous translation from one language to another is improved, the production, support, and use cost of such devices at the global level will drop significantly, language barriers will disappear, customers will be more remote from banks, and traditional banking services will finally become the category of commodity services. This process is especially specific to Europe, where, thanks to the *PSD2* initiative, customers will be free to choose any bank or provider to conduct their operations almost online, and modern banking systems will be able to understand them in any languages. This trend will gradually lead to the disappearance of traditional bank branches, as we know them – with queues, offices, cash, etc.

The AI application to solve cyber resilience problems will significantly improve the quality and validity of decisions made for the business renewal and continuation in emergency situations.

Blockchain technology:
The fifth trend is distributed computing technology, i.e., *blockchain* [186]. Today, there are several attempts to introduce the blockchain and cryptocurrency in banking. Truth be told, most of the cases did not go beyond the pilot stage. This technology is interesting as one of the prospects for developing payment systems and transfer systems in real time. Certainly, in addition to aspects of the technology and protocols implementation, it is

Cyber Resilience is a business priority that supports "continuous availability" that allows companies to meet their business outcome objectives

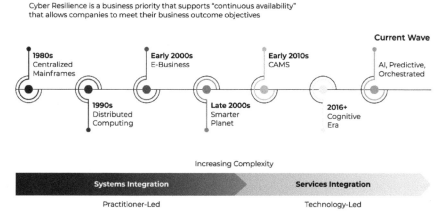

Figure 1.64 From backup and disaster recovery to intelligent cyber resilience services, IBM.

necessary to consider the extent to which technology affects the change in business processes. Any technology must live in the context of the environment in which it is going to be placed: as far as it complies with the bank security requirements, with the operation formalization, how the controller looks at it, etc. That is, even at the experiment level, the whole set of associated factors before recommending the payment infrastructure transfer to new technology. Banks began active work with blockchain in 2014. At the same time, the early prototype systems were on *Bitcoin,* then on *Ripple.* One of the Bitcoin prototypes was developed to exchange fiat currencies using crypto tokens.

Another experiment implemented on *Bitcoin* is an electronic contract for making a transaction. The seller offered the goods by signing the contract with an electronic key. The buyer deposited money in the bank and signed the transaction with his electronic key. The bank acted as an arbitrator for the payment and certified the money deposit fact with its electronic key. As a result, the money was transferred to the seller upon transaction completion, and the goods were transferred to the buyer. Moreover, all transaction stages were reflected in the *blockchain*.

Currently, the technologically advanced companies are actively interested in blockchain technology and ready to apply it everywhere as the unified accounting registers, reporting, voting systems, polls, notarial documents, goods clearance (for example, at the level of the customs documents), government funding, and multilateral services. However, there are few real examples confirming the technology feasibility, but there are a lot more discussions about its application prospects. Working groups are being created everywhere to study blockchain technology, including solving the cyber resilience problems, and lawmakers are preparing its legal base.

Robotics technology:
The sixth trend is robotics, which, according to *Karl Frey* and *Michael Osbourne* of *Oxford University*, means *"a system or tasks automation of such a level when the human labor need disappears and it is replaced with its automated version"* [187]. The robotics scientific basis is a *neurobiology and bioinformatics, AI and high-performance computing, neuromorphic and cognomorphic technologies, genetic algorithms, neuro-engineering,* and other disciplines. In particular, the current focus is to design interfaces between the virtual and physical worlds (*Virtual-to-Physical/V2P or Online-to-Off line/O2O*), which are able to multiply the new possibilities created by the AI development.

According to *ISO 8373:2012*, a robot is understood as a drive mechanism, programmed in two or more axes, having a certain autonomy degree, moving inside its working environment, and performing tasks intended for it. Also, a robot can be called any device (mechanism) that performs the actions intended for it, which simultaneously meets three conditions:

- *SENSE*: perceive the world around with sensors, microphones, cameras (all areas of the electromagnetic spectrum), various electromechanical sensors, etc., can act in this role;
- *THINK:* understand the surrounding physical world and build behavioral models to perform the actions intended for it;
- *ACT:* to influence the physical world in one way or another.

There are two main areas of robotics development: industrial and service robotics.

The *industrial robotics* is growing (on average by **15%** per year) mainly due to the rapid Chinese economical robotization. The industrial robotics market growth rate is ahead of the global *GDP* growth rate: between 2011 and 2016, the average annual growth in sales of industrial robots was **12%**. In 2016, 294,000 industrial robots were sold, and the total market volume reached $13.1 billion (including software and integration services, the market exceeds **$40 billion**).

The *service robotics* automates primarily service economy processes, which is a significant part of the global economy. For this reason, the service robotics in comparison with industrial one shows an even greater growth (at the level of **25%** per year) with the relatively smaller absolute figures, in comparison with the industrial one. For example, in 2015, 48,000 professional service robots were sold, and in 2016, this number increased by **24%**, to **59,000**. The total market volume of professional service robots reached $4.7 *billion*. It is significant that trends in the robotics development in Russia repeat western. Russia has good potential in the field of service robotics: the industrial and service robotics ratio in the country is 1:10. A number of companies, for example, *ExoAtlet* and *CyberTech Labs*, have entered the world market and successfully compete with the foreign robot manufacturers. Most of all, in absolute terms, robots are sold for logistics (*~25,000 units*), military applications (*~11,000*), for commercial spaces (*~7000*), and field works and exoskeletons (*~6000 each*).

Among the priority areas of service robotics development are:

- logistic systems (include indoor logistics, unmanned and airborne delivery vehicles outdoors, etc.);
- robots for customer service;
- industrial exoskeletons;
- robots for household tasks (personal assistants).

It is expected that investments in robots for logistics can quickly pay for themselves, assuming that they will be used 24 hours a day. According to *IFR* estimates, for the United States, investments are compensated for an average of two to three years, given the 15-year robots lifespan. According to a *McKinsey* study, there is an increase (*7%–10%* in developed countries and *300%* in developing countries) mainly due to *B2C* in the field of unmanned delivery vehicles outside the premises. Today, the unmanned aerial vehicles and the autonomous platforms for the goods delivery in the city are the most widespread.

The investments relevance in the robot development for customer service is determined by the following factors:

- mass service personalization strategy;
- human resource cost;
- digitalization and competition level in the customer service market;
- machine learning and robotics application to analyze customer behavior;
- changes in personnel functions;
- well-known initiatives and results on robot promotion in the service sector, especially in Japan and China.

The main difficulties in this robot implementation are inflated customer expectations and AI capabilities exaggeration. In addition to market challenges, the industry also faces technological challenges. Most of these problems are due to the interaction between robots and humans as well as security and standards.

The main exoskeleton tasks include the human capabilities expansion in the field of defense, rescue, and in emergency situations. As a rule, a mechanized exoskeleton is understood to be some active mechanical device with pronounced anthropomorphic properties, suitable for the size of the operator who wears it and coordinated by the operator movements. The general robotics development features should be noted, that are represented in the personalization (customization) of mass service and production, giving rise to all the new organization forms of customer service and production, as well as reducing the overall robot implementation costs in the industry, which leads to an increase in profitability of the latter and lowering the entry threshold into the industry. According to Barclays Research, the average cost of work performed by the robot is 6 euros per hour. Similar work performed by humans is estimated differently in different countries and regions: in Germany – 40, in the USA – 12, in Eastern Europe – 11, and in China – 9. At the same time, the more successful organization forms differ not in the number of robots introduced but the optimal interaction scheme between a robot and a human in a maximally non-deterministic environment. Machine cloud learning makes robots suitable for executing out of the box tasks, without using the costly labor of a programmer/setup engineer team. Reducing the cost allowed small and medium businesses, who previously could not hire expensive engineers for integration and maintenance, to begin the robot implementation. According to *Barclays Research*, by 2022, the average collaborative robot cost will be less than **$20,000**.

Technologies and products are emerging in which Big Data analysis and machine learning help to improve the robot performance. For example, using Amazon *AWS Greengrass service*, a robot can learn how to perform a task based on the other robot experience in a best way. There are companies that place robots on the customer premises for free, and free and take money only for the time the robot works (*robot-as-a-service or pay-as-you-go*). By 2019, **30%** of professional service robots will work on this model (forecast made by *International Data Corporation (IDC)*).

Robotics development is influenced by such technologies as *AR/VR*. Here, *AR* can be used, for example, to see how robots will look and work indoors, as well as to set up and repair with a contextual hint to the service technician. *VR* can be used as a simulator, for example, paired with machine learning, in order to practice the robot control not in a costly physical environment but in a cheaper virtual one. Robots are already integrated with the IoT and additive technologies in the context of creating *"factories of the future."* At the same time, the cybersecurity importance cannot be overestimated when robots are introduced into life-critical and mission-critical processes.

Thus, the modern robots are some physical models of living organisms, designed to carry out a number of industrial and other operations as per a predetermined algorithm in accordance with changes in environmental parameters.

There are three robot generations:

- with program control;
- with the feedback presence;
- integral robots.

Within the framework of which there are the following types of robots: production (for example, a robot-manipulator), transport (delivery robot), accompanying (a robot-consultant), research (robot-laboratory), specialized (delivery drones, collection robot, and robotic security guards), etc. Currently, there are about **50** robot manipulator types, more than **200** unmanned aerial vehicle (drones) types, more than **80** android types, more than **1000** dual-purpose robots, etc., in the world.

Industry 4.0 has demanded from robots a number of abilities that exceed human ones, namely:

- Fast self-development and evolution through deep self-learning, semantic understanding, and generating new knowledge about the subject area based on models and methods of AI.
- Autonomous behavior and independent decision-making based on the complex integration of the macroenvironment and microenvironment factors and parameters variety.
- Group behavior support and integration into a mixed hybrid environment based on group unions of robots and humans. According to world robot statistics, the global market for robotization is more than **75 billion**. At the same time, the annual world sale growth of industrial robots amounts to **15%** and in quantitative terms up to **400,000** robots per year.

What is going to happen in 10–15 years? According to experts, there comes a new era of the so-called *cyborgization and hybridization*, as the next robotization stages. Multirobot systems will learn to make decisions and respond to changing environmental conditions, including in the under destructive intruder cyber-attacks. Man will become part of the robotic and, later, the hybrid world. It is the direction where research and development, creative team *R&D* from world technologically developed countries, primarily *Japan, the USA, South Korea, China, Germany, and Russia*, are conducted.

1.5.3   Mathematical Problem Definition

We introduce the following concepts:

- cyber system;
- cyber system behavior;
- cyber system mission;
- cyber system behavior disturbance;
- cyber system state.

These concepts are among the primary [188], undefined concepts and are used in the following sense.

Primary concepts:
A *cyber system* is understood as a certain set of hardware and software components of a critically important information infrastructure with communications on control and data between them, designed to perform the required functions.

The *cyber system behavior* is understood as some algorithm introduction and implementation for the system functioning in time. At the same time, the targeted corrective actions are allowed ensuring the system behavior cyber resilience.

The *cyber system mission* is called the mission; corrective measures are cyber disturbance detection and neutralization. In other words, a cyber system is designed for a specific purpose and may have some protective mechanism, customizable or adjustable means to ensure the cyber resilience.

A *cyber system behavior disturbance* is a single or multiple acts of an external or internal destructive impact of the internal and/or external environment on the system.

The disturbance leads to a change in the cyber system functioning parameters and prevents or makes the system purpose difficult.

A disturbance combination **forms a disturbance set.**
The *cyber system state* is a certain set of numerical parameter characteristics of the system functioning in space.

The numerical process characteristics depend on the functioning conditions of the cyber system, disturbances, and corrective actions to detect and neutralize the disturbances and, in general, from the time.

The set of all corrective actions for detecting and neutralizing disturbances is called the *corrective action set*; the set of all digital platform
behavior system states is called the *state set*.

Thus, we will assume that without disturbances, as well as the corrective measures for the disturbance detection and neutralization, the cyber
system is in an operational state and meets some intended purpose.

As a disturbance result, the cyber system transits into a new state; this
may not meet its intended purpose.

In such cases, the two main tasks appear:

1. detection of the disturbance fact and, possibly, changes made to the
   normal cyber system functioning process;
2. setting the optimal (cyber-resilient) in a certain sense (based on a
   given priority functional) organization of the cyber system behavior
   to bring the cyber system to an operating state (including redesigning
   and/or restarting the system, if this solution is considered the best).

On the basis of the introduced concepts, we will reveal the content of elementary, complex, and disturbed computationcomputations in terms of
dynamic R. E. Kalman interrelationships [189].

Disturbed machine computation:
Further, we will use the term "*elementary cyber system behavior*," considering the structure, which input receives some input value at certain points
in time and from which some output value is derived at certain points in
time. The above concept of the elementary cyber system behavior as a system Σ includes an auxiliary time point set T. At each time point $t \in T$, the
system Σ receives some input value $u(t)$ and generates some output value
$y(t)$. In this case, the input variable values are selected from some fixed set
U, i.e., at any time moment t, the symbol $u(t)$ belongs to U. The system
input value segment is a function of the form $\omega: (t1, t2) \rightarrow U$ and belongs
to some class Ω. The output variable value $y(t)$ belongs to some fixed set Y.
The output values segment represents a function of the form $\gamma: (t2, t3) \rightarrow Y$.

The *complex cyber system behavior* is understood as a generalized
structure, the components of which are elementary given system behaviors
with communications on control and data among themselves.

Now we define the concept of the *immunity history (memory)* of the
cyber system behavior to destructive influences. We assume that under
group and mass cyber-attacks, the output variable value of the system Σ
depends both on the source data and the system behavior algorithm and on
the immunity *history (memory) destructive influences*. In other words, the

disturbed cyber system behavior is a structure in which the current output variable value of the Σ system depends on the Σ system state with an accumulated immunity *history (memory) to destructive disturbances*. In this case, we will assume that the internal Σ system state set allows containing information about the Σ system *immunity history (memory)*.

Let us note that the considered content of the disturbed cyber system behavior allows describing some "*dynamic*" self-recovery behavior system of the above system under disturbances, if knowledge of the $x(t1)$ state and the restored computation segment $\omega = \omega^{(t1,t2)}$ is a necessary and sufficient condition to determine the state $x(t2) = \varphi(t2;t1,x(t1),\omega)$, where $t1 < t2$. Here, the time point set T is orderly, i.e., it defines the time direction.

Disturbances characteristics:
Let us reveal the characteristic features of single, group, and mass *Industry 4.0* cyber system disturbances using the following definitions.

> **_Definition 1:_** The dynamic self-recovery cyber system behavior system under group and mass cyber-attacks Σ is called *stationary (constant)* if and only if:
>
> (a) T is an additive group (according to the usual operation of adding real numbers);
> (b) Ω is closed according to the shift operator z^{τ}: $\omega \to \omega'$, defined by the relation: $\omega'(t) = \omega(t + \tau)$ for all $\tau, t \in T$;
> (c) $\varphi(t; \tau, x, \omega) = \varphi(t + s; \tau + s, x, z^{s\omega})$ for all $s \in T$;
> (d) the mapping $\eta(t,): X \to Y$ does not depend on t.
>
> **_Definition 2:_** A dynamic system of self-recovery cyber system behavior under group and mass cyber-attacks Σ is called a system *with continuous time* if and only if T coincides with a set of real numbers and is called a system *with discrete time* if and only if T is an *integer set*. Here, the difference between systems with continuous and discrete time is insignificant and, mainly, the mathematical convenience of the development of the appropriate behavior models of the cyber systems under group and mass disturbances determines the choice between them. The systems of self-recovery cyber system behavior under group and mass cyber-attacks with continuous time correspond to classical continuous models, and the mentioned systems with discrete time correspond to discrete behavior models. An important cyber system complexity measure in group and mass cyber-attacks is its state space structure.
>
> **_Definition 3:_** The dynamic system of cyber system behavior in group and mass cyber-attacks Σ is called *finite-dimensional* if and only if X is a

finite-dimensional linear space. Moreover, *dim* $\Sigma = dimX_{\Sigma}$. A system Σ is called *finite* if and only if the set X is *finite*. Finally, a system Σ is called a *finite automaton* if and only if all the sets X, U, and Y are *finite* and, in addition, the *system is stationary* and *with discrete time*.

The finite dimensionality assumption of the given system is essential to obtain specific numerical results.

Definition 4: A dynamic system of cyber system behavior in group and mass cyber-attacks Σ is called *linear* if and only if:

(a) spaces X, U, Ω, Y, and G are vector spaces (over a given *arbitrary field K*);
(b) mapping φ $(t; \tau, \cdot, \cdot): X \times \Omega \rightarrow X$ is *K-linear* for all t and τ;
(c) mapping η $(t, \cdot): X \rightarrow Y$ is *K-linear* for any t.

If it is necessary to use the mathematical apparatus of differential and integral calculus, it is required that some assumptions about continuity are included in the system Σ definition. For this, it is necessary to assume that the various sets *(T, X, U, Ω, Y, G)* are the topological spaces and that the mappings φ and η are continuous with respect to the corresponding *(Tikhonov) topology*.

Definition 5. The dynamic system of cyber system behavior in group and mass cyber-attacks Σ is called *smooth* if and only if:

(a) $T = R$ is a set of real numbers (with the usual topology);
(b) X and Ω are topological spaces;
(c) transition mapping φ has the property that $(\tau, x, \omega) \mapsto \varphi(\cdot; \tau, x, \omega)$ defines a continuous mapping $T \times X \times \Omega \mapsto C^1(T \rightarrow X)$.

For any given initial state (τ, x) and an input action segment $\omega^{(\tau,t_3]}$ of system Σ, the system $\gamma^{(\tau,t_3]}$ reaction is specified, i.e., the mapping is given: $f_{\tau, x} : \omega^{(\tau,t_3]} \rightarrow \gamma^{(\tau,t_3]}$.

Here, the output variable value at time $t \in (\tau, t_1]$ is determined from the relation: $f_{\tau, x}(\omega^{(\tau,t_3]})(t) = \eta(t, \varphi(t; \tau, x, \omega))$.

Definition 6. The *dynamic system of cyber system behavior under group and mass cyber-attacks* Σ (in terms of its external behavior) is the following mathematical concept:

(a) Sets that satisfy the properties discussed above are given.
(b) A set that indexes a function family: $F = \{f_{\alpha}: T \times \Omega \rightarrow Y, \alpha \in A\}$, is defined, where each family F element is written explicitly as $f_{\alpha}(t, \omega) = y(t)$, i.e., it is the output value for the input effect ω obtained

in *experiment* α. Each f_α is called an input–output mapping and has the following properties:

1. (*The time direction*) There is a mapping $\iota: A \to T$, then $f_\alpha(t, \omega)$ such that $f_\alpha(t, \omega)$ is defined for all $t \geq \iota(\alpha)$.
2. (*Causality*) Let $t \in T$ and $\tau < t$. If $\omega, \omega' \in \Omega$, and $\omega_{(\tau,\,t]} = \omega'_{(\tau,\,t]}$, then $f_\alpha(t, \omega) = f\alpha(t, \omega')$, for all α for which $\tau = \iota(\alpha)$.

Cyber resilience hypervisor model:
Let us define a hypervisor model (an abstract converter) of the cyber system behavior under the group and mass cyber-attacks as follows.

Definition 7. The *abstract mapping of the cyber system behavior under group and mass cyber-attacks* Σ is a complex mathematical concept defined by the following axioms:

(a) T time points set, X computation states set, the instantaneous values set of U input variables, $\Omega = \{\omega: T \to U\}$ set of acceptable input variables, the instantaneous value set of output variables Y and $G = \{\gamma: T \to Y\}$ set of acceptable output values are given.

(b) (*Time direction*) set Y is some ordered subset of the real number set.

(c) The input variable set Ω satisfies the following conditions:

1. (*Non-trivial*) The set Ω is not empty.
2. (*Input variable articulation*) Let us call the segment of input action $\omega = \omega^{(t_1,t_2]}$ for $\omega \in \Omega$, the restriction ω to $(t_1, t_2] \cap T$. Then if $\omega, \omega' \in \Omega$ and $t_1 < t_2 < t_3$, then there is $\omega'' \in \Omega$, that $\omega''^{(t_1,t_2]} = \omega^{(t_1,t_2]}$ and $\omega''^{(t_2,t_3]} = \omega'^{(t_{21},t_3]}$.

(d) There is a *state transition function* $\varphi: T \times T \times X \times \Omega \to X$, the values of which are the states $x(t) = \varphi(t; \tau, x, \omega) \in X$, in which the system turns out to be at time $\tau \in T$ if at the initial time $\tau \in T$ it was in the initial state $x = x(\tau) \in X$ and if its input received the input value $\omega \in \Omega$. The function φ has the following properties:

1. (*Time direction*) The function φ is defined for all $t \geq \tau$ and is not necessarily defined for all $t < \tau$.
2. (*Consistency*) The equality $\varphi(t; t, x, \omega) = x$ holds for any $t \in T$, any $x \in X$, and any $\omega \in \Omega$
3. (*Semigroup property*) For any $t_1 < t_2 < t_3$ and any $x \in X$ and $\omega \in \Omega$, we have $\varphi(t_3; t_1, x, \omega) = \varphi(t_3; t_2, \varphi(t_2; t_1, x, \omega), \omega)$.
4. (*Causality*) If $\omega, \omega'' \in \Omega\Omega$ and $\omega_{(\tau,t]} = \omega'_{(\tau,t]}$, then $\varphi(t; \tau, x, \omega) = \varphi(t; \tau, x, \omega')$.

(e) The output mapping η: $T \times X \rightarrow Y$ is given, which defines the output values $y(t) = \eta(t, x(t))$. The mapping $(\tau, t] \rightarrow Y$, defined by the relation $\sigma \mapsto \eta\ (\sigma, \varphi(\sigma; \tau, x, \omega)), \sigma \in (\tau, t])$, is called an input variable segment, i.e., the restriction $\gamma_{(\tau, t]}$ of some $\gamma \in G$ on $(\tau, t]$.

Additionally, the pair (τ, x), where $\tau \in T$ and $x \in X$, is called the event (or phase) of the system Σ, and the set $T \in X$ is called the system Σ event space (or phase space). The transition function of the states φ (or its graph in the event space) is called a trajectory or a solution curve. Here, the input action, or control ω, transfers, translates, changes, and converts the state x (or the event (τ, x)) to the state φ $(t; \tau, x, \omega)$ (or the event $(t, \varphi\ (t; \tau, x, \omega))$). The cyber system behavior motion is understood as the function of states φ.

Definition 8. In a more general form, the *abstract converter model of the cyber system behavior under disturbances* ℜ with discrete time, *m* inputs, and *p* outputs over the field of integers *K* is a *complex object* (ℵ, ℘, ◊), where the mappings ℵ:*l→l*, ℘:*Kᵐ→l*, ◊:*l→Kᵖ* are core abstract *K –homomorphisms*, *l* is some *abstract vector space* that is above *K*. The *space dimension l(dim I)* determines the *system dimension* ℜ*(dim*ℜ*)*.

It is significant that the chosen representation allows formulating and proving statements confirming the fundamental existence of the desired solution.

Cyber resilience control:
Based on the given definitions, let us reveal the ideology essence of the cyber system behavior with a memory for forming immunity to destructive group and mass disturbances as follows.

Definition 9. The *cyber system behavior with memory* is called the complex mathematical concept of the dynamical system Σ, defined by the following axioms:

(a) A time point set *T*, a set of computational states *X* under intruder cyber-attacks, an instantaneous value set of standard and destructive input actions *U*, a set of acceptable input effects $\Omega = \{\omega: T \rightarrow U\}$, an instantaneous value set of output values *Y*, and a set of output values of the reconstructed computationcomputations $G = \{\gamma: T \rightarrow Y\}$.
(b) (*Time direction*) set *Y* is some ordered subset of the real number set.
(c) The set of acceptable input actions Ω satisfies the following conditions:
 1. (*Non-trivial*) The set Ω is not empty.

2. (*Input variable articulation*) Let us call the segment of input action $\omega = \omega^{(t_1, t_2]}$ for $\omega \in \Omega$ the restriction of ω on $(t_1, t_2] \cap T$. Then if ω, $\omega' \in \Omega$ and $t_1 < t_2 < t_3$, then there is $\omega'' \in \Omega$, that $\omega''^{(t_1, t_2]} = \omega^{(t_1, t_2]}$ and $\omega''^{(t_2, t_3]} = \omega'^{(t_2, t_3]}$.

(d) There is a *state transition function* φ: $T \times T \times X \times \Omega \to X$, the values of which are the states $x(t) = \varphi(t; \tau, x, \omega) \in X$, in which the system is at time $t \in T$, if at the initial time $\tau \in T$ it was in the initial state $x = x(\tau) \in X$ and if it was influenced by the input action $\omega \in \Omega$. The function φ has the following properties:

1. (*Time direction*) The function φ is defined for all $t \geq \tau$ and is not necessarily defined for all $t < \tau^4$.
2. (*Consistency*) The equality $\varphi(t; t, x, \omega) = x$ holds for any $t \in T$, any $x \in X$, and any $\omega \in \Omega$.
3. (*Semigroup property*) For any $t_1 < t_2 < t_3$ and any $x \in X$ and $\omega \in \Omega$, we have $\varphi(t_3; t_1, x, \omega) = \varphi(t_3; t_2, \varphi(t_2; t_1, x, \omega), \omega)$.
4. (*Causality*) If ω, $\omega'' \in \Omega$ and $\omega_{(\tau, t]} = \omega'_{(\tau, t]}$, then $\varphi(t; \tau, x, \omega) = \varphi(t; \tau, x, \omega')$.

(e) An output mapping η: $T \times X \to Y$ is specified, which defines the output values $y(t) = \eta(t, x(t))$ as a self-recovery result. The mapping $(\tau, t] \to Y$, defined by the relation $\sigma \mapsto \eta(\sigma, \varphi(\sigma; \tau, x, \omega))$, $\sigma \in (\tau, t])$, is called a segment of the input variable, i.e., the restriction $\gamma_{(\tau, t]}$ of some $\gamma \in G$ on $(\tau, t]$.

Additionally, we introduce the following terms. A pair (τ, x), where $\tau \in T$ and $x \in X$, is called the system Σ *event*, and the set $T \in X$ is called the system Σ event space (or *phase space*). The transition function of states φ (or its graph in the event space) is called the *trajectory of the cyber system self-recovery behavior*. We assume that the input action, or the self-recovery control ω, transforms the state x (or the event (τ, x)) into the state $\varphi (t; \tau, x, \omega)$ or in the event.

The above concept definition of the cyber system self-recovery behavior is still quite general and is caused by the need to develop common terminology, explore, and clarify basic concepts. Further definition specification is presented below.

Behavior simulation in disturbances:

Imagine the cyber system behavior under the disturbances by the vector field in the phase space. Here the phase space point defines the above system state. The vector attached at this point indicates the system state change rate. The points at which this vector is zero reflect equilibrium states, i.e., at these points, the system state does not change in time. The

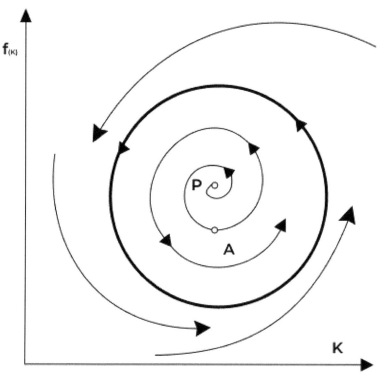

Figure 1.65 Cyber system phase behavior.

steady-state modes are represented by a closed curve, the so-called limit cycle on the phase plane (Figure 1.65).

Earlier *V. I. Arnold* [190] showed that only two main options of restructuring the phase portrait on the plane are possible (Figure 1.66).

1. When a parameter is changed from an equilibrium position, a limit cycle is born. *Equilibrium stability* goes to the *cycle*; the very same equilibrium becomes unstable.
2. In the equilibrium position, an unstable limit cycle dies; the equilibrium position attraction domain decreases to zero with it, after which the cycle disappears, and its instability is transferred to the equilibrium state.

The catastrophe theory begins with the works of *R. Tom and V. I. Arnold* [191] and allows analyzing jump transitions, discontinuities, and sudden qualitative changes in the cyber system behavior in response to a smooth change in external conditions that have some common features. It uses the *"bifurcation"* concept, which is defined as forking and is used in a

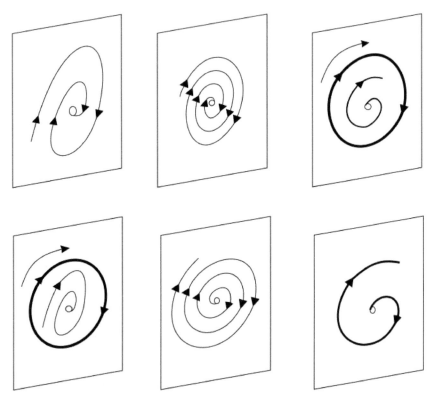

Figure 1.66 Cycle generation bifurcation.

broad sense to denote possible changes in the system functioning when the parameters on which they depend change. A *bifurcation* set is a boundary separating the space domains of control parameters with a qualitatively different system behavior under study.

In order to study the jump transitions in the cyber system behavior, we study the critical points $u \in R^n$ of smooth real functions $f{:}R^n{\to}R$, where the derivative vanishes: $\partial f/\partial_{xi}|_u = 0$, $I = 1$, n. The importance of such a study is explained by the following statement: if some system properties are described by a function f that has the potential energy meaning, then of all possible displacements, there will be real ones for which f has a *minimum* (*the Lagrange fundamental theorem* says that the *minimum of the full potential system energy is sufficient for stability*).

The most common types of critical points for a smooth function are local maxima, minima, and inflexion points (Figure 1.67).

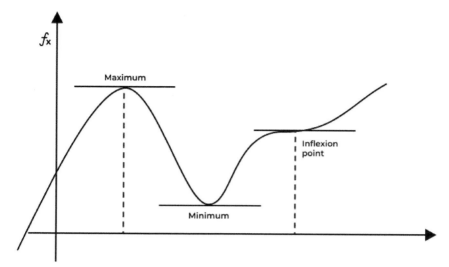

Figure 1.67   Critical points representation when $n = 1$.

In general case, in the *catastrophe theory* (Figure 1.68), the following technique is applied to study the cyber system features: first, the function f is decomposed into a *Taylor series* and then it is required to find a segment of this series that adequately describes the system properties near the critical point for a given number of control parameters. The computation-computations are carried out by correctly neglecting some *Taylor series* members and leaving others that are the *"most important."*

Rene Tom, in his works, pointed out the importance of the *structural stability* requirements or insensitivity to small disturbances. The *"structural stability"* concept was first introduced into the differential equation theory by *A. A. Andronov* and *L. S. Pontryagin* in 1937 under the name *"system robustness."*

A function f is considered structurally stable if for all sufficiently small smooth functions p, the critical points f and $(f + p)$ are of the same type. For example, for the function $f(x) = x^2$ and $p = 2\varepsilon x$, where ε is a small constant, the disturbed function takes the form: $f(x) = x^2 + 2\varepsilon x = (x + \varepsilon)^2 - \varepsilon^2$, i.e., the critical point has shifted (the shift magnitude depends on ε) but has not changed its type.

In the work of *V. A. Ostreykovsky* [192], it is shown that the higher the degree of n, the worse x^n behaves: a disturbance $f(x) = x^5$ can lead to four critical points (*two maxima and two minima*), and this does not depend on how small the disturbance is (Figure 1.69).

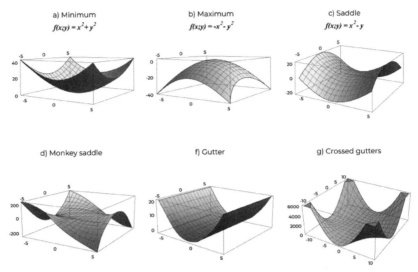

a) Minimum
$f(x;y) = x^2 + y^2$

b) Maximum
$f(x;y) = -x^2 - y^2$

c) Saddle
$f(x;y) = x^2 - y$

d) Monkey saddle

f) Gutter

g) Crossed gutters

Figure 1.68   Critical points representation when $n = 2$.

As a result, the catastrophe theory allows studying the *Industry 4.0* cyber system behavior dynamics under disturbances, like the disturbance simulation in living nature. In particular, to put forward and prove the *hypothesis* that under mass disturbances, the cyber system is in stable equilibrium if the potential function has a strict local minimum.

If certain values of these factors are exceeded, the cyber system will smoothly change its state if the critical point is not degenerate.

When a load increses, the critical point will first degenerate and then, being structurally unstable, will be separated into non-degenerate or disappear. At the same time, the cyber system behavior program will jump into a new state (abrupt stability, destruction, critical changes in structure and behavior, etc.).

The cyber resilience control system image:
In order to design a cyber resilience control system, we use the theory of multilevel hierarchical systems *(M. Mesarovic, D. Mako, and I. Takahara)* [193]. In this case, we will distinguish the following hierarchy types: *"echelon," "layer,"* and *"stratum"* (Figure 1.70).

Here, the main strata are as follows:

- *Stratum 1* is a monitoring of group and mass cyber-attacks and an immunity accumulation: the intruder simulation in the exposure types; modeling of the disturbance dynamics representation and the

A) $f(x) = x^3$ function behavior under disturbance

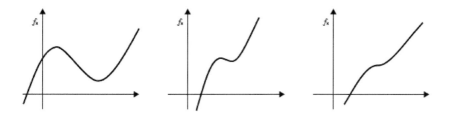

B) $f(x) = x^4$ function behavior under disturbance

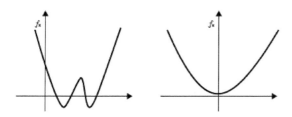

C) $f(x) = x^5$ function behavior under disturbance

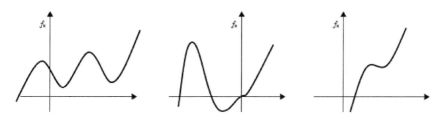

Figure 1.69 Function behavior under disturbance.

scenario definition to return the cyber system behavior to the equilibrium (stable) state; *macro model (program)* development of the system self-recovery under disturbances (*E*).

- *Stratum 2* is a development and verification of the cyber system self-recovery program at the micro level: development of the *micromodel (program)* of the system self-recovery under disturbances; modeling by means of denotational, axiomatic, and operational semantics to prove the partial correctness of the system recovery plans (*D*).

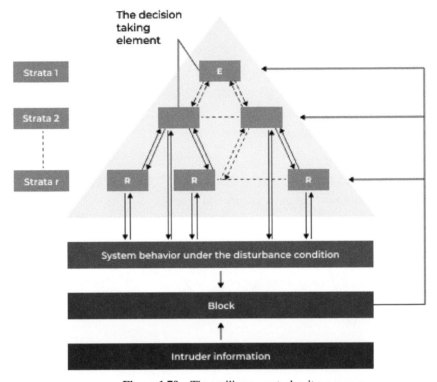

Figure 1.70   The resilience control unit.

- *Stratum 3* is a *self-recovery* of the disturbed cyber system behavior when solving target problems at the micro level: output of operational standards for recovery; model development for their presentation; recovery plan development and execution. Here, (*R*) corresponds to the hierarchy levels of the given organization system.

Let us note that a certain step of some micro- and macro-program self-usable translator (or intellectual controller, or hypervisor) to recover the cyber system behavior under disturbances is consistently implemented here.

A possible algorithm fragment of the named system recovery is shown in Figure 1.71.

Here, $S^k = (S^k_1, S^k_2, ..., S^k_p; t)$ is a state *vector of the cyber system behavior*; $Z(t) = (z_1, z_2, ..., z_m; t)$ are the *parameters of the intruder actions*; $X(t) = (x_1, x_2, ..., x_n; t)$ are the *controlled parameters*; $V(R,C)$ are the *control actions*, where *R* is a set of *accumulated immunities to exposure*; *C* is a variety of *cyber behavior purposes*.

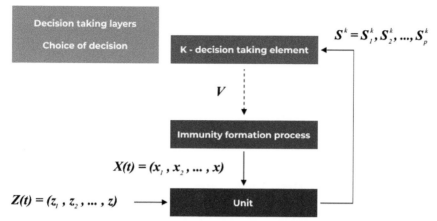

Figure 1.71 Cyber system self-recovery algorithm fragment.

The decision on the cyber system behavior self-recovery under disturbances is made based on the information (S) on the system state, the immunity presence to disturbances R, and considering the system functioning purposes C. The indicators S are formed based on the parameters X, which are input, intermediate, and output data. The attacker influence parameters Z are understood as values that are weakly dependent (not dependent) on the system ensuring the required cyber resilience.

Intermediate research results:

- The *Industry 4.0* cyber system behavior analysis under growing threats to cybersecurity makes it possible to present the above systems as a *dynamic system*, provided that knowledge of the previous system state and the recovered system operation segment is a necessary and sufficient condition to determine the next observed state. It also implies that the time point set is ordered, i.e., it defines the time direction.
- The *selected abstract translator representation* of the cyber system behavior with a memory based on the identified dynamic interrelations allows formulating and proving statements confirming the fundamental solution existence to self-recovery programs of the *Industry 4.0* cyber systems behavior under group and mass perturbations.
- The analysis shows the possibility of the *catastrophe theory* application to analyze the *Industry 4.0* cyber system behavior dynamics under disturbances by analogy with the disturbance simulation in wildlife. It is shown that under mass disturbances, the cyber system

is in stable equilibrium if the potential function has a strict local minimum. If certain values of these factors are exceeded, the system will smoothly change its state if the critical point is not degenerate. With a certain increase in the load, the critical point will first degenerate and then, being structurally unstable, will decay into non-degenerate or disappear. At the same time, the observed cyber system will abruptly move into a new state (*loss of cyber resilience, destruction, critical changes in structure and behavior, irreversible critical state*, etc.).

- The level and hierarchy analysis of the cyber-resilient memory control system made it possible to identify the following strata: *monitoring of group and mass cyber-attacks and immunity accumulation; self-recovery program development and verification of the disturbed system behavior; recovery, which achieves cyber system self-recovery when solving the target problems.*

2

BCM Best Practice

Currently, a new generation of standards in *business continuity management has appeared in different countries.* These documents present best practices for *business continuity management* and *disaster recovery management* of organizations' infrastructure in emergency situations. Here is an incomplete list of them:

- ISO 22300 family standards prepared (and revised) by the ISO/TC 292 – *Security and resilience* technical committee of the International Standardization Organization (ISO)[34]. ISO 22300:2018 – Security and resilience – Vocabulary, ISO 22301:2019 – Security and resilience – Business continuity management systems – requirements, ISO 22313:2020 – Security and resilience – Business continuity management systems – Guidance on the use of ISO 22301, ISO/TS 22317:2015 – Societal security – Business continuity management systems – Guidelines for business impact analysis (BIA), ISO/TS 22318:2015 – Societal security – Business continuity management systems – Guidelines for supply chain continuity, ISO 22320:2018 – Security and resilience – Emergency management – Guidelines for incident management, ISO/TS 22330:2018 – Security and resilience – Business continuity management systems – Guidelines for people aspects of business continuity, ISO/TS 22331:2018 – Security and resilience – Business continuity management systems – Guidelines for business continuity strategy, etc.
- Knowledge base (*The Good Practice Guidelines (GPG) 2018 Edition) Business Continuity Institute (BCI)*[35], *The Professional Practices for Business Continuity Management 2017 Edition) Disaster Recovery Institute International (DRI)*[36] *(SANS)*[37].

---

[34] https://www.iso.org/committee/5259148.html
[35] www.thebci.org
[36] www.drii.org
[37] www.sans.org

- U.S. standard NFPA 1600 – *Standard on Continuity, Emergency and Crisis Management, Edition 2019,* ASIS ORM.1-2017 – *Security and Resilience in Organizations and Their Supply Chains* (ANSI/ ASIS ORM.1-2017; revision, consolidation, redesignation of ASIS SPC.1-2009 and ASIS/BSI BCM.01-2010) Section IX – U.S. PL 110-53 – *Voluntary certification against yet to be announced standards* (2007), NIST SP800-34 – *Contingency planning guide for information technology, NYSE Rule 446 – Business continuity and contingency (SR-NYSE-2002-35), Canada* standard CSA Z1600-2017 – *Emergency and Continuity Management program.*
- British standards *BS 25999-1: 2006 – Business continuity management – Code of practice* (replaced by *ISO 22313:2020*), *BS 25999-2:200 – Business continuity management – Specification* (replaced by *ISO 22301:2019*), specifications *PAS 200:2011 – Crisis management – Guidance and good practice, PAS 56:2003 – Guide to Business Continuity Management, PASManagement,* PAS *77:2006 – IT Service Continuity Management. Code of Practice, etc.*
- Libraries *COBIT 2019, ITIL V4, RESILIA 2015, MOF* 4.0 in business continuity, *ISO/IEC 27001:2013 (A. 17) – Information technology – Security techniques – Code of practice for information security management.*
- Standards of Australia and New Zealand *AS/NZS 5050:2010 – Business continuity – Managing disruption-related risk, HB 292:2006 – a practitioner's guide to business continuity management, HB 221:2004 – Business continuity management, etc.*
- *SS 540:2008 – Singapore standard for business continuity management (BCM)* (TR19:2005 – *Business continuity management (BCM) & technical reference).*
- Standard of Japan – *Business continuity guidelines (2011),* Bank of Japan documents – *Sound practices on business continuity management of financial institutions in preparation for disruption of operational sites (2002), Business continuity planning at financial institutions (2003), Results of a questionnaire study on business continuity management (2003).*
- Israel standard SI 24001:2007 – *Security & continuity management systems – Requirements and guidance for use of the standards institution of Israel (SII).*
- Document of the Basel Committee on banking supervision – *High Level Principles for Business Continuity (2006),* Section 8.11 of the Bank of Russia *standard*-STO BR IBBS-1.0-2008 – *Ensuring*

information security of organizations of the banking system of the Russian Federation. General provisions, Paragraph 3.7 of the Bank of Russia Regulations of December 16, 2003 N 242-P *"on the organization of internal control in credit organizations and banking groups,"* Instruction No. 2194-U *dated March 5, 2009 "on amendments to the Regulations of the Bank of Russia" dated December 16, 2003 N 242-P,* etc.

It is noteworthy that most of them were created based on the analysis and generalization of the best ways to manage business continuity, tested by various specialists in different countries in practice. The listed standards are *de facto* standards and are of a recommendatory nature, instead of mandatory *de jure* standards. However, the standards are not the results of scientific research but based on real practice. However, this does not diminish their practical significance. The fact is that creating the enterprise standards for business continuity management, in practice, turns out to be more expensive and less successful. In addition, creating own closed standards can cause problems with the compatibility of rules and regulations in working with partners and clients. Therefore, the knowledge and ability to use *de facto* standards in practice is undoubtedly useful for most organizations.

According to the recommendations of the mentioned *BCM* standards, the development of a proper business continuity management program, *enterprise continuity program*, may look like this. The first two steps are devoted to identifying the goals and objectives of the organization's business, which are projected into the area of business continuity management. In this period, the design team considers business requirements as following the planned development strategy, profit, the preservation of the positive image and reputation, regulatory compliance, etc., based on business requirements and requirements for continuity and availability of *IT services* (for example, best practices *COBIT 2019 and RESILIA 2015*). In the third step, all business-critical *IT services* are identified and ranked (using the recommendations of the *ITIL V4* library). Risk assessment, *RA* and business impact analysis, *BIA* are performed. At the same time, owners of critical business processes and supporting *IT services* are involved in obtaining objective assessments. In the fifth step, each *IT service* is examined in detail to determine the assets necessary for the services to function. The goal of the sixth step is to clarify the list of threats and possible risks of interruption for each *IT service*. At the same time, certain preventive and remedial measures are reasonably selected for each identified risk, including detailed controls (for example, in accordance with the recommendations of *COBIT 2019 and RESILIA 2015*). The seventh step is to determine the required recovery time-*RTO* and *RPO* return checkpoints for *IT services*. At the eighth step, disaster recovery measures are specified for each *IT service* (based on the recommendations of *COBIT 2019* and *ISO 22301:2019*). The ninth step is to develop a strategy first, and then appropriate business continuity plans and procedures. The tenth step is to develop the testing plans and schedules. The results of this step are used to keep business continuity plans and procedures up-to-date and to review and update these documents in a timely manner. The eleventh step is to support the overall business continuity management program, including enhancing a culture of *BCM*, the launch of appropriate training programs, and awareness-raising.

2.1   The International ISO 22301:2019 Standard

ISO 22301:2019 *"Security and resilience – Business continuity management systems – Requirements"* (hereinafter referred to as ISO 22301) was prepared to replace ISO 22301:2012 *"Societal security – Business continuity management systems – Requirements"* (which, in turn, by 2014, replaced part 2 of the well-known British standard – *Part 2, "BS 25999-2: 2007-Standard Specification for Business Continuity Management"*) [194, 195, 196, 197, 198, 199, 200].

The ISO 22301:2019 standard is intended for certification of *business continuity management systems* (BCMS) organizations (Figure 2.1) operating internationally. Here, BCMS refers to a system that consists of the following key components:

  a. policy;
  b. competent people with defined responsibilities;
  c. management processes relating to:
    • policy;
    • planning;
    • implementation and operation;
    • performance assessment;
    • management review;
    • continual improvement;
  d. documented information supporting operational control and enabling performance evaluation.

Let us look at the history of the appearance and development of the ISO 22301 standard in more detail.

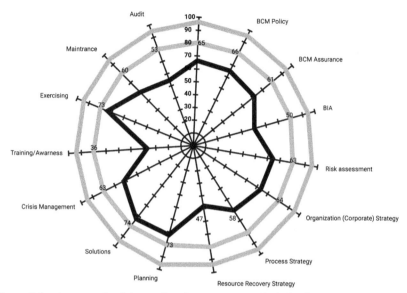

Figure 2.1 An example of the conformity assessment of a BCMS to the requirements of ISO 22301:2019.

2.1.1 First Version of the Standard

In April 2006, a meeting of the *International Organization for Standardization* (ISO) was held in Florence (Italy) under the motto *"Emergency Preparedness"*[38] (Figure 2.2).

This session considered, among other things, the differences between the national standards of *Australia, Israel, Japan, Russia, Singapore, UK,* and *USA* in the field of business continuity management (BCM). As a result, the *ISO/PAS 22399:2007 "Societal security – Guideline for incident preparedness and operational continuity management"* was prepared and recommended for use in all countries of the world (*GOST R 53647.4-2011/ ISO/PAS 22399:2007 "Business continuity management. Guidelines for incident preparedness and business continuity"*). However, most countries continued to use their national *business continuity management standards*.

The following work on BCM was developed by three technical committees of the *International Organization for Standardization* (ISO):

- *ISO/TC 223 "Societal security"* (2001–2014);

---

[38] https://www.iso.org/

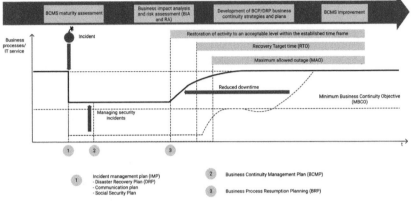

Figure 2.2   BCMS readiness for security incidents.

- *ISO/TC 247 "Fraud countermeasures and controls"* (2009–2014);
- *ISO/TC 284 "Management system for quality of PSC operations"* (2013–2014).

Thus, *ISO/TC 223* has prepared **38** international standards, among which the ISO 22300 family of standards has become more famous:[39]

- ISO 22300:2012 *"Societal security – Terminology"*;
- ISO 22301:2012 *"Societal security – Business continuity management systems – Requirements"*;
- ISO 22313:2012 *"Societal security – Business continuity management systems – Guidance"*;
- ISO 22320:2011 *"Societal security – Emergency management – Requirements for incident response"*;
- ISO 22398:2013 *"Societal security – Guidelines for exercises"*;
- ISO/TR 22312:2011 *"Societal security – Technological capabilities."*
- ISO/TS 17021-6:2014 *"Conformity assessment – Requirements for bodies providing audit and certification of management systems – Part 6: Competence requirements for auditing and certification of business continuity management systems (Joint project with ISO/CASCO)."*

The first version of ISO 22301–ISO 22301:2012 *"Societal security – Business continuity management systems – Requirements,"* was successfully used for certification of more than 100 business continuity management systems, BCMS in 120 countries (Figure 2.3). Significantly, ISO 22301:2012 was initially harmonized with other well-known international

---

[39] https://www.iso.org/committee/5259148.html

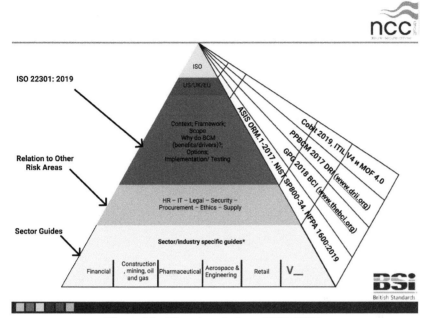

Figure 2.3   The role and place of ISO 22301.

standards, such as ISO 9001 *"Quality management systems,"* ISO 14001 *"Environmental management systems,"* ISO/IEC 27001 *"Information security management systems,"* ISO/IEC 20000-1 *"Information technology-Service management,"* and ISO 28000 *"Specification for security management systems for the supply chain"* (Tables 2.1 and 2.2 and Figure 2.4).

2.1.2   Second Version of Standard

Since 2014, the new ISO/TC 292 – *Security and resilience* technical committee has continued to improve the standards of the ISO 223001 family (Figure 2.5)[40].

In April 2020, the ISO/TC 292 – *Security and resilience* committee has prepared **38** standards and **20** standards that are in the works. At the same time, the key business continuity management standards, *Business continuity management*, were updated:

- **ISO 22300:2012** *"Societal security – Terminology" (replaced by 2018 edition);*

Table 2.1   Harmonization of ISO 22301:201   with other ISO standards.

Requirements	I ISO 90001:2008	ISO 14001:2004	ISO 20000:2011	ISO 22301:2012	ISO 27001:2005
Objectives of the management system	5.4.1	4.3.3	4.5.2	6.2	4.2.1
Policy of the management system	5.3	4.2	4.1.2	5.3	4.2.1
Management commitment	5.1	4.4.1	4.1	5.2	5
Documentation requirements	4.2	4.4	4.3	7.5	4.3
Internal audit	8.2.2	4.5.5	4.5.4.2	9.2	5
Continual improvement	8.5.1	4.5.3	4.5.5	10	6
Improvement	5.6	4.6	4.5.4.3	9.3	7

- **ISO 22301:2012** *"Societal security – Business continuity management systems – Requirements"* (replaced by 2019 edition);
- **ISO 22313:2012** *"Societal security – Business continuity management systems – Guidance"* (replaced by 2020 edition);
- **ISO/TR 22312:2011** *"Societal security – Technological capabilities"* (withdrawn 2017);
- **ISO 22320:2011** *"Societal security – Emergency management – Requirements for incident response"* (replaced by 2018 edition);
- **ISO 22317:2015** *"Societal security – Business continuity management systems – Guidelines for business impact analysis (BIA)"*;
- **ISO/TR 22351:2015** *"Societal security – Emergency management – Message structure for interoperability."*

In particular, the updated second version of the ISO 22301 standard – ISO 22301:2019 *"Security and resilience – Business continuity management systems – Requirements"* was released, which was amended as follows (Figures 2.6 and 2.7):

- changed the order of requirements in paragraph 8, removed repetitions, simplified, and unified terminology;
- removed links to determining the risk assessment;
- the introductory guide has been moved to ISO 22313 – Guide to business continuity management;

Table 2.2 Comparison of ISO 22301:201 with other well-known standards.

BCM Element	ISO 22301	ASIS/BSI BCM,01-2010 ASIS	SPC,1:2009 BS	BS 25999:2	NFPA 1600:2010
Understanding the organization	Section 4.1	N/A	N/A	Section 4.1	N/A
Need and expectations of interested	Section 4.1	N/A	N/A	Section 4.1	Chapter 4.5
Scope	Section 4.3	Section 1	Section 1	Section 3.2.1	Chapter 5.3
BCMS	Section 4.4	Section 4	Section 4	Section 4	Annex D
Management commitment	Section 5.2	Not explicit	Not explicit	Not explicit	Chapter 4.1
Policy	Section 5.3	Section4.3	Section4.2.1	Section 3.2.2	Chapter 4
Roles and Responsibilities	Section 5.4	Section 4.5.2	Section 4.4.1	Section 3.2.4	Chapter 6.6
Planning	Section 6	Section 4.4	Section 4.3	Section 3	Chapter 5
Resources	Section 7.1	Section 4.5.1	Section 4.4.1	Section 4.3	Chapter 6.1
Competence	Section 7.2	Section 4.5.3	Section 4.4.2	Section 3.2.4	Chapter 6.11
Awareness	Section 7.3	Section 4.5.3	Section 4.4.2	Section 3.2.4	Chapter 6.11
Communication	Section 7.4	Section 4.5.7	Section 4.4.3	Section 4.3.3	Chapter 6.8
Documented Information	Section 7.5	Section 4.6.4	Section 4.5.4	Section 3.4.2	Chapter 4.8
Business Impact Analysis	Section 8.2.2	Section 4.4.1.1	Section 4.3.1	Section 4.4.1	Chapter 5.5
Risk Analysis	Section 8.2.3	Section 4.4.1.2	Section 4.3.1	Section 4.1.2	Chapter 5.4
BC Strategies	Section 8.3	Section 4.3	Section 4.2	Section 4.2	Chapter 5
Business continuity procedures	Section 8.4	Section 4.5.6.2	Section 4.3	Section 4.3.3	Chapter 6.7
Testing and Exercising	Section 8.5	Section 4.6.2.2	Section 4.5.2.2	Section 4.4	Chapter 7
Monitoring and Measurement	Section 9.1	Section 4.6.1	Section 4.5.1	Section 4.4	Chapter 7.1
Internal audit	Section 9.2	Section4.6.5	Section 4.5.5	Section 5.11	Chapter 8.1
Management review	Section 9.3	Section 4.7.4	Section 4.6.5	Section 5.2	NVA
Improvement	Section 10	Section 4.7.4	Section 4.6.5	Section 6.2	Chapter 8
Auditing	Section 9.2	Section 4.6.5	Section 4.5.5	Section 5.1	Chapter 8.1

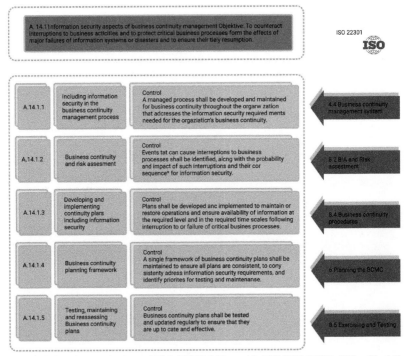

Figure 2.4   Harmonization of ISO 22301:201   with ISO/IEC 27001:200   (A. 14).

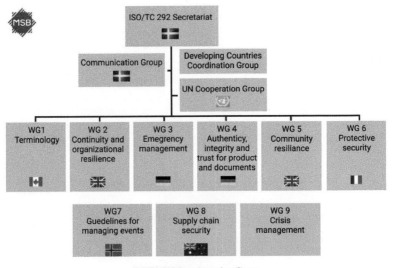

Figure 2.5   The structure of ISO technical committee ISO/TC 292.

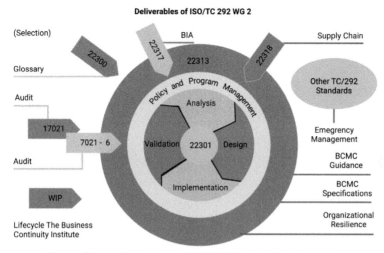

Figure 2.6 Deliverables of ISO/TC 292 in the field of BCM.

- more attention is paid to planning changes in the business continuity management system;
- reduced the number of prescriptive procedures and documentation requirements;
- the business continuity strategy is clarified: "business continuity strategy and solutions";

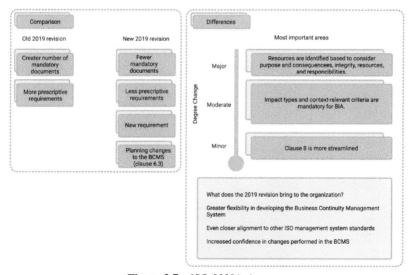

Figure 2.7 ISO 22301 changes.

- business continuity plans provide clear guidance on helping teams and people fix violations.

ISO 22301:2019 *"Security and resilience – Business continuity management systems – Requirements"* consists of the following sections:

Clause 1: Scope;
Clause 2: Normative references;
Clause 3: Terms and definitions;
Clause 4: Context of the organization;
Clause 5: Leadership;
Clause 6: Planning;
Clause 7: Support;
Clause 8: Operation;
Clause 9: Performance evaluation;
Clause 10: Improvement.

ISO 223301:2019 aims to improve business continuity management systems (BCMS) of organizations based on the well-known *Walter Andrew Shewhart (1891–1967)* and *William Edwards Deming (1900–1993)* *"Plan-Do-Check-Act"* (PDCA) or *"Plan-Do-Study-Act"* (PDSA) lifecycle model (Figure 2.8). Here, sections 4–10 of ISO 223301:2019 are arranged as follows (Figure 2.9):

- <u>Clause 4</u> is a component of **P**lan. It introduces requirements necessary to establish the context of the BCMS as it applies to the organization as well as needs, requirements, and scope.
- <u>Clause 5</u> is a component of **P**lan. It summarizes the requirements specific to top management's role in the BCMS and how leadership articulates its expectations to the organization via a policy statement.

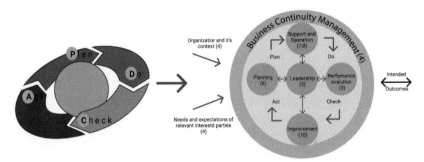

Figure 2.8   The PDCA model of ISO 22301:2012.

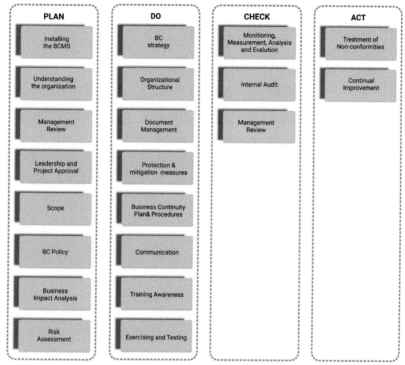

Figure 2.9   The content of PDCA model.

- Clause 6 is a component of **Plan**. It describes requirements as it relates to establishing strategic objectives and guiding principles for the BCMS as a whole.
- Clause 7 is a component of **Plan**. It supports BCMS operations as they relate to establishing competence and communication on a recurring/as-needed basis with interested parties, while document-ing, controlling, maintaining, and retaining required documented information.
- Clause 8 is a component of **Do**. It defines business continuity needs, determines how to address them, and develops the procedures to manage the organization during a disruption.
- Clause 9 is a component of **Check**. It summarizes requirements nec-essary to measure business continuity performance, BCMS compli-ance with this document, and management review.
- Clause 10 is a component of **Act**. It identifies and acts on BCMS non-conformance and continual improvement through corrective action.

According to ISO 223301:2019, a possible BCMS certification algorithm for an organization can be represented as (Figure 2.10) follows.

Stage 1. Implementation of the management system: Before being audited, a management system must be in operation for some time. Usually, the minimum time required by the certification bodies is 3 months.

Stage 2. Internal audit and review by top management: Before a management system can be certified, it must have had at least one internal audit report and one management review.

Stage 3. Selection of the certification body (registrar): Each organization can select the certification body (registrar) of its choice.

Stage 4. Pre-assessment audit (optional): An organization can choose to do a pre-audit to identify any possible gap between its current management system and the requirements of the standard.

Stage 5. Stage 1 audit: A conformity review of the design of the management system. The main objective is to verify that the management system is designed to meet the requirements of the standard(s) and the objectives of the organization. It is recommended that at least some portion of the Stage 1 audit be performed on-site at the organization's premises.

Stage 6. Stage 2 audit (on-site visit): The Stage 2 audit objective is to evaluate whether the declared management system conforms to all requirements of the standard, is actually being implemented in the organization, and can support the organization in achieving its objectives. Stage 2 takes place at the site(s) of the organization's sites(s) where the management system is implemented.

Stage 7. Follow-up audit (optional): If the auditee has non-conformities that require additional audit before being certified, the auditor will perform a follow-up visit to validate only the action plans linked to the non-conformities (usually one day).

Stage 8. Confirmation of registration: If the organization is compliant with the conditions of the standard, the Registrar confirms the registration and publishes the certificate.

Stage 9. Continual improvement and surveillance audits: Once an organization is registered, surveillance activities are conducted by the Certification Body to ensure that the management system still complies with the standard. The surveillance activities must include on-site visits (at least one/year) that allow verifying the conformity of the certified client's management system and can also include: investigations following a complaint, review of a website, a written request for follow-up, etc.

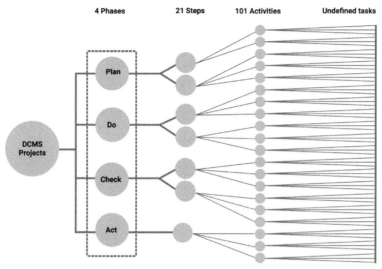

Figure 2.10   Possible algorithm for improving BCMS.

Note that the following guidelines and technical reports of the British Standards Institute (BSI)[41] can provide some methodological assistance in preparing for certification of BCMS organizations to the requirements of ISO 223301:2019 (Table 2.3):

- BCI Horizon Scan 2020 – BCI Horizon Scan 2020 landscape and modern cyber threats research[42].
- Introducing ISO 22301 Business Continuity Management. Minimize risk and protect your business-ISO 22301-client's Guide-Contains information about the capabilities of the business continuity management system, BCMS[43].
- ISO 22301 Self-assessment questionnaire-ISO 22301-Checklist for self-checking the organization's readiness to implement a business continuity management system, BCMS[44].
- ISO 22301 – Business continuity management – Your implementation guide – ISO 22301[45].

---

[41] https://www.bsigroup.com

[42] https://www.bsigroup.com/ru-RU/ISO-22301/BCI-Horizon-Scan-Report/

[43] https://www.bsigroup.com/globalassets/localfiles/ru-ru/documents/iso-22301/iso-22301-intro-4-pager-2016.pdf

[44] https://www.bsigroup.com/globalassets/localfiles/ru-ru/documents/iso-22301/iso-22301-self-assessment-form-2016.pdf

[45] https://www.bsigroup.com/globalassets/localfiles/ru-ru/documents/iso-22301/iso-22301-implementation-guide-2016.pdf

Table 2.3   Audit of an BCMS organization.

Exercise Type	What is it?	Benefit	Disadvantage
Checklist	Distribute plans for review	Ensures plan addresses all activities	Does not address effectiveness
Structured Walkthrough	Thorough look at each step of the BCP	Ensures planned activities are accurately described in the BCP	Low value in proving response capabilities
Simulation	Scenario to enact recovery proceures	Practice session	When subsets are very different
Parallel	Full test, but primary processing does not stop	Ensures hight level of reliability without interrupting normal operations	Expensive as all personnel is involved
Full interruption	Disaster is replicated to the point of ceasing normal operations	Most reliable test of BCP	Risky

- ISO 22301 – Features and advantages of standard ISO 22301[46].
- Business continuity management – our top ten tips – Business continuity management – Ten main recommendations for preparing and responding to failures using the ISO 22301 standard[47].
- Measurement matters. The role of metrics in ISO 22301 – A BSI whitepaper for business[48].
- Beyond recovery – The broader benefits of Business Continuity Management – A BSI whitepaper for business[49].

Organizations certified to ISO 22301:2012 are expected to switch to ISO 22301:2019 within three years. According to the decision of the certification bodies, after October 30, 2022, ISO 22301:2012 certificates become invalid.

---

[46] https://www.bsigroup.com/globalassets/localfiles/ru-ru/documents/iso-22301/iso-22301-features-benefits-2016.pdf

[47] https://www.bsigroup.com/globalassets/localfiles/ru-ru/documents/iso-22301/bcm-top-tips-uk-en.pdf

[48] https://www.bsigroup.com/globalassets/localfiles/ru-ru/documents/iso-22301/bsi-bcm-metrics-whitepaper-uk-en.pdf

[49] https://www.bsigroup.com/globalassets/localfiles/ru-ru/documents/iso-22301/iso-22301-whitepaper-beyond-recovery-uk-en.pdf

Figure 2.11   The development of *Good Practice Guidelines 2010–2018.*

2.2 BCI Practice

In 1994, the Business Continuity Institute (BCI), a non-profit organization, was founded in the United Kingdom[50]. Currently, BCI unites more than 8.5 thousand certified business continuity specialists from more than 100 countries.

2.2.1 Activity Directions

The main activities of the British Continuity Institute (BCI) (Figures 2.11–2.13) refer to the dissemination and promotion of best practices in business continuity management and disaster recovery in emergency situations.

- *BCI International Symposium* is an international forum of specialists in business continuity management, where ideas are exchanged on topical issues and further steps are planned for the development of business continuity management practices.
- *Business Continuity EXPO* is a major international exhibition where system integrators, software and hardware manufacturers, and consulting companies annually present their achievements and share best practices in the field of BCM (at the same time, thematic conferences on BCM issues are held).
- *Business Continuity Awareness Weeks* – regularly held events around the world, during which presentations, guidelines, and reference literature on BCM issues are available on the institute's website to attract

---

[50] https://www.thebci.org/.

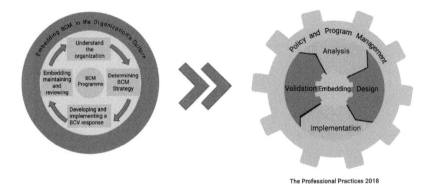

The Professional Practices 2018

Figure 2.12   Business continuity management (BCM) lifecycle.

the attention of specialists and managers of government and commercial organizations.

• *The Business Continuity Awards* is an annual event that celebrates outstanding achievements in the field of BCM in several categories.

2.2.2 Main Results

The main results of BCI are the following:

• *2002–2006:* A document containing the best practices of business continuity management, *Good Practice Guidelines 2002*, has been developed. The document was approved by the UK government authorities for use in public and commercial organizations and was

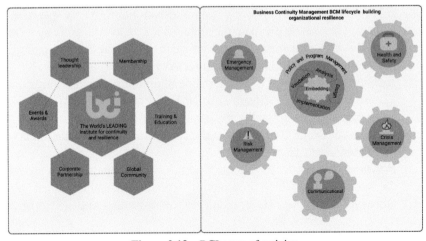

Figure 2.13   BCI areas of activity.

distributed free of charge. The Publicly Available Specification, PAS 56:2003, is based on *Good Practice Guidelines 2002*. *PAS 56 has been further developed and the business continuity standard BS 25999 has been prepared – Part 1. Code of best practices-BS25999-1: 2005. Code of Practice.* An updated version of *Good Practice Guidelines 2005*, which developed a curriculum for training of specialists in the field of business continuity management, organized didactic education fundamentals of business continuity management. Participation in BCI training programs was considered, when passing certification.

- *2007–2011:* The *Good Practice Guidelines 2007* version is being released. The audit part of *BS 25999 Part II has been prepared. System specifications BCM-BS 25999-2: 2007. Specification.* Based on BS 25999 was improved the *ISO/PAS 22399:2007* standard *"Societal security – Guideline for incident preparedness and operational continuity management."* Methodological assistance was provided to Australia, New Zealand, Singapore, and China in developing national standards for risk management and business continuity. Was continued the work on harmonization of approaches to business continuity in the UK and the US. The system of certification of specialists in the field of business continuity management based on the knowledge of the ten certification standards for professionals in the field of business continuity was improved – *Certification Standards for Business Continuity Professionals*. The following levels of training of specialists in the field of business continuity are highlighted:
  - ○ *ASSOCIATE (AMBCI)* – a novice specialist who is familiar with the listed business continuity management standards;
  - ○ *SPECIALIST (SBCI)* – a specialist with at least two years of experience in business continuity management who knows and can use in practice at least four standards;
  - ○ *MEMBER (MBCI)* – a practitioner with at least two years of experience in business continuity management who uses all ten standards in practice;
  - ○ *FELLOW (FBCI)* – a senior management practitioner with at least five years of experience in business continuity management who is well-versed in and able to apply all ten standards in practice.

An updated version of *Good Practice Guidelines 2010* has been prepared.

- *2012–2020. 2008.* Based on part 2 of BS 25999, the international standard ISO 22301:2012 has been prepared. The system of training specialists in business continuity management has been updated and the levels have been highlighted: *Affiliate/Student, CBCI, AMBCI, MBCI, FBCI* (Table 2.4). Updated versions of *Good Practice Guidelines* 2013 and 2018 have been prepared. Work on the development and improvement of the ISO 22301:2012 standard has continued, and the ISO 22301:2019 version has been prepared; #A478A9 #DFCFE0.

2.3   DRI Practice

In 1988, a nonprofit organization, the International Disaster Recovery Institute International, DRII, was established at the University of Washington (www.drii.org), which includes three branches: DRI Canada (Toronto), DRI Asia (Singapore), and DRI Japan (Tokyo). Currently, DRI unites more than 15,000 certified business continuity specialists from more than 100 countries.

2.3.1   Direction of Activity

The main activities of the DRII Institute include:

- preparation and dissemination of the so-called general body of knowledge for *Professional Practices for Business Continuity Management (2017 Edition)*;
- initial training of specialists in the field of ensuring the smooth operation of the organization in the event of a disaster;
- advanced training of specialists in disaster recovery;
- expertise of relevant standards and regulations in the field of disaster recovery, etc.

The above-mentioned DRII body of knowledge includes the following main topics.

1. *Program initiation and management:*
   - establish the need for a business continuity program;
   - obtain support and funding for the business continuity program;
   - build the organizational framework to support the business continuity program;
   - introduce key concepts, such as program management, risk awareness, identification of critical functions/processes, recovery strategies, training and awareness, and exercising/testing.
2. *Risk assessment:*
   - identify risks that can adversely affect an entity's resources or image;

Table 2.4 Professional training levels in BCI.

Grades	Criteria	Experience	Benefits	Suggested development
FBCI Can reformulate or develop original thinking in this subject and/or specialist understanding to solve complex problems	**FBCI** +Must have held MBCI for at least 5+ years +Must have made significant contribution to BC community and be able to demonstrate this activity	**FBCI** 10 years' experience in BC across all six professional practices	**FBCI** All benefits listed below PLUS: +Prestige = knowing that has achieved the highest possible grade available	**FBCI** +20 hours of CPD each year +Volunteering +Speaking engagements +Research contributions +Awards submission +Content production
MBCI Comprehensive understanding of the subject and can evaluate information that might be complex and/or can apply this in day-to-day work to solve complex problems	**MBCI** +20 hours of CPD each year +Hold one of the following: • CBCI • Business continuity management BCI diploma • AMBCI	**MBCI** 3+ experience in BC across all six professional practices	**MBCI** All benefits listed below PLUS: +may act as a Mentor	**MBCI** +Professional practice courses +Business continuity management BCI diploma +Volunteering +Speaking engagements +Research contributions +Awards submission +Content production
AMBCI Foundational understanding with an awareness of different approaches and/or apply this in day-to-day work to carry well-defined tasks	**AMBCI** +20 hours of CPD each year +Hold one of the following: • CBCI • Business Continuity Management BCI Diploma	**AMBCI** 1+ years' experience in BC across all six professional practices or 2+ years in one or two professional practices	**AMBCI** All benefits listed below PLUS: +Full access to BCI research +15% discount on selected BCI products and events +Free subscription to the printed copy of Continuity&Resilience magazine	**AMBCI** +Professional practice courses +Business continuity management BCI Diploma +Volunteering +Speaking engagements +Awards submission +Content production +Attend events and grow personal network

Continued

Table 2.4 Continued

Grades	Criteria	Experience	Benefits	Suggested development
CBCI Demonstrates knowledge of the BCI Good Practice Guidelines; level may be upgraded upon submitting evidence of professional experience	**CBCI** +20 hours of CPD each year +Pass the CBCI exam	**CBCI** Good knowledge of the BCI's Good Practice Guidelines	**CBCI** All benefits listed below PLUS: +Eligible for BCI CPD program +10% discount on selected BCI products and events +Internationally recognized credential +Access to Regus facilities and other value added benefits	**CBCI** +Professional practice courses +Attend events and grow personal network
Affiliate/student As a non-certified member. Evidence of knowledge and professional experience has yet to be supplied	**Affiliate/student** +Affiliate – an individual who has an active interest in BC or who is considering pursuing a career as a BC professional +Student – an individual who is currently undertaking study on either a full time or part time course that is related to BC or resilience	**Affiliate/student** No knowledge check or information about level of experience	**Affiliate/student** +Join BCI Mentoring program as a member +Free GPG download +5% discount on selected BCI products and events +Access BCI membership social media groups +Claim third party discounts on conferences, etc. +Free digital Continuity&Resilience magazine +Online access to the Continuity and Resilience Review	**Affiliate/student** +Introduction to business continuity course +Good Practice Guidelines training course (Certification training) + Business continuity management BCI diploma +E-learning course – introduction to business continuity

- assess risks to determine the potential impacts to the entity, enabling the entity to determine the most effective use of resources to reduce these potential impacts.
3. *Business impact analysis:*
   - identify and prioritize the entity's functions and processes in order to ascertain which ones will have the greatest impact should they not be available;
   - assess the resources required to support the business impact analysis process;
   - analyze the findings to ascertain any gaps between the entity's requirements and its ability to deliver those requirements.
4. *Business continuity strategies:*
   - select cost-effective strategies to reduce deficiencies as identified during the risk assessment and business impact analysis processes.
5. *Incident response:*
   - develop and assist with the implementation of an incident management system that defines organizational roles, lines of authority, and succession of authority;
   - define requirements to develop and implement the entity's incident response plan;
   - ensure that incident response is coordinated with outside organizations in a timely and effective manner when appropriate.
6. *Plan development and implementation:*
   - document plans to be used during an incident that will enable the entity to continue to function.
7. *Awareness and training programs:*
   - establish and maintain training and awareness programs that result in personnel being able to respond to incidents in a calm and efficient manner.
8. *Business continuity plan exercise, assessment, and maintenance:*
   - establish an exercise, assessment, and maintenance program to maintain a state of readiness.
9. *Crisis communications:*
   - provide a framework for developing a crisis communications plan;
   - ensure that the crisis communications plan will provide for timely, effective communication with internal and external parties.
10. *Coordination with external agencies:*
    - establish policies and procedures to coordinate incident response activities with public entities.

Table 2.5   Main areas of BCM certification

Evaluation result	Stage 1	Stage 2	Stage 3	Stage 4	Stage 5
1. Initiation and management of Business continuity projects.	◇		✓		★
2. Business Impact Analysis	◇	✓		★	
3. Risk Evaluation and Control	◇		✓	★	
4. Developing Business Continuity Management Strategies	◇		✓	★	
5. Emergency Response and operations	◇	✓		★	
6. Developing and Implementing Business Continuity and Crisis Management Plans	◇		✓		★
7. Awareness and Training Programs	◇	✓			★
8. Maintaining and Exercising Business Continuity and Crisis Managements Plans	◇	✓			★
9 . Crisis Communications (employees of the company, contractors, partners, media, etc.)	◇	✓		★	
10. Coordination with External Agencies	◇	✓		★	
Current position ✓			**Final position** ★		

Note that the DRII Institute maintains working relations with the BCI Institute. Thus, the general body of knowledge *DRII-Professional Practices for Business Continuity Management (2017)* correlates *with* the best practices of business continuity management BCI – *Good Practice Guidelines 2018 Edition*. This body of knowledge includes topics similar to *Certification Standards for Business Continuity Professionals* (Table 2.5).

Based on professional practices, DRII developed and improved the appropriate certification of specialists (Table 2.6).

The mentioned DRII certification system provides the following basic training levels for BCM specialists:

• *Associate Business Continuity Planner (ABCP)* – a novice specialist who is getting acquainted with professional practices in the field of business continuity management;

Table 2.6 Structure of the DRII specialist certification system.

Continuity	Get certified. Prove your expertise.	Business Continuity(BCLE 2000)Business Continuity Review(BCP 501)	ABCP CFCP CBCP CBCV
Advanced continuity	Become an MBCP, DRI's highest level of certification.	Continuity Master's Case Review(BCP 601)	MBCP
Risk Management	Create a robust risk management program.	Risk Management (RMLE 2000)Risk Management Review(RMP501)	ARMP CRMP
Audit	Add a skill with our ANSI and CQI IRCA approved, case study-based course.	Continuity Audit NFPA 1600(BCLE AUD NFPE) Continuity Audit: ISO 22301(BCLE AUD ISO)	CBCA CBCLA
Healthcare Continuity	Protect service delivery, safeguard records, maintain compliance.	Healthcare Continuity(HCLE 2000) Healthcare Continuity Review(HCP 501)	AHPCP CHPCP
Public Sector	Protect your agency' mission-essential functions.	Public Sector Continuity (GCLE 2000) Public Sector Continuity Review	APSCP CPSCP
Cyber Resilience	Master the elements of cyber resilience by integration business continuity and cyber security.	Cyber Resilience (CRLE 2000) Cyber Resilience Review(CRP 501)	ACRP CCRP5

- *Certified Functional Continuity Professional (CFCP)* – certified functional specialist with experience in business continuity management;
- *Certified Business Continuity Professional (CBCP)* – certified specialist with experience in business continuity management;
- *Master Business Continuity Professional (MBCP)* – a specialist in business continuity management, who has the necessary knowledge and is able to correctly use this knowledge in practice.
- Efforts to unify business continuity approaches in the United Kingdom and the United States continued.

2.3.2 Features of the Approach

The DRII model (Figure 2.14) for emergency planning includes the following steps:

- *Project initiation – project initiation phase:*
  ○ clarification of the problem;

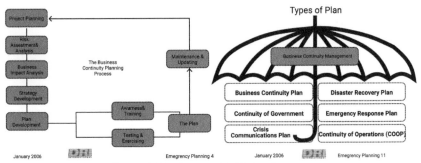

Figure 2.14   Stages of BCP development.

  o defining goals and objectives – requirements analysis;
  o defining assumptions and terms used;
  o defining the scope and cost of the project;
  o establishing the project steering committee;
  o defining business continuity policies.
 • *Requirements analysis – functional requirements phase:*
  o risk assessment and management, RA;
  o business impact assessment, BIA;
  o developing alternative BC strategies;
  o cost analysis (ABC) of BC strategies;
  o determining the budget for the business continuity management program.
 • *Development of the BCP and DRP plan – design and development phase:*
  o defining the goals and objectives of the plan;
  o clarification of recovery goals and tasks;
  o determining the content and structure of the plan;
  o develop a plan and the necessary action scenarios;
  o the procedure for putting the plan into effect;
  o creating a backup office;
  o personnel management program;
  o computation of acceptable data loss;
  o plan administration.
 • *Implementation of the BCP and DRP plan – implementation phase:*
  o the definition of priority actions in emergency situations;
  o defining the operating procedure of the anti-crisis center;
  o distribution of permissions and responsibilities;
  o checking the effectiveness of business continuity management;
  o details of the procedures for actions in emergency situations;
  o specifying the required resources;

○ verification of suppliers' contractual obligations.
* *Testing the BCP and DRP plan – testing and exercising phase:*
  ○ defining goals and objectives;
  ○ development of necessary testing scenarios;
  ○ evaluating the adequacy of test plans;
  ○ training and implementation of the BCM awareness program;
* *Maintenance and support of the BCP and DRP – maintenance and updating phase plan:*
  ○ planning deadlines and the required budget;
  ○ support of the necessary software;
  ○ revision of criteria for ensuring business continuity;
  ○ audit of the BCP and DRP plan;
  ○ organization of familiarization with the mentioned plan.
* Execution phase (if required)

2.4   SANS Institute Practice

Experts of the institute of system administrators and security administrators – *SANS Institute*[51] – distinguish between a business continuity plan (BCP) and a disaster recovery plan (*DRP*). At the same time, the first differs from the second, playing a major role in the *business continuity management* program, BCM organization (Figure 2.15).

According to *SANS Institute*, the main stages of the BCP and DRP plan development lifecycle are shown in Figure 2.16.

Let us comment on the marked stages in more detail.

2.4.1   BCP Development

For a successful start of the project, there is a need to get support from the organization's management.

Getting management support:
Receiving management support guarantees sufficient funding for the project as well as some assistance in coordinating and managing the project.

Getting to know the history of incidents:
Analysis and discussion of the factography of incidents in the field of business continuity, BC, will timely identify existing problems in this area and further focus on solving the most pressing issues of business continuity management.

---

[51] www.sans.org

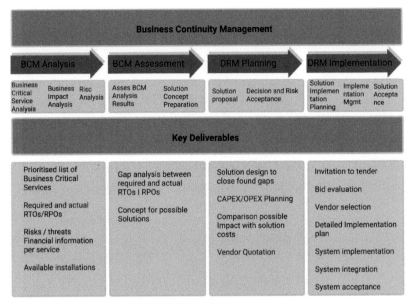

Figure 2.15   ECP components.

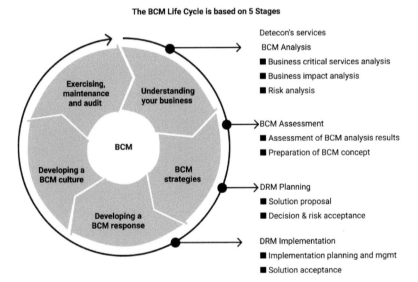

Figure 2.16   The BCM lifecycle.

Requirements analysis:
Before developing a BCP/DRP plan, it is recommended to define business requirements, for example:
- minimize the indicators of continuity of business processes;
- minimize the time to resume critical processes and operations;
- minimize the financial losses in case of emergencies;
- maintain the organization's reputation and positive image in an emergency;
- compliance with international and national regulatory requirements, etc.

The assigning of the responsible person:
It is recommended to appoint a single responsible person who will be responsible for the development and implementation of BCP/DRP plans. A responsible person is a Project Manager for the development and implementation of BCP/DRP plans. Its responsibilities may include:
- defining project goals and objectives;
- creating a project office;
- project risk identification and control;
- identify and control critical factors for project success;
- develop the project charter;
- project management;
- control the budget allocation;
- participate in the project management committee;
- organizing the required interaction;
- manage project changes;
- timely presentation of project results to the organization's management;
- maintain and support the project.

The establishment of the project office:
One has to create a project office. It is recommended to include representatives from all major divisions of the organization in the project office to successfully complete the project. For example, such representatives can be employees of the departments of financial management, accounting, production, sales, development, information technology, logistics, marketing, security, human resources, legal department, and so on. There is a sample of the role distribution among all project participants, which is as follows:

- Project Manager – manages the progress of the project and coordinates the activities of various groups.
- Management group – approves the project, allocates the budget, and sets requirements.

- HR team – manages the hiring of temporary staff to support business operations, if necessary.
- Information group – interacts with the media in case of emergencies.
- Legal group – deals with resolving legal issues in the event of emergencies.
- A group of IT security ensures the confidentiality, integrity, and availability of machine data in the course of the project.
- Physical security group – provides physical security of assets in case of emergencies.
- Operation group – responsible for the operation of technologies during a crisis.
- Emergency response team – puts business continuity and disaster recovery plans in place in the event of an emergency.
- Damage assessment group – determines the type and nature of damage caused by an accident.
- Backup group – organizes backup, storage, and recovery of information assets.
- Backup office group – responsible for organizing a working backup office in case of an emergency to resume business.
- Repair group – responsible for repairing assets that have failed as a result of an accident.

Computation of the required BCM costs:
It is necessary to calculate the direct and indirect costs of the BCM program in order to make adequate decisions in the field of business continuity management. Various methods can be used for this purpose, such as TCO and TVO Gartner methods.

Risk analysis (RA):
Risk analysis will assess the completeness and sufficiency of organizational and technical countermeasures of an organization to counter threats to business continuity. In addition, risk analysis further will make cost-effective decisions on business continuity management (Figure 2.17).

Ranking critical business processes:
During interviews with key employees of the organization, it is recommended to identify and rank the critical business processes of the organization. At the same time, for each business process, clearly define the assets involved, residual risks, and preventive and corrective countermeasures (controls). The result of this ranking is a map of the organization's ranked critical business processes.

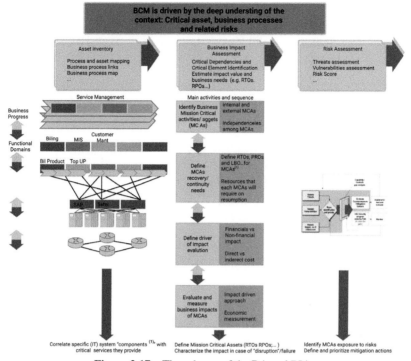

Figure 2.17 The scheme of the RA and BIA.

Involvement of external experts:
It is recommended to consider the involvement of external experts to assess the risk assessment decisions made in specific subject areas.

Familiarization with available statistics:
Up-to-date data on existing threats, vulnerabilities, and risks is collected from reliable sources (reports from analytical agencies, international institutions, professional communities, statistics from incident response centers, etc.). In particular, the subscription to the following newsletters might be beneficial to get up-to-date information: SANS Institute[52], SANS Incident Website,[53] the Incident Response Center, and the CERT Coordination Center (CERT/CC)[54].

- According to SANS, reliable sources of information include:
- SANS Institute;

---

[52] www.sans.org
[53] www.incidents.org
[54] www.cert.org

- Federal Emergency Management Agency (FEMA);
- Office of Emergency Services (OES) USA;
- Occupational Safety and Health Administration (OSHA), etc.

Selecting a strategy for managing residual risks:
At this step, one has to decide about what to do with the remaining risks: accept, transfer, minimize, manage, etc.

Determination of the MTD and RTO:
The maximum tolerable downtime (MTD) is defined as the maximum period of time during which regular business processes can be performed without critical consequences for the business.

The restore time objective (RTO) is defined as the required recovery time for each business process.

Identifying priorities for recovery of business processes:
This step sets priorities for restoring critical business processes in the event of emergencies.

Checking the availability of reserve areas:
The reserve areas (premises) and the corresponding infrastructure (means of computer technology, communication systems, telecommunications, office equipment, backup equipment, backup power supply, air conditioning, fire extinguishing, etc.) are checked for compliance with the requirements for ensuring business continuity of the organization.

Presentation to management:
Based on the results of the assessment and analysis of residual business continuity risks, a presentation is prepared for the organization's management, which helps to coordinate and obtain the necessary approval for the proposed BCP/DRP initiatives.

Business impact analysis (BIA):
The planning stages:
Having received the required support from management, one needs to form a project team, set goals and objectives, and start implementing the stage. At this stage, it is possible to attract an external team of performers. It is also possible to use the appropriate automated tools for developing BCP/DRP plans.

Data collection and classification:
At this step, there is a need to decide on the method of data collection and processing (questionnaires, interviews, round tables, etc.) and to set

Table 2.7   Example of asset classification.

Asset	Description	Responsible	Criticality level	Exposure level
B2B information and analytical and trading operating system	Market analysis of products, services and technologies for the development	A.petrov	Second	First
Automated payment system	The company's main AS	Ivanov	First	second

classification criteria for the information collected, for example, as shown in Tables 2.7 and 2.8.

Data analysis:
Data analysis will evaluate the consequences of undesirable events for critical business processes and set a guideline recovery time for these processes.

Documenting BIA results:
The results of the BIA as well as supporting information (executive summaries, schedules, questionnaires, interview notes, recommendations received, etc.) should be saved and documented in the relevant report on the executed work.

The presentation of the work results:
At this step, the main work results are presented in a visual form to the organization management. If necessary, the BIA main results are protected.

Table 2.8   Possible criteria for criticality and impact.

Criticality level 1: immediate recovery	Criticality level 2:recovery in 12 hours	Criticality level 3:recovery in 24 hours	Criticality level 1:recovery in 7 days	Criticality level 2:recovery in 14 days
This is most critical for maintaining your business. Immediate recovery is required to prevent significant losses or loss of the organization's market value	Relatively critical for maintaining business. Further downtime may result in a critical level of 1.	Important for Business of organization	Important for effective business activity of organization	

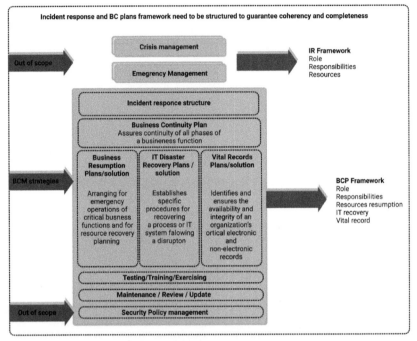

Figure 2.18   BCP project details.

Defining BC strategies:

Once critical business processes and assets have been identified and the RTO and RPO have been established, BC strategies must be defined (Figures 2.18 and 2.19) to develop a plan for responding to undesirable events and combining the procedures required to recover from an incident.

It is useful to work out in detail the content and structure of the BCP and DRP plan, containing at least the following:

- contact details of the person responsible for the BCP and DRP plan;
- results of ranking of business processes and related assets by the degree of importance and the amount of risk (the amount of financial losses in the event of business interruption);
- composition, authority, and responsibility of the management team as well as functional disaster recovery teams;
- issues of mobilization of required personnel;
- basic contact information of officials (full name, position, office, home and mobile phone numbers, home address, etc.) and ways to notify them in an emergency;

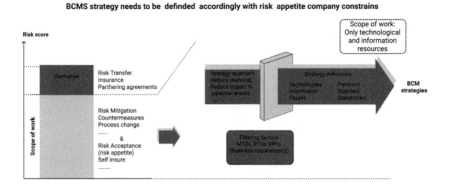

Figure 2.19   Defining BC strategies.

- issues of information interaction both inside and outside the organization (with the main suppliers, partners and clients, state control structures, law enforcement agencies, mass media, etc.);
- the compliance with critical business processes;
- the accommodation of the required personnel in alternative areas;
- a list of resources that are minimal for disaster recovery, including archived information, primary documents, forms, templates, etc.;
- disaster recovery requirements for core business processes and supporting infrastructure, including priorities and recovery time;
- disaster recovery procedures for functional architecture;
- procedure for reviewing and maintaining the BCP and DRP plan.

Defining recovery strategies:
Selection of the typical recovery strategies:

- "hot" room;
- "warm" room;
- "cold" room;
- fault-tolerant solutions;
- mutual assistance agreements;
- backup data center;
- technical support and support by the supplier;
- combined approach.

Preparing contact information:
Prepare the list of contact information to quickly contact the necessary employees of the organization in case of emergencies. The list should

be stored in an easily accessible place (if possible, in the organization's on-call service). The following data must be included in the contact list:

- management of the organization;
- technical service personnel;
- the medical officers and psychologists on the staff;
- suppliers of process equipment and software;
- state bodies and law enforcement agencies;
- representatives of the mass media.

Inventory of funds:
The inventory list should include the following:

- inventory of communications, fire-fighting, and video surveillance equipment;
- list of employees with phone numbers and home addresses;
- inventory of premises and production areas;
- the inventory documentation;
- inventory of information system components;
- inventory of industrial equipment;
- inventory of external storage;
- inventory of reserve funds;
- technical passports of premises, etc.

Preparing service level agreements (SLA) with suppliers:
SLAs will help to prepare in advance and anticipate possible obligations and actions in case of emergencies.

Preparing alternative deliveries:
It is recommended that a list of alternative service providers and solutions be prepared in advance in the event that regular suppliers are not able to make the required deliveries or such deliveries will be difficult.

2.4.2 BCP Testing

Testing the BCP and DRP plan is the next important step in the development lifecycle of the mentioned plan (Figures 2.20 and 2.21).

The definition of the goals and objectives of testing:
The goals and objectives of testing may include:

- getting evidence of the plan's operability;
- verification of the reliability and sufficiency of methodological support for business continuity;

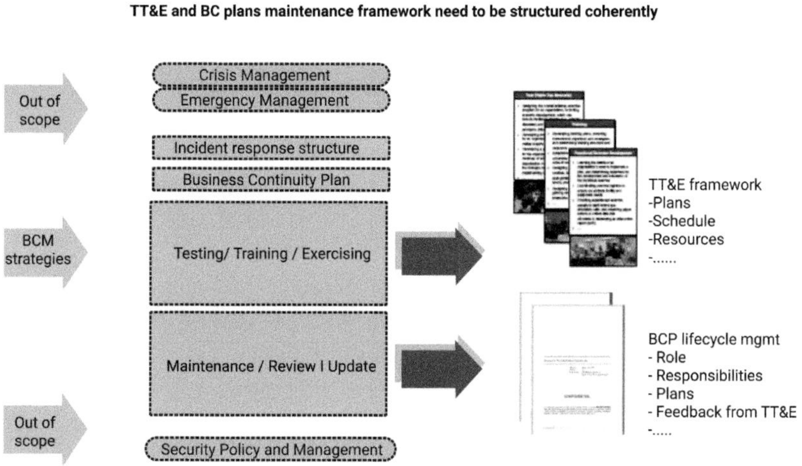

TT&E and BC plans maintenance framework need to be structured coherently

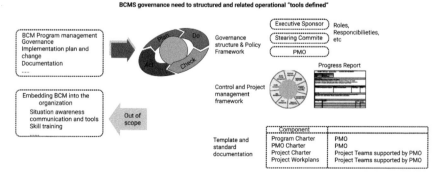

Figure 2.20   BCP/DRP testing.

- checking the adequacy and completeness of business continuity tools;
- determining the plan's shortcomings;
- getting the necessary skills and knowledge emergency response;
- keeping the plan up-to-date, etc.

Defining the required resources:
Depending on the testing methodology, certain resources may be required. Here, the following must be ensured:

- whether the necessary equipment is available on demand;
- how quickly the required equipment is delivered to the reserve areas;
- whether the staff and their competence are sufficient to work on the reserve areas.

Figure 2.21   BCMS maturity assessment.

Selecting employees for BCP and DRP testing:
This step selects the employees involved in the BCP and DRP testing process. Then, the job responsibilities are defined for each group and individual employees.

Defining the testing methodology:
The following options of testing methods are proposed, which differ in the degree of approximation of the simulated situation to real events:

- full-scale testing – affects most structural divisions of the organization and is carried out in the fullest possible testing volume;
- test based on "checklists" – the method involves sending out pre-prepared survey questionnaires to the organization's divisions, containing a questionnaire with clarifying questions about the BCM program, collecting, summarizing, and analyzing the received data;
- simulation test – the test simulates actions in the event of an emergency and begins only from the moment of actual relocation to an alternative location and installation of the relocated equipment;
- "parallel" test – performed simultaneously on the actual functioning infrastructure of the organization and on alternative areas; any discrepancy or difference between the actual functioning systems and alternative ones is recorded and taken into account;
- "complete violation" test – in this test, regular operations are completely stopped and business processes start running on an alternative area with pre-prepared infrastructure.

Evaluating the test effectiveness:
To determine the effectiveness of the BCP/DRP test plan, the test results are compared with the expected test results. In case of deviation, the test plan is adjusted.

Test plan administration:
The administration of the test plan consists of assigning a responsible person to coordinate testing procedures and document test results.

Presenting test results to management:
The test results are clearly presented to the organization's management for review and approval.

Support and maintenance of the BCP and DRP test plan up-to-date:
The BCP and DRP test plan should be regularly reviewed and updated based on the test results obtained to keep it up-to-date.

Support for BCP and DRP plans:

As a rule, the BCP and DRP plan is updated annually in the normal operation of the organization. However, in some of the cases listed below, the plan may be updated in a shorter time frame:

- regular changes in the structure of the organization and business infrastructure;
- emergence of new connections and business dependencies on modern technologies;
- restructuring of the organization – mergers, acquisitions, transfer to new areas, or temporary termination of production;
- changes in the organization's operating activities;
- increased influence of regulators, partners, and clients on the organization's BCM program;
- financial and other losses incurred by the organization during the implementation of previous BCP and DRP plans;
- long-term business interruption in the event of recent emergency events;
- increased threat factors (increased probability of an undesirable event or business impact).

Tracking test results:
When updating the BCP and DRP plan, one must consider the latest test results.

Updating the BCP and DRP plan when the infrastructure changes:
If the infrastructure changes, the plan must be updated to review the new infrastructure components (Figure 2.22). For example, if an organization has launched a new wireless network, the relevant items in the BCP and

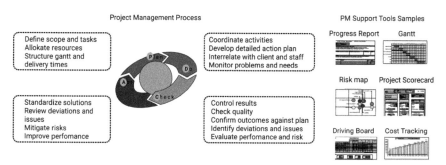

Figure 2.22 ECP development project management.

DRP plan should be revised to include disaster recovery actions for added components of the critical infrastructure components.

Updating the BCP and DRP plan when changing the composition and structure of business processes:
If there is a change in the composition and structure of critical business processes, the BCP and DRP plan must be updated.

Approving and implementing BCP and DRP plans:
Once a plan is developed and tested, it must be reviewed and approved by management. This means that the plan meets the goals and objectives of the organization's business and management authorizes its implementation.

Evaluating the adequacy of the BCP and DRP plan:
First, the organization's management makes sure that the BCP and DRP plan meets the goals and objectives of the business. Then the necessary policies, procedures, and responsibilities for implementing the organization's BCM program are declared. However, if the organization's business depends on the services of external partners and contractors, the adequacy of the partners' and contractors' continuity plans is assessed in terms of compatibility with the organization's BCP and DRP plan.

The acquisition of the necessary skills:
It is necessary to develop plausible business interruption scenarios and verify the effectiveness of the approved BCP and DRP plan. At the same time, it is desirable to work out the action situations in various situations. For example, in the event of hacker attacks, virus attacks, terrorist attacks, equipment failures, fire, other natural disasters, etc.

A detailed work operation analysis:
As a result of checking the effectiveness of the BCP and DRP plan, it is necessary to hold working meetings with a detailed work operation analysis to find out the reasons for non-fulfillment or late fulfillment of business continuity tasks. Here it is necessary to draw up a protocol of each working meeting with an indication of the issues under consideration, the appointment of responsible persons and the definition of further steps to improve the BCP and DRP plan.

Preparing for an external certification audit:
As a rule, the organization is checked by external auditors once a year. At the same time, the auditors identify the existing risks of the BCM program and make recommendations for proper business continuity management. It is recommended that these comments be considered when preparing and passing the certification audit of the organization.

2.5 AS/NZS 5050:2010 Standard

The Australian National Audit Agency has prepared a number of regulatory documents for the certification audit of business continuity management programs[55]; the main ones are:

- Standard *AS/NZS 5050 Business continuity – Managing disruption – Related risk. 5050.1 Part 1: Specification*.
- Standard *AS/NZS 5050 Business continuity – Managing disruption – Related risk. 5050.2 Part 2: Practice*.
- Standard *HB 292:2006 – A Practitioners Guide to Business Continuity Management*.
- *Better Practice Guide Business Continuity Management – Keeping the wheels in motion. 2000*. This guide consists of two parts. The first part contains the main terms and definitions in the field of business continuity management. The second part contains recommendations for developing BCM strategies, plans, and procedures.
- Report No. 53 2002-03 – Business Support Process Audit – Business Continuity Management Follow-on Audit, 2002.
- Business Continuity Management and Emergency Management in Centrelink, the Auditor-General Audit Report No. 9 2003-04 Performance Audit.
- Business Continuity Management Guide, ANAO Report No. 6 2014-15, etc.

2.5.1 Basic Recommendations

AS/NZS 5050:2010 (HB 292:2006) considers business continuity management as part of a more general risk management process. This considers both internal and external risks that are beyond the direct control of the organization. For example, these may be internal risks at the strategic level that affect the organization as a whole or operational risks that locally affect the organization's business processes. It is also considered that the consequences of risks could be different and lead to financial losses, legal harassment, loss of image and reputation, harassment by regulatory authorities, business interruption, etc.

Here, business continuity refers to the property of maintaining the availability of critical business process and supporting assets for the stable functioning of the organization in emergency situations. At the same time, the traditional top-down approach to risk management from business

---

[55]https://www.anao.gov.au/sites/default/files/ANAO_Report_2014-2015_06.pdf

goals and objectives to business processes and assets, discussed in detail in the *Australian and New Zealand Standard on Risk Management (AS/NZ 4360:1999)*, is supplemented by a bottom-up approach to business continuity management itself. Figure 2.23 shows the main features of risk management and business continuity.

It is assumed that risk management determines organizational and technical countermeasures that will be aimed primarily at preventing events that interrupt business (Table 2.9). Business continuity management is considered as a component of risk management and will determine economically justified countermeasures taken in the event of business interruptions (Table 2.9). Essentially, BCM concerns actual events – the implementation of threats to business continuity – and the actions to be taken in response to these events. In this sense, BCM complements the risk management process, which mostly concerns the probability of implementing threats of business interruption and the choice of preventive measures to protect against such events. For example, using passwords to access an organization's information systems is a preventative measure to protect against unauthorized access. Whereas, viewing the computer log of attempts to access the information system is a corrective measure that will detect the fact of unauthorized access. In the context of BCM, an organization starts with the assumption that preventive security measures did not help (or were not used) and business interruption occurred. In this case, the organization must respond to the designated events in proportion to

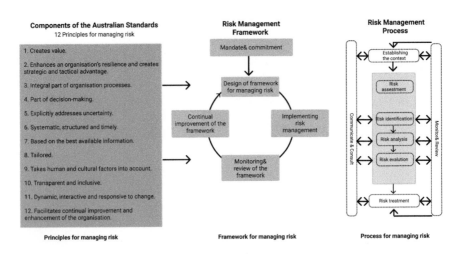

* AS/NZS ISO 31000: 2009 refers to 11 principles. The 12th principle comes from the Business Continuity standart AS/NZS 5050:2010, which is aligned with AS/NZS ISO 31000

Figure 2.23   ANAO recommendations on risk management and business continuity.

their significance – questions of the probability of threats and the original cause (vulnerability of the system), therefore, are no longer considered. Otherwise, if the organization does not take adequate response actions in a timely manner, its inaction leads to serious negative consequences for a period exceeding the maximum allowed time for business interruption.

Table 2.9 Risk management principles.

Risk management principles:
Governance Institute's Risk Management Framework applies the principles for managing risk set out in the Australian Standards (AS/NZS ISO 31000:2009 and AS/NZS 5050: 2010) as shown below.

No.	Principles for managing risk	How Governance Institute applies these principles
1)	Creates value	Risk management adds value to and improves the success rate of Governance Institute's advocacy and educational and training offerings and helps maintain the viability and quality of its business.
2)	Enhances an organization's resilience and creates strategic and tactical advantage	The process for managing disruption-related risk involves anticipating rapid change, operating in non-routine modes, and adapting to a changing environment within the context of our objectives. The experience of doing this enhances our adaptive capacity in that Governance Institute has developed a business continuity program to manage the impact of changes.
3)	Integral part of organization processes	Risk management is applied to Governance Institute's processes starting with strategic planning, through budgeting, to operations, and is embedded in staff KPI's and performance appraisals.
4)	Part of decision-making	Governance Institute's decision-making is done against the background of known documented risks in a risk register. The risk register, with mitigation strategies, is updated where necessary as decisions are made.
5)	Explicitly addresses uncertainty	Governance Institute addresses uncertainty in the analysis and evaluation phases of its risk management process.
6)	Systematic, structured, and timely	Governance Institute has documented its Risk Management Policy and Process and established a Risk Register. Management addresses risk in its decision-making and incorporates risk management into its performance appraisal. The RAF Committee periodically reviews risk management and reports to the Board.

Continued

Table 2.9   Continued

No.	Principles for managing risk	How Governance Institute applies these principles
7)	Based on the best available information	Governance Institute uses historical data, experience, stakeholder feedback, observation, forecasts, and expert judgment in the identification, analysis, and evaluation phases of the Risk Management Process.
8)	Tailored	Governance Institute's Risk Management Framework has been developed specifically for Governance Institute. Risks have been categorized by risk type as well as by responsible manager.
9)	Takes human and cultural factors into account	Governance Institute's Risk Management Framework has been developed to be easily understood by all levels of management within Governance Institute and to be a benchmark in Governance Institute's promotion of effective governance.
10)	Transparent and inclusive	Governance Institute's Risk Management Framework is widely available to all levels of management and the Board of Governance Institute. All managers are responsible for the management of risks within their area of responsibility and are evaluated accordingly against their KPI's. Risk is monitored and reviewed by the RAF Committee and the Board.
11)	Dynamic, iterative, and responsive to change	Governance Institute's Risk Register and risk mitigation strategies are regularly reviewed by responsible managers and through the performance appraisal process. Risk is reported up to the RAF Committee and Board, and feedback is provided at each management level. The Policy, Process and Register are subject to annual review by the RAF Committee.
12)	Facilitates continual improvement and enhancement of the organization*	The Risk Management Framework provides the assurance and confidence that management and the Board need to develop Governance Institute's future strategies and improve the operation of its business.

*AS/NZS ISO 31000:2009 refers to 11 principles. The 12th principle comes from the Business Continuity Standard AS/NZS 5050:2010, which is aligned with AS/NZS ISO 31000.

Table 2.10 Risk assessment methodology.

Setting the overall risk rating

The matrix used then categorizes risk into four levels as follows:
E: extreme
H: high
M: moderate
L: low

Likelihood	Consequences				
	1. Insignificant	2. Minor	3. Moderate	4. Major	5. Catastrophic
A. (Almost certain)	M	H	H	E	E
B. (Likely)	M	M	H	H	E
C. (Possible	L	M	H	H	H
D. (Unlikely)	L	L	M	M	H
E. (Rarely)	L	L	M	M	H

Risk control effectiveness
This is achieved by linking each mitigating strategy control using the following ratings:

Control rating	Definition
Excellent	• Highly dependable risk control process and procedures in place that can be relied upon to prevent risk materializing • 90%–100% effective
Good	In most circumstances, control will be effective to prevent risk event occurring or to mitigate risk in the event it does occur. • 80%–90% effective
Satisfactory	Control is in a place and works most of the time. Risks will be controlled most of the time. • 50%–70% effective
Poor	Control in place, however, is considered to be generally unreliable or relatively ineffective. No guarantee risk will be controlled: • 20%–50 % effective
Unsatisfactory	Control is totally ineffective. Risk will not be controlled. • Less than 20% effective

Figure 2.23 shows the business interruption risk management algorithm, which consists of the following four steps:

- the definition of the characteristics of the business;
- residual risk assessment;
- implementing the required countermeasures;
- evaluating the effectiveness of countermeasures taken.

2.5.2 The Application Specifics

The AS/NZS 5050 standard (HB 292:2006) emphasizes that the main purpose of BCM is to keep up-to-date a sufficient number of resources necessary for the stable functioning of the organization in emergency situations. This BCM representation differs significantly from the concept of disaster recovery (DR), which has been closely, if not exclusively, associated with information technology. Here, the concept of continuity has been strengthened by shifting the focus to the organization as a whole and not just on technological aspects.

The AS/NZS 5050 standard (HB 292:2006) distinguishes the following main stages of an enterprise business continuity management program, the features of which are described in detail in Table 2.11:

- run the project;
- identification of key business processes;
- business impact assessment, BIA;
- developing business continuity measures;
- implementation of business continuity measures;
- BCP testing and support.

Among the listed stages, a fairly important place is given to the stage of business impact analysis, BIA. Let us comment on this stage in more detail.

At the BIA stage, the main focus is on assessing possible losses resulting from business process failures. In this case, various categories of losses are considered – for example, exceeding the standard level of operating costs, penalties as a result of violation of contractual obligations, reduced return on investment relative to the planned level, loss of business reputation of the company, and so on up to a decrease in the market value of the company.

The source map of the organization's key business processes is required for proper analysis of the BIA. Further, for each business process, various types of violations of functioning are identified, which potentially lead to losses. Based on the map of key business processes, it is possible to build an analytical model that links various violations in the functioning of

Table 2.11   Features of the BCM enterprise program.

1) Project launch • Documentation of project boundaries, goals, and objectives • Approval of the project charter • Establishment of a steering committee • Approval of the budget and timeline of the project	Project charter	Requires management involvement
2) Identification of key business processes • Identifying key business processes • Ranking of key business processes • Defining the scope and mutual dependencies for each business process • Determining the required resources to perform business processes	Map of ranked critical business processes	List of critical business processes • Methods of business process ranking • Ranked list of business processes
3) Business impact analysis, BIA • Identifying interview participants • Planning and conducting interviews • Documenting interview results • Determining the maximum allowable business interruption time, MAO • Analysis of the impact on the operating and financial performance of the organization	Values of maximum acceptable time (MAE) during which business can be continued	• Reduce the impact of the emergency situation • Acceptable process and service interrupt recovery
4) Development of business continuity measures • Identification and evaluation of possible measures • Selection of alternative activities and resources	Residual risk management plan for measures	Contents • Title sheet • Terms and definitions • Procedure for registering incidents • Contingency plans • Description of services • Background material • Contact details • Assumptions and limitations

Continued

Table 2.11   Continued

5) Implementation of business continuity measures	• Contractual commitments	
• Identify the recovery team • Documentation of necessary actions and steps • Determination of escalation of events • Obtaining contact lists • Preparation of business continuity plan, BCP	• Procedures review regulations business continuity plan	
6) Testing and plan support • Development of test plans • Checking test plans • Determination of required reserves • Reserve management	BCP testing plan	BCP testing interval • Annually

business processes with the category and scale of losses as a result of such violations. Depending on the scale, losses can be estimated quantitatively (in monetary terms) or qualitatively (according to a specially developed qualitative scale). According to the results of possible losses, an assessment model will evaluate the criticality of business processes as a whole and assess the criticality of various kinds of malfunctions with reference to the scale of losses.

Together with the analysis of the criticality of business processes and the dependence of the scale of losses from violations of business processes, it is recommended to analyze information services with reference to business processes and information flows. For example, the analysis of an enterprise accounting system, consolidated reporting system, or business intelligence system based on a data warehouse, information data, information portal, enterprise email, network printing service, and so on could be performed. At the same time, a deeper level of detail is recommended since, for example, the enterprise accounting system actually provides several services (accounting support, human resource management support, support for material and technical accounting, accounting, etc.) that are involved in various ways in the company's business processes. During the analysis of information services, these services are identified, their use within the company's business processes and possible violations in the functioning of the services are analyzed, and a preliminary assessment of the significance of information services from the company business point of view is performed.

The business impact analysis is recommended to be completed by building a model of cause-and-effect relationships between the functioning of business processes, information processes, information services, and information flows. This model will give assessment for each class of services based on information about the criticality of business processes and information flows as well as the scale of possible losses:

- criticality of the service from the company's business point of view;
- possible losses for the business depending on the disruption in the operation of the service and the recovery time;
- costs associated with the increased availability of the service.

Thus, possible results of the business impact analysis may include the following:

- Results of the business processes analysis and information flows:
  - map of the company's business processes and information flows with the necessary level of detail;
  - map of possible violations of the company's business processes and information flows;
  - analytical model for assessing losses as a result of violations of the company's business processes and information flows (quantitative and/or qualitative assessments);
  - resulting assessments of the criticality of the company's business processes and information flows;
  - resulting assessment of the criticality of various types of violations of the functioning of business processes and information flows of the company with reference to the scale of possible losses.
  - Results of analysis of information services:
  - list of information services under consideration with brief descriptions;
  - map of the use of information services by the company's business processes;
  - map of possible violations of information services;
  - preliminary assessment of the criticality of information services for business.
  - Results of the analysis of the impact of information services on business:
  - analytical model for estimating losses as a result of an information service malfunction;
  - resulting assessments of the criticality of information services themselves and various types of violations of their functioning;
  - preliminary estimates of economically justified expenses for increasing the level of information services accessibility.

2.6   Risk Management Practices

2.6.1   ISO 31000 Family of Standards

The ISO 31000 family of standards are prepared by the ISO/TC 262 *"Risk management"* technical committee of the international organization for standardization (ISO)[56] to manage various types of organization risks, including business interruption risks.

The **ISO 31000** family includes three current standards:

- **ISO 31000:2018** *"Risk management – Guidelines,"* Risk Management. *Principles and guidelines*;
- **ISO/TR 31004:2013** *"Risk management – Guidance for the implementation of ISO 31000,"* Risk management. *ISO 31000 Implementation guide;*
- **IEC 31010:2019** *"Risk management – Risk assessment techniques," Risk Management. Risk assessment methodology.*

And six **ISO 31000** family standards are in the works:

- **IWA 31** *"Risk management – Guidelines on using ISO 31000 in management systems"*;
- **ISO/FDIS 31022** *"Risk management – Guidelines for the management of legal risk,"* Risk management. Guidelines for the legislative risk management;
- **ISO/CD 31030** *"Risk management – Managing travel risks – Guidance for organizations,"* Risk management. Guidelines for the legislative risk management. Tourism risk management. Guide for organizations;
- **ISO/AWI 31050** *"Guidance for managing emerging risks to enhance resilience,"* Risk management – Guidelines for improving the organization resilience;
- **ISO/CD 31070** *"Risk management – Guidelines on core concepts,"* Risk management. Guidelines on basic risk management concepts;
- **ISO/AWI 31073** *"Risk management – Vocabulary,"* Risk management. Terms dictionary.

In February 2018, ISO 31000:2009 was replaced with an updated version of ISO 31000:2018 (Figure 2.24). The standard recommends that government and commercial entities integrate the risk management process into the overall management system of the organization. ISO 31000:2018

---

[56] https://www.iso.org/committee/5259148.html

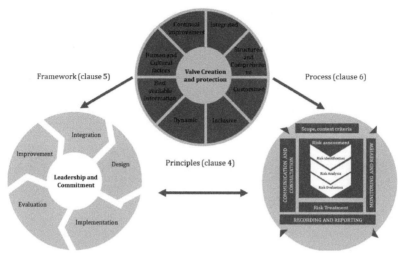

Figure 2.24   ISO 31000:2018 recommendations for risk management.

offers the following principles, framework, and key processes for managing the organization's risks.

The main principles of ISO 31000:2018 include:

a. *Integrated*. Risk management is an integral part of all organizational activities.

b. *Structured and comprehensive*. A structured and comprehensive approach to risk management contributes to consistent and comparable results.

c. *Customized*. The risk management framework and process are customized and proportionate to the organization's external and internal context related to its objectives.

d. *Inclusive*. Appropriate and timely involvement of stakeholders enables their knowledge, views, and perceptions to be considered. This results in improved awareness and informed risk management.

e. *Dynamic*. Risks can emerge, change, or disappear as an organization's external and internal context changes. Risk management anticipates, detects, acknowledges, and responds to those changes and events in an appropriate and timely manner.

f. *Best available information*. The inputs to risk management are based on historical and current information as well as on future expectations. Risk management explicitly considers any limitations and uncertainties associated with such information and expectations. Information should be timely, clear, and available to relevant stakeholders.

g. *Human and cultural factors.* Human behavior and culture significantly influence all aspects of risk management at each level and stage.

h. *Continual improvement.* Risk management is continually improved through learning and experience.

The ISO 31000:2018 Framework includes *leadership and responsibility; integration; development; implementation; monitoring; improvement.* At the same time, the new stage is the introduction of risk management in the organization, which implies:

- developing an appropriate plan including time and resources;
- identifying where, when, and how different types of decisions are made across the organization and by whom;
- modifying the applicable decision-making processes where necessary;
- ensuring that the organization's arrangements for managing risk are clearly understood and practiced.

According to ISO 31000:2018, the main stages of an organization's risk management are (Figure 2.24):

- communication and consultation;
- scope, context, and risk criteria;
- risk assessment (identification, analysis, and evaluation);
- risk treatment (selection of risk treatment options, preparing and implementing risk treatment plans, etc.);
- monitoring and review;
- recording and reporting.

Here, more than **30** methods from the ISO/IEC 31010 standard could be used for risk assessment; in particular, brainstorming, Delphi method, preliminary hazard analysis, *HAZOP, HACCP, FMEA, FTA,* decision tree, *SWIFT* technique, *Monte Carlo method,* and other well-known risk management methods.

2.6.2   Managing Cyber Risks

An approach to a cyber resilience (and cybersecurity) on the basis of cyber risks management ISO/IEC 27005 (2018), ISO/IEC 22301 (2012), and NIST SP 800-160 (2018) is relatively recent. This approach has been attended by the driving forces from the two most advanced groups of regulatory documents (Figure 2.25):

- Standards and recommendations for IT risks management within the entire company (there is a large group of standards among which the

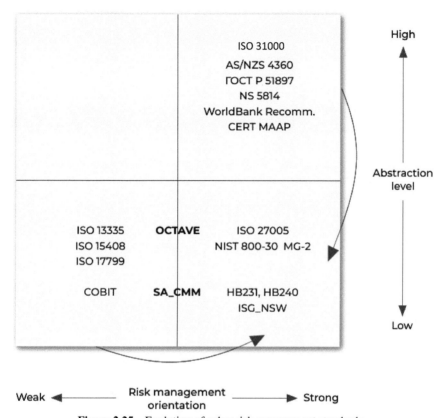

Figure 2.25   Evolution of cyber risk management standards.

Australian standard is clearly distinguished AS/NZS 4360 (2004). First of all, due to the wide practical application and, second, due to the fact that several related and developing documents, such as HandBook 231 "Information security risk management guidelines," HandBook 240 "Guidelines for managing risk in outsourcing," and "Information Security Risk Management Guideline for NSW Goverment," are directly related to risk management in IT). Currently, the standard AS/NZS 4360 (2004) is the basis of the international standard ISO 31000:2018 – Risk management – Guidelines.

- Set of standards ISO/IEC 27000 (ISO/IEC 17799:2000), ISO/IEC 15408, and ISO/IEC 22300 (BS 25999:2006) for cybersecurity management and business continuity, respectively.

As a result, these driving forces have generated several national guidelines and standards, which have been recognized by the international community as the best cyber risk management practices. Thus, the following publications have achieved the certain prominence:

- "Specific NIST publication (American Standards Institute) NIST SP 800-30 Guide for Conducting Risk Assessments[57];
- Guidelines and the method "OCTAVE – Operationally Critical Threat, Asset and Vulnerability Evaluation" of the initiative group of specialists from the Carnegie Mellon University (Software Engineering Institute (SEI) at Carnegie Mellon University)[58];
- Canadian guide to security risk management for information technology systems (Security Risk Management for Information Technology Systems(MG-2) (having the status of recommendations for government agencies)[59];
- Highly specialized IT risk management recommendations, including the Australian security outsourcing guidelines, Risk management guidelines for software acquisition capability at Carnegie Mellon University (SA-CMM) (Software Acquisition Capability Maturity Model)[60], etc.

The main goals of the specialized standards and risk management guidelines are the following:

- Filling a void between the senior management, operating with terms of business processes, business continuity and sustainability, and the technicians operating with terms of vulnerabilities and technical/organizational security tools.
- Specifying the adequate organizational and technical tools required for proper cybersecurity (Figures 2.26 and 2.27).
- *Cyber risk* – a combination of an event probability and its consequences.
- *Vulnerability* – an error or a defect of organizational processes, structure, or implementation of technical means which can cause (accidentally or deliberately) violation of cyber resilience (cybersecurity).
- *Threat* – a potential for implementation of a certain vulnerability.
- *Impact* – a degree and form of damage caused to the asset by the vulnerability realization.
- *Cyber risk management* – a set of the coordinated events to manage the company (both the components of informational infrastructure

---

[57] https://csrc.nist.gov/publications/detail/sp/800-30/rev-1/final
[58] https://resources.sei.cmu.edu/library/asset-view.cfm?assetid=13473
[59] http://www.cse-cst.gc.ca/en/services/publications/itsg/MG-2.html
[60] https://resources.sei.cmu.edu/asset_files/TechnicalReport/2002_005_001_14036.pdf

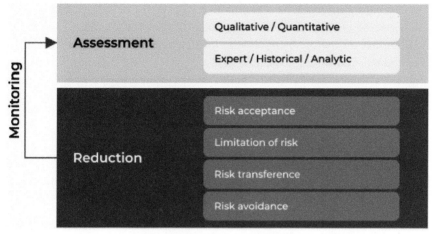

Figure 2.26 Risk management

Selecting Tools and Services for ITRM

Risk Management Tool Service

• **ERM**	Operational Risk Management
• **IT**	IT Risk Management
	GRC
	IT Risk Assessment
	Control Assessment
	Business Impact Assessment
• **Business Continuity**	Business Impact Assessment
	Risk Assessment
• **Security**	Risk Assessment
	Business Impact Assessment
	Threat Assessment
	Vulnerability Assessment
	Control Assessment

Figure 2.27 Selecting tools and services for ITRM.

and means of resiliency and cybersecurity, and the entire vertical of management in general) in order to minimize the overall cyber risk.

• *Cyber risk assessment* (the resulting measure of probability and impact can be expressed either qualitatively – **3/4/5** degrees, or quantitatively – the probability is in average expected frequency of the event in a given time interval (month/year), and the impact is in monetary terms). One of the main results of the cyber risk assessment process is their prioritization, according to the degree of potential impact on the company assets. In this case, the assessment of cyber risks is carried by means of:

o Expert assessment (directly (explicitly) or indirectly –with the use of special software and hardware, based on some knowledge about the dependence of any cyber risk measure on the observed conditions).

o Historical data about possible vulnerability and an impact caused by its implementation (the method's drawbacks are the need for a large amount of historical data (and for some threats, it may not simply exist) and the impossibility of the accurate assessment of the trend in the event of changing environment, which happens almost in all areas of cybersecurity).

o Analytical approaches (which are mostly in academic research), for example, the plotting of weighted transition graphs to determine the impact size from the vulnerability implementation.

• *Measures aimed at countering cyber risk* (reducing the overall risk of a company), including:

o Passive actions:
  o cyber risk acceptance (the acceptance of the observed cyber risk level without any countermeasures);
  o cyber risk avoidance (the decision to transform the activity, which will entail this level of cyber risk).

o Active actions:
  o limitation or reduction of a specific cyber risk (consists of a set of organizational and technical measures which we have used to interpret as measures to ensure information security);
  o risk transference (insurance) is still rather a rare procedure which is gradually gaining a recognition.

• *Set of measures for internal audit as well as internal and external status monitoring of the cyber resilience (cybersecurity).* First, they check the quality of measure implementation to reduce cyber risks, their adequacy, the achievement of a target function through the internal alterations in the company and then assess the changing external surroundings (the new types of threats and new ways of implementing the already known threats). In all the circumstances, in case of a significant discrepancy between the current surrounding and implemented measures, the monitoring subsystem shall initiate partial or full review of the company's cyber risk policy.

For example, the following approaches could be used for the primary analysis of cyber risks:

- *consequences and probabilities matrix computation*;
- structured what-if technique (SWIFT);
- root causes assessment (RCA);
- business impact analysis (BIA);
- failure mode and effects analysis (FMEA);
- layers of protection analysis (LOPA);
- event tree analysis (ETA);
- cause-consequence analysis;
- human reliability assessment (HRA);
- sneak analysis (SA), etc.

For a more profound analysis of cyber risks, the following techniques can be applied:

- *Delphi technique*;
- checklists-based method;
- brainstorming technique;
- method of partially structured or structured interview;
- preliminary hazard analysis (PHA);
- methods of Bayes network-based analysis;
- Monte Carlo technique and others.

In order to develop models of cyber threats, the following methods can be applied:

- *methods of expert assessment*;
- methods of mathematical statistics;
- Markov technique;
- event-logical approach;
- failure mode, effects and critical analysis (FMECA);
- fault tree analysis (FTA);
- event tree analysis (ETA), etc.

Let us proceed with the special issues of cyber risk management on the examples of the mainstream standards and guidelines [201].

2.6.3 The NIST SP 800-30 Standard

Guidelines NIST SP 800-30 Guide for Conducting Risk Assessments – American National Standards Institute: this is the most neutral standard, describing all the stages of IT risk management processes[61]. The above-mentioned guidelines became widely known and subsequently formed the basis of the international standard ISO/IEC 27005:2018 Information technology – Security techniques – Information security risk management[62].

Figure 2.28 illustrates the key milestones of the IT risk assessment process recommended by the NIST SP 800-30:

- classification of systems and services;
- identification of threats;
- identification of vulnerabilities;
- management system analysis;
- defining the probabilities of vulnerability;
- influence quantity analysis;
- risk estimation;
- recommendations;
- documents drawing up.

Traditionally, the standard considers three classes of threat sources:

- Natural (e.g., earthquake, flooding, etc.).
- Human factor:
  ○ indeliberate;
  ○ deliberate.
- Technical (voltage loss, flooding, fire, etc.).

The definition of vulnerability probabilities is made on the basis of qualitative criteria:

- High: the motivation and vulnerability are present, and security tools are weak.
- Middle: the motivation and vulnerability are present, and security tools are strong.
- Low: the motivation is weak, the vulnerability is insignificant, or security tools are rather strong.

Determination of influence quantity can be either qualitative or quantitative. For example, the computation of IT risk can be performed using a matrix, as it is shown in Table 2.12, which also illustrates the quantitative version of the influence quantity.

---

[61] https://csrc.nist.gov/publications/detail/sp/800-30/rev-1/final
[62] https://www.iso.org/standard/75281.html

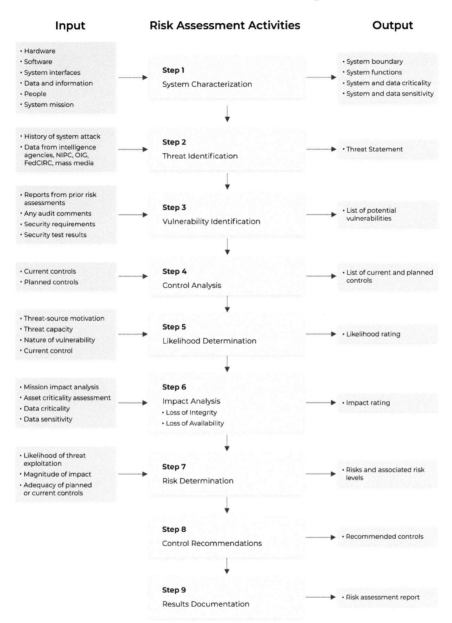

Figure 2.28   IT risk assessment.

Table 2.12   Cyber risks computation example.

Threat probability	Cyber risk level		
	Low (10)	Middle (50)	High (100)
High (1.0)	Low $10 \times 1.0 = 10$	Middle $50 \times 1.0 = 50$	High $100 \times 1.0 = 100$
Middle (0.5)	Low $10 \times 0.5 = 5$	Middle $50 \times 0.5 = 25$	Middle $100 \times 0.5 = 50$
Low (0.1)	Low $10 \times 0.1 = 1$	Low $50 \times 0.1 = 5$	Low $100 \times 0.1 = 10$

Figure 2.29 illustrates the possible stages of the IT risk reducing process:

- prioritization of activities (risks);
- countermeasure optimization;
- assessment "cost-effectiveness";
- tool selection;
- responsibility assignment;
- countermeasure integration plan development;
- instrumental assign.

"Cost-effectiveness" assessment is performed by means of the costs comparing (for the implementation of a particular tool) with IT risks, in case of refusal to implement these tools. Herewith, the principle of minimal sufficiency of security tools is formulated.

2.6.4   OCTAVE Methodology

"OCTAVE – Operationally Critical Threat, Asset and Vulnerability Evaluation" method was developed by the Software Engineering Institute (SEI) at Carnegie Mellon University on behalf of the Department of Defense[63]. At the same time, for an academic community, the tendency toward a higher level of abstraction and universalism has been expressed both in the external document's structure and in their internal one. First of all, the document itself is divided into three parts: OCTAVE-criteria (Figure 2.30), which contains the most abstract requirements and recommendations, and two documents describing options for implementation of these criteria for large companies (the so-called OCTAVE-method) and for small companies (the OCTAVE-S-method).

---

[63] https://resources.sei.cmu.edu/library/asset-view.cfm?assetid=13473

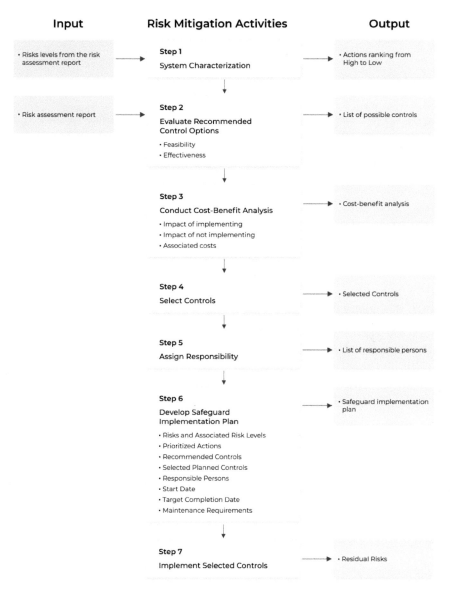

Figure 2.29 IT risk estimator.

OCTAVE-criteria are first formulated as 10 principles of assessment and cyber risks management, which lead to 15 basic requirements (attributes) for the conducted processes. Then three main stages (phases) are formulated:

- threats profiling for assets;

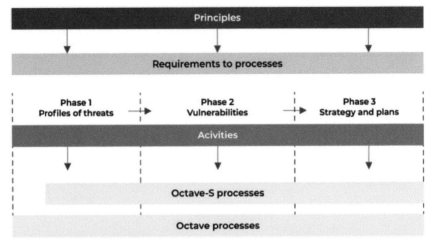

Figure 2.30   OCTAVE-method structure.

- identification of informational structure vulnerabilities;
- development of cybersecurity strategies and plans.

In these phases, there are 16 activities that show themselves (variously in the companies of different size) in the actual processes of IT risk assessment and management. Definition and description of the processes are methodically documented in OCTAVE.

It must be noted that these recommendations carry out the entire process of assessment and management of IT risks, proceeding from the classification of assets. Thus, at the end of each phase, the main documents are the profiles of threats/vulnerabilities/cyber risks/residual IT risks for each type of asset.

OCTAVE guidelines pay great attention to the composition and content of output documents for each of the stages of IT risk analysis and management.

2.6.5   MG-2 Lifecycle

Guidelines for IT risk management of the Canadian government MG-2 (A Guide to Security Risk Management for Information Technology Systems) have been created on the basis of three previously existing government documents: Security policy, Guidelines for assessing and countering IT risks, and Guidelines for certification and accreditation of systems[64].

---

[64] http://www.cse-cst.gc.ca/en/services/publications/itsg/MG-2.html

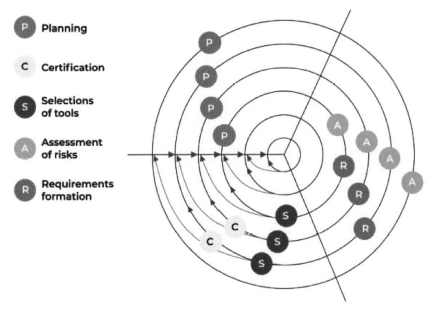

Figure 2.31 Lifecycle model of cyber risks management, MG-2.

The aforementioned guidelines formulate the main stages of IT risk management. Thus, their statement is based on the lifecycle of an information system (the spiral line, each turn of which requires the performance of certain processes from the methodological set of IT risks management – Figure 2.31):

- Planning:
  - data acquisition and system description;
  - determination of acceptable risk levels.
- Preparation for the assessment and analysis of IT risks:
  - identification of assets;
  - drawing up an "Assets Classifier" indicating the security requirements.
- IT risk assessment:
  - identification of threats, vulnerabilities, and existing security tools;
  - identification of the possible threat influence quantity on the asset.
- Decision-making on mitigation, avoidance, transfer, or acceptance of IT risk.
- Development of asset security requirements (in case of the accepted decision to mitigate IT risk).

- Selection of measures and security tools (organizational and/or technical).
- Implementation of security and safety systems in general.
- Certification.
- Accreditation.
- Functioning support.
- Proposal generation for improvements.

Particular attention is paid to the "Asset Classifier," which provides detailed descriptions of assets, their roles and values, and security requirements.

2.6.6   COBIT 2019 Standard

In the COBIT 2019 – Control Objectives for Information and Related Technology, the process of IT risks assessment and management is included in the first part of the guideline "Planning"[65]. Table 2.13 displays the recommended steps of the process as well as the possible assignment of responsibility.

The quality metrics of the IT risks management process according to COBIT 2019 are:

- share of IT budget, which is spending on the procedures of risks assessment and management;
- percent of IT objectives of the companies covered by risks management procedures;
- share of potential threats for which the risk assessment has been carried out in full and the risk management plan has been developed;
- percent of incidents in the sphere of information security, which, at the time of occurrence, has not been considered by the process of risk assessment, etc.

2.6.7   SA-CMM Maturity Model

In the Software Acquisition Capability Maturity Model (SA-CMM) of the Carnegie Mellon University (Software Engineering Institute (SEI) at Carnegie Mellon University – CMU/SEI), the issues of IT risks assessment and management appear at the third level (of five levels traditionally). This means that the IT risk management process is properly established and documented. However, metrics and measures of IT risk management required for measuring, comparing, and optimizing of processes are missing.

---

[65] http://www.isaca.org/Knowledge-Center/COBIT/Pages/Overview.aspx

Table 2.13   COBIT 2019 guidelines for IT risks management.

Functions and tasks	Executive director	Finance director	Directors	IT director	Lead managers	Responsible for operation	Chief network architect	Development director	Director of administrators	Project managers	Responsible for monitoring and risk management
Defining task of a risk management program	+	+	+	+	+	+					+
Linking the program to a business goal		+	+	+	+	+					+
Defining and ranking the business processes				+	+	+					+
Defining the IT tasks and objectives					+		+	+	+		+
Importance ranking of IT tasks	+				+		+	+	+		+
Defining of information security risks	+			+	+	+	+	+	+		+

Continued

Table 2.13   Continued

Functions and tasks	Executive director	Finance director	Directors	IT director	Lead managers	Responsible for operation	Chief network architect	Development director	Director of administrators	Project managers	Responsible for monitoring and risk management
Defining of a risk management program	+			+	+	+	+	+	+		+
Developing a plan for IT controls	+	+	+	+	+	+	+	+	+		+
Residual risk assessment		+	+		+	+	+	+	+		+
Program maintenance and risk monitoring	+	+	+	+	+	+	+	+	+	+	+

The above-mentioned guidelines[66] enumerate the following activities in IT risks management for acquiring software:

- risk identification;
- constraint identification, default assumptions, and guidance documents at all levels, under which the risk management will be implemented;
- creation of communication infrastructure for the data exchange in the course of risks management;
- personnel training (qualifications corresponding to the risks management plan);
- acquisition, analysis, and reporting on data, including historical data, which is required for making scheduled decisions.

The considered methods of IT risks management allow us to work on:

- formation of requirements to the risks management system;
- developing of methods of analysis and risks management;
- formation of requirements to the risks management software;
- identification of threats and vulnerabilities of information infrastructure;
- residual IT risk assessment.

The results of such works include the following:

- procedure of risks analysis and management;
- rationale for choosing software for risks analysis;
- risks management strategy;
- risks management guideline;
- risks analysis software standard;
- report on residual risks assessment, etc.

Let us note that the above-mentioned guidelines and standards exclude a number of important details, which shall be specified in practice. Specification of these details depends on the overall level of stability of the company's business culture as well as the specifics of its activities. As can be seen from the facts mentioned above, it is impossible to submit some universal enterprise procedure of cyber risks management. In each particular case, the general methods of IT risks management will have to be adapted to specifics of enterprise activity.

---

[66] https://resources.sei.cmu.edu/asset_files/TechnicalReport/2002_005_001_14036.pdf

2.7   Business Process Description Practices

Effective construction of a proper business continuity management program, *ECP*, is impossible without a clear definition and description of the business processes of the enterprise. Various international and European organizations, *ISO, ITU, ETSI, TMF, OASIS*, etc., deal with business-oriented representation of the company's activities. The main efforts of these organizations are focused on the development of appropriate models, methods, and standards for modeling and supporting the main business and production processes; for example, the processes of production management, development and sales management, and management of the enterprise. At the same time, interest arises in the so-called reference approaches to modeling and supporting the main business and production processes, which are applicable for most enterprises.

The modeled process is usually understood as a certain stable sequence of actions related to the production and economic activities of the enterprise and focused on creating a new value chain. At the same time, the process can include a whole hierarchy of interrelated functional actions that implement one (or more) business goal of the enterprise; for example, management and analysis of product output or resource support for the implementation of services, etc. This structural and functional complexity of business and production processes determines the multicomponent nature of the corresponding OSS/BSS modeling and support systems (Figure 2.32).

Historically, the first tools for modeling business and production processes of an enterprise were special CASE-tools (in practice, the capabilities of *MS Power Point* and *Visio* were not enough), such as *CASE – Computer Aided Software/System Engineering, ERWin (CA), S-Designer (Sybase), Rational Rose (IBM), ARIS (Scheer AG)*, and *ORACLE Designer (Oracle Corporation)*. The advantages of the listed CASE-tools include the following:

- built-in capabilities for detailed decomposition of the studied processes into steps and procedures;
- mechanisms for visual representation of processes in the form of graphical diagrams;
- readiness to set and solve various problems of analysis and synthesis in the field of modeling and support of the main business and production processes of the enterprise.

Further, it turned out that following only the recommendations of the universal open system interaction model (*OSI*) and creating the primary regulation of enterprise processes is not enough to support these processes. Moreover, it was necessary to take into account the goals and objectives of

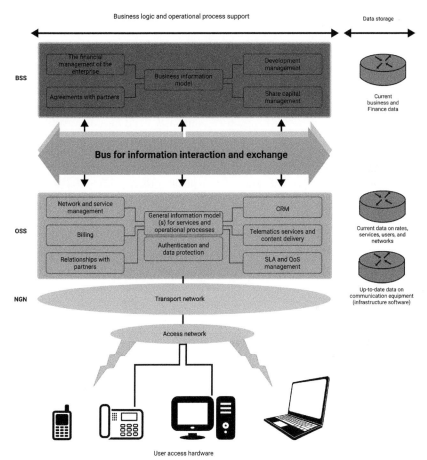

Business logic and operational process support

Data storage

Bus for information interaction and exchange

BSS

The financial management of the enterprise

Business information model

Development management

Agreements with partners

Share capital management

Current business and Finance data

OSS

Network and service management

General information model (s) for services and operational processes

CRM

Billing

Telematics services and content delivery

Authentication and data protection

Relationships with partners

SLA and QoS management

Current data on rates, services, users, and networks

Up-to-date data on communication equipment (infrastructure software)

NGN

Transport network

Access network

User access hardware

Figure 2.32   Possible scheme of the business support model.

the business (the mission and strategy of the business), the requirements of customers and partners, the results of the competitive environment analysis, and the so-called benchmarking analysis (benchmark comparison). As a result, the evolution of methods and tools for modeling and supporting business and production processes has led to the creation and dissemination of comprehensive reference models and tools for business support, which will be discussed later.

2.7.1   Process Modeling

As a rule, modeling of business and production processes of an enterprise begins with the analysis and generalization of the results of previously completed or ongoing projects that affect the description of these processes. These projects most often include:

Implementation of the ERP meta system for decision-making process performed by the Manager. It describes the processes of sampling, analysis, generalization, and presentation of analytical reports for decision-making by the heads of the company's structural divisions. A specialized DataMining VisualLogic tool is usually used as a description tool.

- Implementation of the ERP system. Formalizing key organizational processes for implementing ERP class systems. Visio and Word softwares are usually used as the description tool, and Oracle Tutor is less commonly used
- SOA project. The processes of the financial block, IT and IS services, and the technical block are described. It is usually used by Visio software and Excel files. It can be loaded into the Oracle – Internal Control Management (ICM) module.
- Activity on automation of IT department processes. It describes the processes of interaction between IT departments and other functional blocks on the issues of operation and maintenance of information systems. It is usually used for Visio and MS Word.
- The establishment of the project office. Project management processes in the company are described. Visio and Word softwares are usually used as the description tool.

As a rule, the analysis of the obtained results indicates the need to aggregate the obtained private representations of the business and production processes of the enterprise into a generalized summary model of processes (Figure 2.33). At the same time, the software for describing these processes will create data models that will be used for automating the enterprise's activities by automation services. In particular, the software must provide interfaces with ERP systems to avoid possible inconsistencies in data formats. For example, for the HR component, to check the compliance of the organizational structure and staffing, for the ICM component, to notify process owners of changes, and so on.

Next step – choosing a reasonably mature methodology for modeling business and production processes of the enterprise.

In general, there are two main approaches to modeling these processes. The first involves creating a new unique model of the company's business processes each time based on certain enterprise principles. The second is focused on the use of well-known methods of process modeling. The advantages of the first so-called "*axiomatic*" approach include creating a

AS IS		TO BE
There is no single SOFTWARE that is not compatible and does not allow you to track changes	➡	Using a single SOFTWARE that allows you to" combine " materials created in different programs and manage business processes, creating a common picture
There are no uniform standards for describing business processes, as a result, processes are difficult to compare (description, policy, procedure, process, regulations, standard, regulation, etc.).	➡	Development and implementation of a single methodology (a single "coordinate system"), common terminology, decomposition levels, and so on).Creating a General business process model based on the reference process model
Inconsistency of functional blocks in the description of processes, as a result, cross-functional differences and difficulties in describing cross-functional processes	➡	Coordination and methodological support in describing cross-functional and functional business processes
There is no procedure for implementing the created regulations	➡	Development of the process of implementing regulations and monitoring implementation
There is no analysis of the effectiveness of business processes There is no timely update of business processes	➡	Setting up a process for monitoring the efficiency and effectiveness of business processes and their subsequent improvement

Figure 2.33 Existing problems with the description of business processes.

unique process model for the company that fully meets its needs. The disadvantages of this approach include the following:

- certain subjectivism: the decomposition and description of processes are performed using expert methods during interviews with senior management and ordinary employees of the company;
- complexity of the submission: the complexity of the selection, submission, and approval process requires a lot of time and high competence of the performer;
- difficulties with decomposition: there are contradictions at every junction of processes;
- difficulties in proving the completeness, reliability, and consistency of the approach: there is no guarantee that a holistic view of the company's business processes will be presented, the processes are

correctly decomposed, and all the necessary links between the processes are established;

- high project risks: the approach seems to be applicable only in small companies, mainly production companies, where the number of business processes is small and all of them are intuitive.

The advantages of the second so-called "*constructive*" approach (if it is proved to be applicable to the company) include:

- using proven best practices in business process modeling;
- quick start of work, the approach assumes a systematic view of the company and an example of decomposition of processes at several levels (as a rule, 3–4), which will coordinate the results, and increases the efficiency of work in general;
- using a unified methodology, data process formats, ready-made templates, and so on;
- clear reduction in the risks and complexity of the project, the presence of a single tool for describing and managing the created process model.

The disadvantages of this approach include difficulties in choosing the appropriate business process model for the company (not only taking into account industry but also local specifics), as well as the need to adapt the model to the company's goals and objectives. However, in practice, the second "*constructive*" approach is usually chosen (Figure 2.34), containing developed functional tools, including special software to support existing (creating schemes and regulations for processes) and future tasks (*BPR*, *ISO*, automation support, etc.).

2.7.2   NGOSS Methodology

Currently, a number of approaches to modeling and supporting business production processes are known. Most of them are based on the application of system analysis and the use of appropriate process modeling tools. One of the most famous approaches is the approach based on the use of the *NGOSS methodology – New Generation Operation Systems and Software*.

The *NGOSS* methodology was developed by the international non-profit organization TeleManagement Forum (TMForum)[67], which studies the development and optimization of the business of Telecom operators. Currently, TMForum unites more than 90% of the world's leading tele-

---

[67] www.tmforum.org

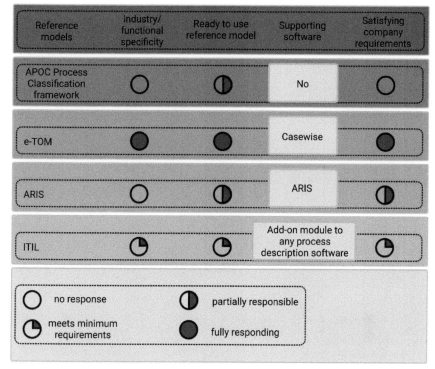

Reference models	Industry/ functional specificity	Ready to use reference model	Supporting software	Satisfying company requirements
APOC Process Classification framework	○	◑	No	○
e-TOM	●	●	Casewise	●
ARIS	○	◑	ARIS	◑
ITIL	◔	◔	Add-on module to any process description software	◔

○ no response ◑ partially responsible
◔ meets minimum requirements ● fully responding

Figure 2.34   An example of the justification of the method for describing business process.

communication companies and includes more than 500 companies in the communications and service industry; in particular, major communication service providers, including MTS and Beeline, hardware and software vendors, software developers, system integrators, etc.

NGOSS is a methodology for implementing, operating, and developing *operational support systems* (*OSS*) and enterprise business support systems (*BSS*). The obvious advantages of this methodology include simultaneous consideration of both technical and non-technical requirements of the business as well as a high level of automation of support for the main processes of the enterprise in heterogeneous IT environments. *NGOSS* includes the extended *enhanced telecom operations* (*eTOM*) business process map, the *shared information data* (*SID*) model, *telecom applications*, and the *technology neutral architecture* (*TNA*) (Figure 2.35). Figure 2.36 shows the structure of the extended telecom operator business process map, the *enhanced telecom operations map*.

At the beginning of 2004, the *eTOM* map (or model) (version GB921 v4. 0) was recognized by the *ITU-T* Institute (the main body for establishing

NGOSS - New Generation OSS

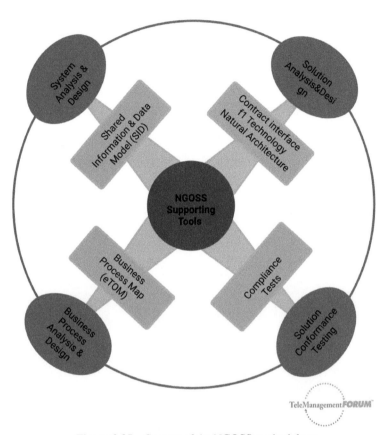

Figure 2.35   Content of the NGOSS methodology.

international recommendations in the field of telecommunications) and offi-
cially recommended for the formal description of the architecture of business
and production processes of telecommunications service operators (recommendations *M. 3050*). Today, the *eTOM* model is a kind of benchmark for
business-oriented representation of communications enterprises. At the same
time, the mentioned model is developing dynamically, considering the needs
of both existing and new organizations participating in the TMForum community. The obvious advantage of the *eTOM* model is that it is not tied to
specific ways of building and technological equipment of the business, and,
therefore, it is applicable to enterprises with different levels of automation.

   Formally, the *eTOM* map is represented as a certain interconnected
hierarchy containing certain sets of business process groups (seven

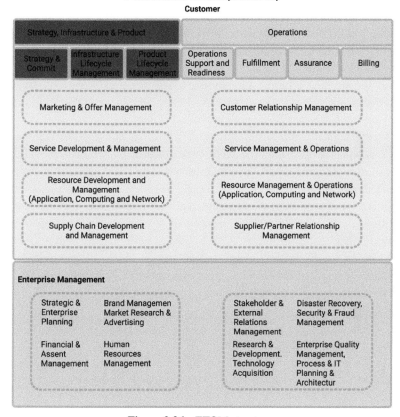

Figure 2.36   ETOM structure.

vertical groups are allocated). Process decomposition starts at the enterprise level and then continues at more detailed lower levels. This defines all the necessary input and output data as well as the relationships between business processes. Conceptually, *eTOM* covers three main process areas:

- decision lifecycle planning and management processes (strategies, infrastructure, and services);
- operational process;
- enterprise management processes.

The "strategies, infrastructures, and services" processes reflect issues of strategic business development, infrastructure, and services lifecycle management. "Operational processes" – issues related to the provision, provision, and billing of services. The "enterprise management" processes cover general issues of supporting the company's activities, such as knowledge management, external relations management, and so on.

The *shared information data (SID) model* describes the operator's activities in terms of: "Customer," "Product," "Tariff," "Service" and "Resource." This model is used for the integration of IT systems and unified presentation of data during the exchange and sharing of information by various divisions of the Telecom operator. The structures of the *SID* data models and the *telecom application map (TAM)* are shown in Figures 2.37 and 2.38.

Technology neutral architecture (TNA) does not depend on specific applications and describes the basic principles of *OSS/BSS* system design and the main technical and design solutions. The structure of this architecture is shown in Figure 2.38.

It is significant that, in practice, various techniques can be used to decompose and correlate the hierarchy of *eTOM* and business processes (Figure 2.39).

In the example shown in Figure 2.40, the business and production process maps consist of two levels: upper and lower. All processes are divided into groups:

- "strategy, infrastructure, and product" – strategic business processes;
- "operational process" – business processes related to the daily activities of the company;

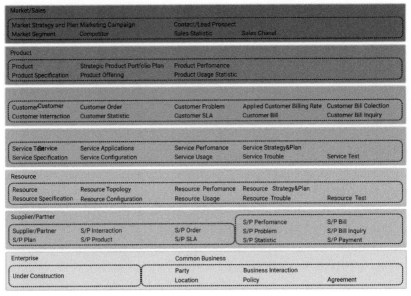

Figure 2.37   Structure of the SID model.

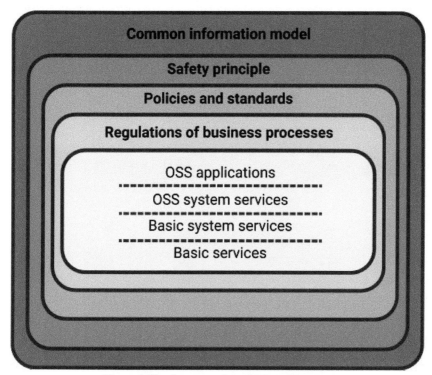

Figure 2.38 Structure of the TNA architecture.

- "company management" – business processes that are responsible for supporting all aspects of the company's services.

Note that in *eTOM*, there are no clear criteria for dividing processes into groups (the specifics of business enterprises may differ). Processes from the "company management" group are auxiliary (these are processes that do not participate in the value chain). Examples of such processes are: "evaluating the company's performance," "putting objects into commercial operation," and "managing physical access to IS."

The most difficult is the division of business and production processes into strategic and operational ones. The features of this division can be demonstrated by the example of the "development of a new product and service" process (Figure 2.40).

In general, the *eTOM* presents the activity of each enterprise as an activity that provides certain services to end users with an acceptable price–quality ratio. The reflection in *eTOM* of such a business-oriented presentation of the enterprise is shown in Figure 2.41.

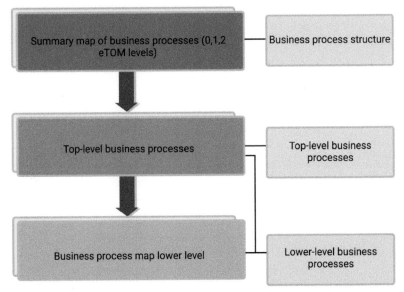

Figure 2.39   Example of matching a business process map and a business process map of an enterprise.

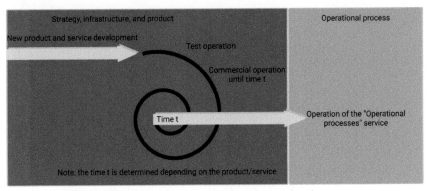

Figure 2.40   Separation of strategic and operational processes.

Here, the processes in the "Execution" area are directed to providing communication services (ordering a service, installing, and configuring it). Processes from the "Provision" area are focused on supporting the provided communication services. The main goal here is to reduce the number of network failures as well as quickly solve emerging technical maintenance problems. Billing processes are designed for timely and error-free payments to both the end user and the intermediate operators involved.

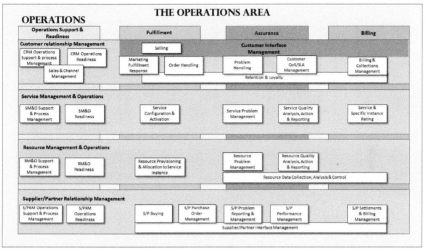

Figure 2.41 Focus on the delivery of services.

Horizontally, the "Operational processes" area is divided into the following processes:

- managing customer relationships;
- managing the service in operational processes;
- managing the service in operational processes;
- managing the supplier/partner relationship.

This division is based on assessments of effective customer service and the development of partnerships. At the same time, such concepts as "product," "service," and "resource" are clearly defined. Here, a product is understood as a tangible and intangible offer to customers, which is additionally characterized by the corresponding parameters of the service provided, orientation to customer segments, as well as certain price offers. A service can be considered as a component of a certain product or products. A resource is a physical and non-physical element used to build a service, for example, elements of a communication network, information systems and technological infrastructure, software, etc.; a possible example of defining the concepts "product," "service," and "resource" for a telecom operator is shown in Figure 2.42.

In the "Company management" area, the following processes are highlighted:

- strategic planning of the company's development;
- financial and asset management;

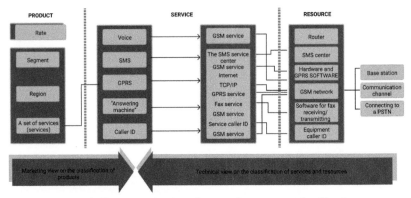

Figure 2.42　Example of product-service-resource classification.

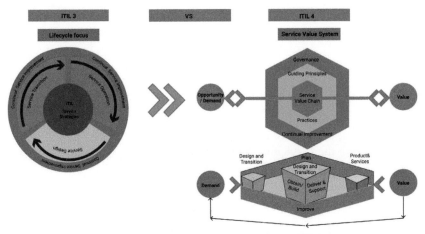

Figure 2.43　Comparison of ITIL V3 and ITIL V4.

- company risk management;
- company performance management;
- IT management;
- managing stakeholder relations and external relations;
- personnel management;
- knowledge and research management.

At the same time, the *Information Technology Infrastructure Library – ITIL* V4 – can be used as the basis for selecting functional subgroups in the "IT Management" area (Figure 2.43), which is currently the *de facto* standard for managing IT resources in companies. The list of *ITIL V4* practices is presented in Table 2.14.

Table 2.14 ITIL practices.

General management practices	Service management practices	Technical management practices
• Architecture management	• Availability management	• **Deployment management**
• **Continual improvement**	• Business analysis	• Infrastructure and platform management
• **Information security management**	• Capacity and performance management	• Software development and management
• Knowledge management	• **Change control**	
• Measurement and reporting	• **Incident management**	
• Organizational change management	• **IT asset management**	
• Portfolio management	• **Monitoring and event management**	
• Project management	• **Problem management**	
• **Relationship management**	• **Release management**	
• Risk management	• Service catalog management	
• Service financial management	• **Service configuration management**	
• Strategy management	• Service continuity management	
• **Supplier management**	• Service design	
• Workforce and talent management	• **Service desk**	
	• **Service level management**	
	• **Service request management**	
	• Service validation and testing	

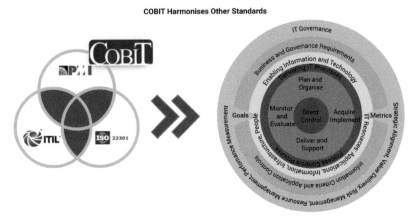

Figure 2.44   Role and place of the COBIT 2019 methodology.

2.8   COBIT Standard® 2019

The standard "*Control Objects for Information and Related Technology (COBIT®)*" is prepared by the *IT Governance Institute (ITGI)*[68] – established in 1998 by the *Information Systems Audit and Control Association (ISACA)*. The current version of the standard is *COBIT® 2019* (Figure 2.44). This standard is also a *de facto* standard and is a description of a structured set of universal IT management tasks that are recommended instead of mandatory requirements in the *de jure standards*. The value of this standard is the business-oriented representation of IT, the use of metrics and maturity models to achieve the goals of IT management, and a clear and unambiguous definition of the responsibility of owners of business processes and IT services.

2.8.1   Description of the DSS04 process

The COBIT® 2019 standard is based on the following statement – in order to provide the information that an organization needs to achieve its business goals, IT resources must be managed by a set of naturally grouped IT processes (Figure 2.45).

In total, 34 IT processes are defined (Figures 2.46 and 2.47 and Table 2.16). Among them, an important place is given to the *DSS04 – Managed Continuity process*.

---

[68] www.itgi.org

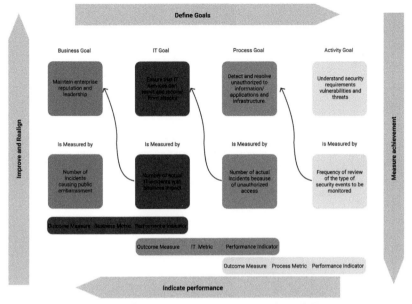

Figure 2.45   The business-oriented view of IT.

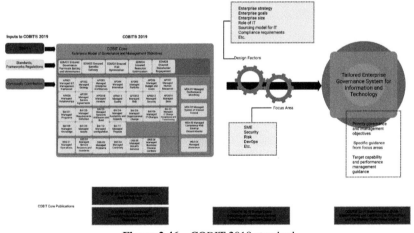

Figure 2.46   COBIT 2019 standard.

The *DSS04 – Managed Continuity* process description contains recommendations for managing business continuity for some IT-dependent company (Table 2.16).

Each of the listed steps in the DSS04 – Managed Continuity process is further explained by a separate table that describes the input and output,

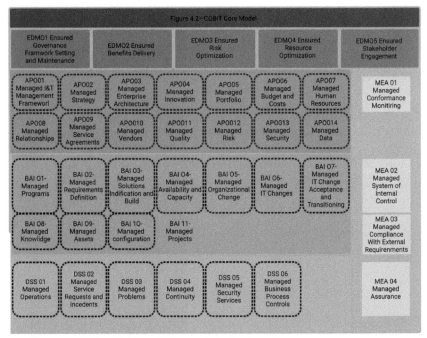

Figure 2.47   COBIT® process 2019.

as well as actions or recommendations (Activities). For example, the DSS04.01 stage of this process is described as follows (Table 2.17).

Significantly, the responsible persons (RACI Chart) and their roles for each of the stages of the DSS04 process are also immediately identified (Table 2.18).

Thus, the *DSS04 – Managed Continuity* process needs to be controlled, meaning certain mechanisms (policies, procedures, methods, and organizational structures) that provide a reasonable assurance that business requirements will be met and undesirable events will be prevented.

The clear advantages of the COBIT® 2019 standard is a clear structure for describing the *DSS04 – Managed Continuity* IT process, including a structure of controls that allow the process to be audited for compliance with the requirements of the standard.

2.8.2   DSS04 Maturity Levels

Note that the COBIT® 2019 standard also addresses the following maturity levels of the *DSS04 – Managed Continuity* process in an organization.

Absent:

There are no signs of implementation of the *DSS04 – Managed Continuity* process in the organization. There is no understanding of the risks of interrupting critical business processes and IT services as well as possible losses for the business. The organization does not recognize the existence of a business continuity problem and accordingly believes that this issue does not deserve the management's attention.

Naive:

The organization recognizes the existence of a business continuity problem and the need to solve it. However, solutions to this problem are not considered. The main attention is paid to the issues of operation and maintenance of information systems. The issue of implementing the *DSS04 – Managed Continuity* process is poorly understood. Clients use their own techniques at their own risk for disaster recovery after IT failures. The IT service's response to major failures is not premeditated. Scheduled shutdowns of computer systems are practiced for preventive maintenance. However, business requirements for availability of critical business processes and IT services are not considered. Responsibility for the *DSS04 – Managed Continuity* process is not formally defined. The powers of officials are limited.

Repeated:

The organization understands the problem of ensuring business continuity. Employees responsible for the *DSS04 – Managed Continuity* process have been assigned. However, the approach to ensuring business continuity is implemented in fragments. As a result, the incoming information about the availability of critical business processes and IT services is inconsistent, incomplete, and does not consider the state of the business. Business continuity indicators are not defined. There are no documented BCP/DRP plans, although there are intentions to develop them. The list of critical business processes and IT services is not complete. In general, the activities for business continuity management are just beginning to be put into practice of the organization. However, the end result is highly dependent on the initiative of individual employees.

Fixed:

The need to act in accordance with the approved principles of the business continuity management is fully realized. A basic set of the business continuity management indicators is being developed. The relationship between results and business continuity indicators is defined, which is documented

Table 2.15   ITIL4® – COBIT® 2019 mapping.

	Architecture management	Continual improvement	Information security management	Knowledge management	Measurement and reporting	Organizational change management	Portfolio management	Project management	Relationship management	Risk management	Service financial management	Strategy management	Supplier management	Workforce and talent management
EDM – Evaluate direct monitor														
EDMO1 – Ensured governance framework setting and maintenance														
EDM02 – Ensured benefits delivery														
EDMO3 – Ensured risk optimization														
EDMO4 – Ensured resource optimization														
EDMOS – Ensured stakeholder engagement														
APO – Align, plan, and organize														
APOO1 – Managed management framework														
APOO2 – Managed strategy													�	
APOO3 – Managed enterprise architecture	▮													

Availability management	Business analysis	Capacity and performance management	Change control	Incident management	Asset management	Monitoring and event management	Problem management	Release management	Service catalog management	Service configuration management	Service continuity management	Service design	Service desk	Service level management	Service request management	Service validation and testing	Deployment management	Infrastructure and platform management	Software development and management

Continued

Table 2.15   Continued

	Architecture management	Continual improvement	Information security management	Knowledge management	Measurement and reporting	Organizational change management	Portfolio management	Project management	Relationship management	Risk management	Service financial management	Strategy management	Supplier management	Workforce and talent management
APOO4 – Managed innovation							X							
APOO5 – Managed portfolio							X							
APO06 – Managed budget and costs											X			
APOO7 – Managed human resources														X
APOO8 – Managed relationships									X					
APOO9 – Managed service agreements														
APO10 – Managed vendors													X	
APO11 – Managed quality														
APO12 – Managed risk										X				
APO13 – Managed security			X											
AP014 – Managed data BAI – Build, acquire, and implement														

Availability management	Business analysis	Capacity and performance management	Change control	Incident management	Asset management	Monitoring and event management	Problem management	Release management	Service catalog management	Service configuration management	Service continuity management	Service design	Service desk	Service level management	Service request management	Service validation and testing	Deployment management	infrastructure and platform management	Software development and management

Continued

Table 2.15   Continued

	Architecture management	Continual improvement	Information security management	Knowledge management	Measurement and reporting	Organizational change management	Portfolio management	Project management	Relationship management	Risk management	Service financial management	Strategy management	Supplier management	Workforce and talent management
BAI – Build, acquire, and implement														
BAI01 – Managed programs							▓							
BAI02 – Managed requirements definition														
BAI03 – Managed solutions identification and build														
BAIO4 – Managed availability and capacity														
BAIO5 – Managed organizational change						▓								
BAI06 – Managed IT changes														
BAI07 – Managed IT change acceptance and transitioning														
BAI08 – Managed knowledge														

	Availability management	Business analysis	Capacity and performance management	Change control	Incident management	Asset management	Monitoring and event management	Problem management	Release management	Service catalog management	Service configuration management	Service continuity management	Service design	Service desk	Service level management	Service request management	Service validation and testing	Deployment management	infrastructure and platform management	Software development and management
													▪		▪					
													▪							▪
	▪		▪																	
			▪																	
									▪								▪			

Continued

Table 2.15   Continued

	Architecture management	Continual improvement	Information security management	Knowledge management	Measurement and reporting	Organizational change management	Portfolio management	Project management	Relationship management	Risk management	Service financial management	Strategy management	Supplier management	Workforce and talent management
BAI09 – Managed asset				▨										
BAI10 – Managed configuration														
BAI11 – Managed projects								▨						
DSS – Deliver, service support														
DSS01 – Managed operations														
DSS02 – Managed service requests and incidents														
DSS03 – Managed problems														
DSS04 – Managed Continuity														
DSS05 – Managed security services			▨											
DSS06 – Managed business process controls														
MEA – Monitor, evaluate, and assess														

Availability management	Business analysis	Capacity and performance management	Change control	Incident management	Asset management	Monitoring and event management	Problem management	Release management	Service catalog management	Service configuration management	Service continuity management	Service design	Service desk	Service level management	Service request management	Service validation and testing	Deployment management	infrastructure and platform management	Software development and management
					▓														
										▓									
▒	▒	▒	▒	▒	▒	▒	▒	▒	▒	▒	▒	▒	▒	▒	▒	▒	▒	▒	▒
							░											▒	
					▓								░		▓				
							▓												
											▓								
▒	▒	▒	▒	▒	▒	▒	▒	▒	▒	▒	▒	▒	▒	▒	▒	▒	▒	▒	▒

Continued

Table 2.15 Continued

	Architecture management	Continual improvement	Information security management	Knowledge management	Measurement and reporting	Organizational change management	Portfolio management	Project management	Relationship management	Risk management	Service financial management	Strategy management	Supplier management	Workforce and talent management
MEA01 – Managed performance and conformance monitoring				▓										
MEA02 – Managed system of internal control														
MEA03 – Managed compliance with external requirements														
MEA04 – Managed assurance														

▓ Well covered: 60 % and above
☐ Partially covered: less than 60 %

Availability management	Business analysis	Capacity and performance management	Change control	Incident management	Asset management	Monitoring and event management	Problem management	Release management	Service catalog management	Service configuration management	Service continuity management	Service design	Service desk	Service level management	Service request management	Service validation and testing	Deployment management	infrastructure and platform management	Software development and management

Table 2.16   Description of the *DSS04 – Managed Continuity* process.

Purpose

Adapt rapidly, continue business operations, and maintain availability of resources and information at a level acceptable to the enterprise in the event of a significant disruption (e.g., threats, opportunities, demands, etc.).

Establish and maintain a plan to enable the business and IT organizations to respond to incidents and quickly adapt to disruptions. This will enable continued operations of critical business processes and required I&T services and maintain availability of resources, assets, and information at a level acceptable to the enterprise.

DSS04.01. Define the business continuity policy, objectives, and scope.

Define business continuity policy and scope, aligned with enterprise and stakeholder objectives, to improve business resilience.

DSS04.02. Maintain business resilience.

Evaluate business resilience options and choose a cost-effective and viable strategy that will ensure enterprise continuity, disaster recovery, and incident response in the face of a disaster or other major incident or disruption.

DSS04.03. Develop and implement a business continuity response.

Develop a business continuity plan (BCP) and disaster recovery plan (DRP) based on the strategy. Document all procedures necessary for the enterprise to continue critical activities in the event of an incident.

DSS04.04. Exercise, test, and review the business continuity plan (BCP) and disaster response plan (DRP).

Test continuity on a regular basis to exercise plans against predetermined outcomes, uphold business resilience, and allow innovative solutions to be developed.

DSS04.05. Review, maintain, and improve the continuity plans.

Conduct a management review of the continuity capability at regular intervals to ensure its continued suitability, adequacy, and effectiveness. Manage changes to the plans in accordance with the change control process to ensure that continuity plans are kept up-to-date and continually reflect actual business requirements.

DSS04.06. Conduct continuity plan training.

Provide all concerned internal and external parties with regular training sessions regarding procedures and their roles and responsibilities in case of disruption.

DSS04.07. Manage backup arrangements.

Maintain availability of business-critical information.

DSS04.08. Conduct post-resumption review.

Assess the adequacy of the business continuity plan (BCP) and disaster response plan (DRP) following successful resumption of business processes and services after a disruption.

Table 2.17 Detailed description of the *DSS04.0* stage.

DSS04 process practices, inputs/outputs, and activities				
Management practice	**Inputs**		**Outputs**	
DSS04.01. Define the business continuity policy, objectives, and scope. Define business continuity policy and scope aligned with enterprise and stakeholder objectives.	**From**	**Description**	**Description**	**To**
	AP009.03	SLAs	Policy and objectives for business continuity	APO01.04
			Disruptive incident scenarios	Internal
			Assessments of current internal continuity capabilities and gaps	Internal
Activities				
1) Identify and outsource business processes and service activities that are critical to the enterprise operations or necessary to meet legal and/or contractual obligations.				
2) Identify key stakeholders and roles and responsibilities for defining and agreeing on continuity policy and scope.				
3) Define and document the agreed-on minimum policy objectives and scope for business continuity and embed the need for continuity planning in the enterprise culture.				
4) Essential supporting business processes and related IT services.				

and implemented in the relevant strategic planning and monitoring processes. BCP/DRP plans and procedures are developed and documented. The responsibility of officials is defined. Regular training of the organization's employees on business continuity and BCM issues is provided. Periodic reporting on testing of BCP/DRP plans has been established. At the initiative of individual employees, the *DSS04 – Managed Continuity* process is being further improved. The organization's management pays attention to the business continuity issues. We rely on solutions, characterized by a high degree of stability and fault tolerance. The organization's critical business processes and IT services are maintained in a healthy state.

Controlled:

There is a full understanding of the problem of ensuring business continuity at all levels of the organization. Possible emergencies have been identified and classified. The classification and ranking of critical business processes and IT services is carried out. Regular training of the organization's employees on business continuity and BCM issues is provided. In

Table 2.18   Assignment of responsible people and defining their roles.

DSS04 RACI Chart									
Key management practice	Board	Chief Executive Officer	Chief Financial Officer	Chief Operating Officer	Business Executives	Business Process Owners	Strategy Executive Committee	Steering (Programs\Projects Committee)	Project Management Office
DSS04.01. Define the business continuity policy and scope.					A	C	R		
DSS04.02. Maintain a continuity strategy.					A	C	R		
DSS04.03. Develop and implement a business continuity response.						I	R		
DSS04.04. Exercise, test, and review the BCP.						I	R		
DSS04.05. Review, maintain, and improve the continuity plan.					A	I	R		
DSS04.06. Conduct continuity plan training.						I	R		
DSS04.07. Manage backup arrangements.									
DSS04.08. Conduct post-resumption review.						C	R		

Value Management Office	Chief Risk Officer	Chief Information Security Officer	Architecture Board	Enterprise Risk Committee	Head Human Resources	Compliance	Audit	Chief Information Officer	Head Architect	Head of Development	Head of IT Operations	Head of IT Administration	Service Manager	Information Security Manager	Business Continuity Manager	Privacy Officer
						C	C	R			R	C	R		R	
	I					C	C	R	R	C	R				R	
					I	C	C	R	C	C	R				A	
					I		R	R		C	R				A	
	I							R		C	R				R	
								R		R	R	R			A	
										C	A				R	
	I							R	C	C	R	R			A	

practice, the skills of the organization's employees' actions in emergency situations are practiced. Service level agreements (SLAs) are defined and maintained. The responsibility of officials for the development and implementation of the *DSS04 – Managed Continuity* process is clearly distributed. This process corresponds to the business strategy and IT strategy. Best practices in business continuity and cyber resilience are being consistently implemented. Business continuity goals and benchmarks may be implemented, but their measurements are not regular.

Optimal:
The organization has a complete understanding of the *DSS04 – Managed Continuity process*. Problems of ensuring business continuity in general are identified and resolved in a timely manner. The prospects for the development of this process in the organization are defined. BCP/DRP integrated business continuity plans and procedures are developed and maintained up-to-date. Training and communication on business continuity issues are maintained at the appropriate level. At the technical level, modern tools are used to ensure business continuity and cyber stability. Business continuity requirements are set out in SLA agreements, among other things. Regular comprehensive checks and testing of BCP/DRP plans are carried out. The results of checks and tests are used to improve BCP/DRP plans. Implementing feedback based on verification data and test results is part of the *DSS04 – Managed Continuity* improvement process. The organization's management has been provided with all necessary guarantees of preparedness for emergency situations. Appropriate cyber-training is being conducted. We measure the continuity of critical business processes and IT services on a regular basis, based on which the *DSS04 – Managed Continuity* process is improved. As a result, the resulting process conforms to the "best practice" in business continuity, BCM.

2.9   ITIL V4 Library

In 1985, the *Central Computer and Telecommunications Agency* (CCTA) commissioned the British government to develop the *IT Infrastructure Library – ITIL* (Figure 2.48). Initially, the CCTA sought to create a universal methodological approach in the field of IT management that does not depend on the use of private information technologies and the specifics of organizations' activities. However, this approach was not a scientific development and did not reflect the requirements of narrowly specialized technical standards.

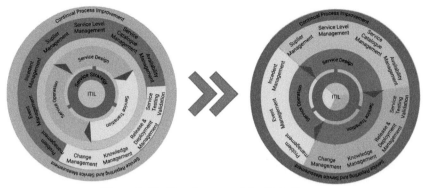

Figure 2.48   Development of the ITIL library.

2.9.1   The ITSCM Process

In 2001, after the merger of CCTA with the State Chamber of Commerce of the United Kingdom, the *Office of Government Commerce (OGC)* became the new owner of the ITIL library. The development of ITIL has continued with OGC's objectives in mind, the main of which is to help British government customers to improve their purchasing activities and improve overall services by maximizing the use of IT.

Currently, the ITIL library is a collection of *"best practices"* in the field of IT management and belongs to the *de facto* standards. Methodological descriptions of ITIL have gained well-deserved popularity, are of a recommendatory descriptive nature, and can be used in most domestic companies. It is worth noting that the development and popularization of ITIL is supported not only by its publisher (OGC) but also by the independent professional society *itSMF (IT Service Management Forum)*, which includes leading vendors Microsoft, Oracle (SUN), HP, IBM, and others.

During its existence, the content and structure of the library changed several times (from a few dozen books to the main five books), but the main ideas and principles were preserved. The ITIL library still conceptually describes all the main stages of the IT services lifecycle:

- incident management;
- problem management;
- configuration management;
- change management;
- release management;
- service level management;
- financial management for IT services;

- capacity management;
- IT service continuity management;
- availability management.

Currently, the third version of the library is available (*ITIL Version 3*); it includes the following five books:

- Service Strategy;
- Service Design;
- Service Transition;
- Service Operation;
- Continual Service Improvement.

Currently, the concepts of IT services in the ITIL library have become familiar to many companies. Ordering and monitoring the quality of provided IT services in accordance with certain SLA agreements have become convenient and efficient for both the customer and the corresponding service provider (IT service provider).

The *ITIL* library value for the proper organization of the enterprise business continuity process is that IT provides a conceptual description of the lifecycle of the *IT service continuity management* (ITSCM) process. The authors of ITIL explain the relevance of this process by the fact that in emergency situations, its use provides businesses with the ability to manage the recovery of their critical IT services, reduce downtime in the operation of IT services, and thereby minimize interruptions of IT-dependent business. Here, an emergency (disaster) is an event that has such a destructive impact on the functioning of the IT service and/or information system that significant efforts are required to restore the normal functioning of the Information and Communication Technologies (ICT) company. In other words, an emergency situation is much more serious than an incident and, in fact, means suspension of business.

The ITIL library states that the main purpose of ITSCM is to support the more general business continuity management process, BCM. This support means that in the event of a business interruption, all the supporting ICT infrastructure and related IT services (including computer systems, networks, applications, data sets, telecommunications, external devices, as well as the support service and service desk) can be restored within a specified period of time after an emergency occurs. And since ITSCM is part of BCM, the scope of the said process should be determined based on the business goals and objectives.

In general, the ITSCM process correctly defined in the organization allows:

- assessment of the consequences of failures and failures of IT services after an emergency situation;
- identification of business-critical IT services for the business that require additional preventive measures;
- computation of the time intervals in which IT services must be restored;
- taking measures to prevent, detect, and counteract the impact of emergencies;
- establishment of a common approach to IT services recovery;
- development and implementation of an appropriate IT disaster recovery plan with sufficient detail to help survive an emergency and restore to normal operation within a given time frame.

Thus, the ITIL library describes the ITSCM IT service continuity management process integrated into BCM's more general business continuity management process. It should be noted here that this ITSCM view was not immediately developed. Since the initial publication of the CCTA Contingency Planning Module, information technology has moved from the category of functionally "self-sufficient" to the category of business-oriented and significantly penetrated into all areas of business. Even such special concepts as business-oriented representation of IT, IT-dependent business, ICT business support, etc. have appeared. The nature of emergency planning has also changed. Whereas previously the traditional planning process was reactive and aimed at eliminating the consequences of emergency situations, now the ITSCM process tends to play a preventive role and aims to prevent interruptions in the business of the organization; in Figure 2.49, the main stages of ITSCM lifecycle (based on OGC model) are presented. Let us take a closer look at the marked milestones.

At the stage of initiation of the ITSCM process, it is recommended to conduct a system analysis of the company's activity and perform the following actions:

- Develop ITSCM policy – the management of the organization should demonstrate its clear interest in the project of implementation of the ITSCM process in the organization. The named project should be initiated by company and company management and supported by it.
- Define ITSCM scope of action – definition of scope of the mentioned process and also some adjacent areas and scopes of action is meant. Here relevant provisions of business strategy and IT strategy, annual reports, audit results, business process descriptions according to ISO 9000, insurance requirements, formalized descriptions of the

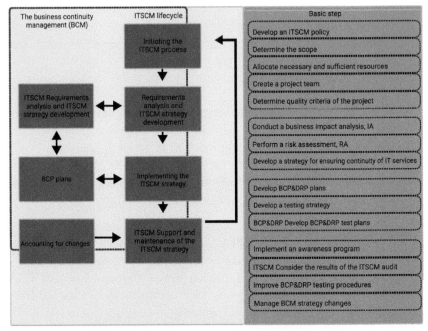

Figure 2.49   ITSCM process lifecycle on OGC model.

IS management system based on ISO/IEC 27001, etc. can be considered. At the same stage, it is also recommended to determine the organizational structure of the management for emergency situations.

- Allocation of necessary and sufficient resources for ITSCM – proper organization of IT services continuity requires significant efforts and costs.
- In order to organize a project team, it is necessary to define project roles, rights, and responsibilities. It is recommended to use mainly known formal methods (for example, PMBOK or PRINCE2) and software tools for project management of ITSCM implementation.
- To define the quality criteria of the carried-out project – for proper organization of process ITSCM, it is recommended from the very beginning to develop corresponding procedures of quality assurance of the project performance.

Analysis of requirements and development of ITSCM strategy:
Business impact assessment, BIA:
Here the main purpose of business impact analysis (BIA) is to determine the impact of IT service downtime (including the cost) caused by ICT

failures and outages. The main reasons for implementing the ITSCM process are the following:

- business process protection;
- rapid restoration of services;
- need to withstand a competition;
- maintaining market positions;
- preservation of profitability;
- protection of the company's reputation, etc.

Then it is necessary to select and, in a certain way, to rank the critical business processes, highlighting and documenting the processes that are subject to the increased requirements for continuity and, consequently, the availability and continuity of IT services supporting them. Further, it is necessary to conduct the ranking of IT systems supporting them (for example, ERP and BI systems, CRM, electronic document management system and office systems, accounting system, email, etc.). Agreements with external IT service providers for emergency services may also be required For other critical IT services, it will be necessary to find a compromise between taking preventive measures and using adequate disaster recovery methods (Figure 2.50).

After the analysis and ranking of IT services, the dependencies between services and IT assets are assessed. The loss of business in the event of a failure or failure of IT services is determined. In this step, it is important to

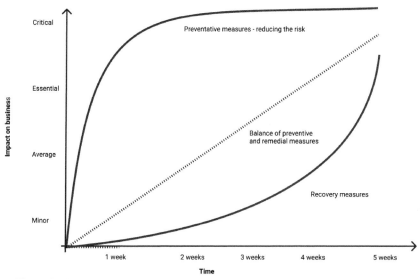

Figure 2.50 The nature of the balance between preventive and recovery measures.

identify the so-called structural shifts in business continuity by correctly assessing the consequences of IT service downtime, scenarios, and their correlation for disaster recovery and business continuation.

Based on the allowable losses, process owners are able to determine the minimum allowable level of IT services as well as calculate the allowable time to restore IT services, RTOs and identify RPO recovery points.

Risk assessment, RA:
The purpose of risk assessment is to determine the risks threatening business continuity (Figure 2.51 and Tables 2.19 and 2.20).

Preventive measures:
Preventive measures are typically defined as measures that reduce the risk of business interruption – in our case, the risk of IT service downtime due to failure or failure. In practice, however, business interruption threats can never be completely eliminated. For this reason, the remaining residual risks must be addressed in the appropriate disaster recovery plan.

Disaster recovery measures:
By restoring ITIL means not only restoring itself but also providing the ICT backup components to ensure the continuity of IT services. At the same time, the application of recovery measures requires the availability of staff with certain qualifications and pre-prepared workstations with redundant ICT components, support services, archives and data warehouses, pre-prepared contracts for third-party services, and so on.

At this stage, probable threats and vulnerabilities are first identified, risks are calculated, and then certain preventive defined countermeasures are taken to manage them. Here, it is recommended to use well-known risk analysis and management methods such as the CCTA Risk Analysis and Management Method (CRAMM). This method allows for the necessary decomposition of business processes and IT services into asset components, including buildings, information systems, applications, data, etc. Assign persons responsible for disaster recovery of IT infrastructure and resumption of activities.;carry out the analysis of probable threats and vulnerabilities. And, finally, calculate residual risks and give recommendations on risk management. After the necessary countermeasures have been taken, it should be determined whether the risks for which disaster recovery countermeasures are required and, in particular, the Contingency Plan, remain.

Developing a strategy to ensure continuity:
Business always strives to find an economically sound balance between the need to manage risks and disaster recovery planning. Here, one should

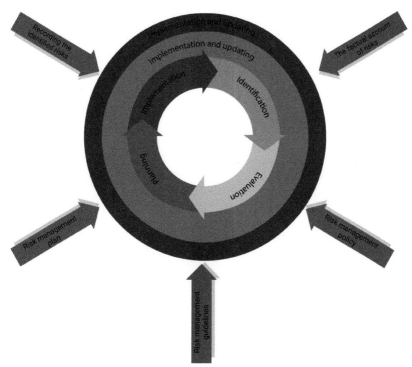

Figure 2.51 ITIL risk management lifecycle.

distinguish between risk management, disaster recovery, and recovery techniques. ITIL recommends referring to the *BCI Good Practice Guidelines* for more information on this issue.

2.9.2 ITSCM Implementation

Process organization and implementation planning:
Once the business continuity strategy has been defined and a reasonable choice of disaster recovery methods has been made, the ITSCM process will be implemented. Detailed plans are developed to use the selected recovery tools. A dedicated workgroup is created, consisting of a Crisis Manager, a Manager, and teams of performers to restore each service. In this case, general organizational issues are first considered:

- emergency response plan;
- damage assessment plan;
- work recovery plan;
- critical data plan;
- management plan for crisis and public relations.

Table 2.19   Example risk profile.

Risks	Threats
Failure of internal IT infrastructure, networks, ATC, etc.	• Fire • Power failure • Vandalism • Inundation • Incidents related to air transport • Natural weather events (hurricanes, etc.) • Disasters (earthquakes, etc.) • Terrorist attack • Sabotage • Serious damage • Damage to the electrical part (for example, as a result of a lightning strike) • Accidental damage • Low-quality software
Failure of external IT systems/servers/networks	• All of the above • Increased load on services • "Service denial" attack • Technological failures (e.g., cryptophagidae)
Data loss	• Technological failures • Human error • Viruses and malware
Failure of network services	• "Service denial" attack • Failure of the service provider's IT systems or networks • Data loss on the service provider's side • Other failures on the service provider's side
Unavailability of key technical personnel	• Strike • Denial of access to premises • Dismissal • Disease • Problems with transport
Failure on the part of external providers of IT services, i.e., IT being outsourced	• Insolvency • Denial of access to premises • Unavailability of the service provider's technical staff • Failure of the provider to meet SLA requirements

The following issues are then elaborated in detail:
- plan for accommodation and service delivery;
- plan for computing systems and LANs;
- telecommunications plan (access organization and communication channels);

Table 2.20 Comparative analysis of possible recovery methods.

	Manual	Urgent	Fast	Average	Consistent
Service support systems	Yes		Yes	Yes	Yes
The company's key mainframe-based systems	Yes			Yes	Yes
Systems for financial control and planning			Yes		Yes
Systems for working with clients and partners		Yes		Yes	Yes

- security plan;
- personnel plan;
- IT operations funding plan and administrative plans.

Reservation and risk residual measures:
The phase consists of putting into practice previously identified preventive measures and recovery methods. The implementation of these preventive measures is usually done with activities within the accessibility management process and may include:

- use of uninterruptible power supplies;
- use of special fault-tolerant computing systems;
- use of remote storage systems and RAID arrays, etc.

For the prompt start of disaster recovery work, it is necessary to make preparation in advance. In particular, it is recommended to conclude special contracts with suppliers in advance for the supply of backup computing equipment. This will allow, if necessary, to promptly update contracts for reserve supply at prices already set in advance. In order to update the conditions and terms of delivery, it is recommended to revise the mentioned contracts every year, as prices and models of technical means may change with time. At adjustment of contracts, it is necessary to consider also base configurations of technical means defined within the limits of process of management of configurations.

Development of recovery plans and procedures:
Detailed disaster recovery plans are recommended to be properly documented and approved by the company management. It is recommended

that all changes to the plans be made in accordance with the established procedures. Here, the marked activities can be carried out jointly as part of the change management process as well as the process of managing the technical equipment configurations.

Recovery plan:

- The structure of the recovery plan may be as follows:
- Introduction – a brief description of the disaster recovery plan;
- Timeframe for reviewing the plan – how and when to review and keep the recovery plan up-to-date;
- Itinerary – a visual and accessible graphical representation of the recovery plan;
- Start point – description of initial conditions for initiating the recovery plan;
- Classification of emergencies – by severity (serious, medium, and acceptable), by duration (months, weeks, and days), and by damage (serious, limited, and minor);
- Special sections – the following six sections are recommended:
  o management issues in emergency situations;
  o supporting IT infrastructure;
  o personnel reserve;
  o security rules and measures;
  o reserved resources for recovery;
  o emergency recovery procedures.

Procedures:
It is recommended that the procedures of the disaster recovery plan should include the installation and testing of ICT components as well as the recovery of applications, databases, etc., in the disaster recovery plan procedures.

Initial test:
The testing step is quite important for the ITSCM process. It is recommended that testing is performed after the disaster recovery plan is developed and adjustments are then made at least once a year. In this case, testing can be conducted without prior notice.

Operations management:
In general, this phase aims to keep the ITSCM process up-to-date. Its main steps include the following steps.

Training and awareness:
The training of staff and their awareness of ITSCM issues are important factors for the successful implementation of this process. Moreover,
they often exceed technical measures in terms of effectiveness. In general, training and awareness programs are aimed at improving the enterprise culture in the area of business continuity and are quite effective in
practice.

Analysis and audit:
To respond in a timely manner to new threats and risks of business interruption, it is necessary to conduct regular audits and check the relevance
of all plans and procedures to ensure the continuity of IT services. In doing
so, the audit should cover all aspects of the ITSCM process. In addition to
regular audits, the audit should be conducted at any significant change in
IT infrastructure, and even more so at changes in IT strategy and business
strategy. Organizations in the process of transformation can implement
a regular program of ITSCM analysis and audit. Identified observations
should be taken into account as part of the change management process.

Testing:
It is recommended that a disaster recovery plan is tested on a regular basis
to ensure its proper appearance. Testing allows weaknesses in the plan to
be identified and timely responses to changes made. It is also a practice
to test changes in the recovery equipment before it is put into commercial
operation.

Change management:
The change management process plays an important role in keeping disaster recovery plans up-to-date. It is recommended to analyze the impact of
each change on the respective disaster recovery plan.

Ensuring safeguards:
Verifying that the quality of the ITSCM process (corresponding plans, policies, procedures, and instructions) meets the goals and objectives of the
business can ensure that the process works.

Additional questions:
The ITIL library draws attention to the fact that the effective maintenance
of the ITSCM process is based on timely informing and reporting to management, identifying critical success factors, and monitoring key quality
indicators.

Reports to management:
In the event of an emergency, management must be informed in good time. It is recommended that a report on the causes and consequences of an emergency situation be prepared and recommendations made to address and prevent such situations in the future. Identified gaps and concerns should be documented and considered in future work.

ITSCM integration with other processes:
The authors of ITIL recommend integrating the ITSCM process with other processes, namely:

- change control;
- incident and problem management;
- availability management;
- service quality management;
- capacity management;
- configuration management;
- information security management.

The input and output data for the listed processes and the place and role of the ITSCM process should be clearly defined.

Critical success factors:
The key success factors include:

- effective configuration management process;
- support for the ITSCM process in the company;
- availability of modern technologies and disaster recovery tools;
- BCM training and awareness for the organization;
- regular, unannounced testing of the disaster recovery plan.

Key quality indicators:
The key quality indicators are:

- number of identified gaps in disaster recovery plans and procedures;
- loss of income of the company as a result of the emergency situation;
- the cost of the business continuity project.

Functional responsibilities and roles:
It is recommended to define several roles and responsibilities of company officials. In addition, a distinction should be made between liability under normal circumstances and liability in emergencies (Table 2.21).

Cost estimates:
The main costs of the ITSCM process include:

Table 2.21   Examples of roles and responsibilities in the ITSCM process.

No.	Role	Responsibility in the normal operation of ICT	Responsibility in an emergency
1)	Board of Directors	• BCM process initiation • Allocation of sufficient resources • Policy development in the field of VSM • Empowering	• Guide actions in emergency situations • Decision-making
2)	Company management	• Managing the ITSCM process • Approving policies, strategies, plans, and procedures • Organization of communications in the company • Initiating an awareness program • Integrating the ITSCM process into the general BCM process	• Coordination and arbitration • Decision-making. Allocating the necessary resources
3)		• Approval of residual risks • Planning the work results • Preparation of agreement projects • Managing testing, evaluation, and reporting	• Managing testing, evaluation, and reporting • Putting in place mechanisms for restoring and ensuring continuity of IT services • Guide teams to ensure continuity • Reporting
4)	Line managers and work team members	• Working out ways to achieve results • Negotiating for the services provided by IT departments • Conducting tests and assessments	• Development and implementation of disaster recovery procedures • The implementation of the recovery plan and reporting

• costs of initiation, development, and implementation of the ITSCM process;
• expenses for support and maintenance of the ITSCM process, including organization of risk management, interaction with other ITIL processes, decomposition, aggregation of ITSCM continuity indicators, etc.

2.10   ISO/IEC 27001:2013 and ISO/IEC 27031:2011 Standards

International Standard *ISO/IEC 27001:2013 – "Information technology – Security techniques – Code of practice for information security management"* BS 7799-1 *"Information security management – Part 1: Code of practice for information security management."*

ISO/IEC 27031:2011, Information technology – *Security techniques – Guidelines for information and communication technology readiness for business continuity*, for IT-dependent government and business organizations. The implementation of the recommendations of this standard makes it possible to develop and implement an enterprise ICT readiness for business continuity (IRBC) program in the context of growing threats to information security (Figure 2.52). Thus, ISO/IEC 27031: 2011 includes and

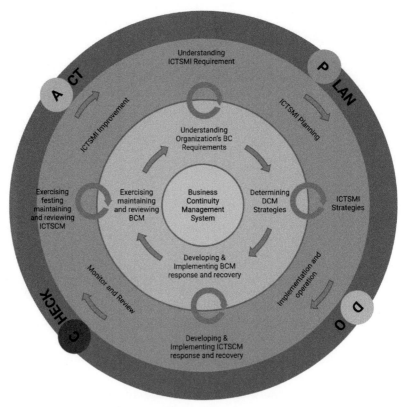

Figure 2.52   The lifecycle of development of the ICT readiness for business continuity program (IRBC).

extends the practice of information security incident management and handling as well as ICT readiness planning and related services.

2.10.1   BCM Aspects

The ISO/IEC 27001:2013 (A. 17, formerly A. 14) and ISO/IEC 27031:2011 standards address the following aspects of business continuity:

- relevance of business continuity management;
- place and role of the information security process;
- business continuity and risk assessment.
- development and implementation of a business continuity plan;
- the structure of the business continuity plan;
- support and maintenance of the business continuity plan.

Let us look at these questions in more detail. The purpose of the business continuity management is to prevent and protect critical business processes of the organization from the possible business interruptions as well as to resume business in the event of an emergency. Here, an emergency is defined as a natural disaster, accident, large-scale equipment failures, cyber-attacks, terrorist acts, etc.

It is recommended for each organization to implement a proper business continuity management process by selecting and using an appropriate combination of preventive and remedial measures (so-called controls). This process will identify critical business processes and their component assets (buildings, infrastructure, equipment, personnel, IT services, etc.). At the same time, the information security requirements should be defined, which should be integrated with other business continuity requirements.

The consequences of incidents, including security breaches, service failures, and service availability violations, should be the subject of a business impact analysis. In order to re-activate business, the appropriate business continuity plans and procedures must be developed and implemented. At the same time, the information security issues should be part of the business continuity process.

The business continuity management process should include an appropriate information security measures, including:

- risk identification and mitigation, in addition to the overall process of risk assessment and management;
- fending off security incidents;
- ensuring the required availability of IT services, etc.

The relationship between business continuity and cybersecurity:
Control:
The organization needs to implement and maintain a proper business continuity management process that considers the information security requirements.

Recommendations:
This process should include all the key elements of business continuity management, including information security issues:

- understanding the risks faced by an organization in terms of probability of occurrence and consequences for the business;
- identification and ranking of critical business processes and supporting assets;
- understanding the consequences of business interruption caused by information security incidents;
- risk management by insuring them as part of the overall business continuity management or operational risk management process;
- identify (if necessary) additional preventive and remedial measures to ensure business continuity;
- determining the necessary and sufficient resources (financial, organizational, technical, etc.) to meet information security requirements;
- ensuring physical security and information security in the organization in various business environments;
- defining and documenting a business continuity plan in accordance with an agreed business continuity strategy and information security issues;
- regularly testing and updating existing plans and processes;
- ensuring that the business continuity management plan and procedures are integrated seamlessly into the organization's structure and processes. Responsibility for implementing the Business continuity management process should be assigned to the responsible persons of the organization with an appropriate authority, such as the information security management committee.

Business continuity and risk assessment:
Control:
One must identify events that may cause business interruptions. Determine the probability of destructive events and their impact on the business as well as possible consequences for the information security management process.

Recommendations:

The choice of the appropriate measures to ensure information security and business continuity should be based on the identification of events (or a sequence of events) that may lead to business interruption; for example, business interruption due to natural disasters, terrorist attacks, floods, fires, equipment failures, operator errors, theft, deliberate actions of an attacker, and so on. After that, one needs to assess the risks of business interruption and calculate the probability of the destructive events and the amount of possible damage.

For the objectivity and completeness of risk assessment, it is necessary to investigate all critical business processes of the organization. At the same time, it is recommended to conduct a risk assessment with the direct participation of business process owners, paying attention to special issues of information security. Here, one needs to link together all the different aspects of risk assessment to get a complete picture of the requirements and challenges in the field of business continuity. Next is computation and ranking the risks by significance in accordance with the goals and objectives of the organization, taking into account such factors as the criticality of business processes, the consequences of processes, the consequences of interruptions, the allowed downtime, recovery priorities, and so on.

Based on the results of the risk assessment, it is necessary to develop an appropriate business continuity strategy that will determine the overall approach to ensuring the business continuity of the organization. The developed business continuity strategy must be approved by the organization's management. In addition, it is necessary to outline a plan for the implementation of this strategy.

2.10.2   BCP Development and Implementation

Control:

It is necessary to develop and implement plans that will allow the organization to continue its business in emergency situations as well as its resumption after an interruption.

Recommendations:

Business continuity plans should reflect the following issues:

- the responsibility of officials;
- estimation of acceptable data and service losses;
- disaster recovery processes;
- support for the availability of services for the required time;
- content of operational procedures for disaster recovery;

- documentation of agreed procedures and processes;
- training and awareness of the organization's employees on BCM issues;
- working out the skills of employees' actions in emergency situations;
- how to test and update business continuity plans.

It is necessary that the business continuity plan corresponds to the goals and objectives of the business. All necessary services and assets must be identified. The issues of attracting the personnel reserve and using the means of reserving and fault tolerance of computing systems as well as the issues of using assets for disaster recovery of information systems are considered. The organization's agreements with third parties in the form of mutual agreements for using IT services or providing commercial services by subscription should be updated.

Because a business continuity plan may contain confidential information, it must be properly protected. The copies of the plan must be stored in a secure, remote location. Copies of the business continuity plan must be updated in a timely manner. Other materials needed to complete the business continuity plan must also be stored in a secure location. If the organization uses remote sites for storing and reserving supporting assets, the level of security and security equipment should be similar to the levels at the organization's head office.

The structure of the business continuity plan:
Control:
To ensure that all types of the business continuity plan are complete and consistent, a single plan structure must be maintained.
Recommendations:
Each business continuity plan should describe the appropriate ways to ensure business continuity, such as ways to ensure availability and security. Each plan must contain information about the plan terms of activation, implementation, and maintenance, contact details of responsible persons, and so on. When the new requirements arise, disaster recovery plans and procedures must be adjusted accordingly. In this case, changes must be made in accordance with the established procedure of the change management process.

It is necessary that a specific employee of the organization is responsible for certain items in the plan. Procedures for disaster recovery, switching to backup data processing modes, and renewal procedures should be the responsibility of the owners and the corresponding business processes. As a rule, service providers are responsible for services provided by third-party organizations.

The business continuity plan should also consider information security requirements:

- conditions for plan initializing and executing (for example, how officials act, how decisions are prepared and made, and so on);
- disaster recovery procedures initiated after incidents that threaten business interruption;
- procedures for switching to standby modes of operation that describe the necessary actions to activate the reserve and restore business processes in a set time frame;
- temporary operating procedures for business recovery;
- business renewal procedures that describe the necessary actions to return to normal operation;
- plan support schedule, which defines the timing and methods of testing the plan, as well as how to update and support the plan;
- training and awareness-raising activities aimed at improving the enterprise culture in the field of business continuity;
- responsibilities of the officials, responsible for the implementation of each item of the plan; if necessary, several responsible persons can be assigned;
- a list of critical assets and resources needed to complete disaster recovery procedures, switch to backup funds, and resume business.

Support and maintenance of a business continuity plan:
Control:
Keeping the business continuity plan up-to-date requires regular testing and revision.

Recommendations:
When testing a business continuity plan, the attention should be paid to checking the knowledge of responsible persons and develop skills for coordinated action in the event of an emergency.

The schedule for testing the business continuity plan must contain the composition and frequency of checking the plan items. However, each item in the plan must be tested.

Recommended ways to test a business continuity plan includes the following:

- simulation testing of possible emergency scenarios (discussion of business recovery measures based on various examples of interruptions);
- simulation (in particular, for training officials to develop skills in emergency situations);

- field testing of recovery procedures (ensuring confidence in the effective recovery of information systems);
- testing recovery procedures on reserved areas (execution of business processes is carried out in parallel with recovery activities);
- testing supplier services (checking services provided by third-party organizations for compliance with the terms of the agreement);
- shakedown run (a comprehensive check of the company's ability to function in a business interruption).

The listed testing methods can be used by each organization and should reflect the actual nature of the company's disaster recovery in emergency situations. The test results should be recorded and used in the future to analyze and improve the business continuity plan.

Responsible persons must be appointed to regularly review and update the business continuity plan. Identified changes in business processes that are not yet reflected in the business continuity plan should be considered by updating the plan accordingly. The change management process should ensure that the business continuity plan is updated and revised.

Examples of changes that should be considered when updating the business continuity plan are changes related to the acquisition of new equipment and system upgrades. Changes must also be considered in a timely manner:

- legislations;
- business strategy;
- location, production capacity, and resources;
- the pool of contractors, suppliers, and main clients;
- processes (including the addition of new and cancelation of existing);
- risks (operational and financial);
- personnel structure;
- addresses or phone numbers of contact persons.

Note that in the previous version of *ISO/IEC 27001* from 2005 (Table 2.22), Annex A. 14 (Annex A. 17 in the current version of the standard), possible functional specifications (controls) for BCM were defined.

2.11   Possible Measures and Metrics

2.11.1   Introducing a Passport System for Programs

Selecting the invariant classification characteristics of the program behavior of some secured infrastructure (in this task, into two classes: correct

Table 2.22   Examples of BCM controls according to ISO/IEC 2700   (A. 1   or A. 17).

Requirement	Standard item	Area of security management	The issue of audit	Results (scale from 1 to 5)
		Audit scope, purpose, and issues		
A 14.1.1	14.1.1	Business continuity management process	• Is there a process for developing and maintaining business continuity? • Business continuity plan that includes: a business continuity plan, regular review and updating of the plan, and the formulation and documentation of a business continuity strategy.	3
A 14.1.2	14.1.2	Business continuity and evaluation risks	• Whether actions that may cause disruption to business processes have been identified, such as equipment accidents, floods, fires, etc. • Whether a risk analysis has been conducted to determine the damage of such impacts. • Whether a strategic plan has been developed based on the analysis. • Risk management that defines the overall approach to business continuity.	3
A 14.1.3	14.1.3	Development and implementation of a business continuity plan	• Whether plans are in place to restore business operations in the event of accidents within the required time frame. • Whether the plan is regularly updated and tested.	2
A 14.1.4	14.1.4	The structure of the business continuity plan	• Is there a single approach to developing a business continuity plan? • Is this infrastructure supported to ensure all plans? • Agreed and defined priorities for testing and support.	2
A 14.1.5	14.1.5	Testing, supporting, and reviewing business continuity plans	• Whether the activation conditions and personal responsibility for each component of the plan are defined. • Is the business continuity plan regularly tested to ensure that it is relevant and effective? • Are there procedures that define how to audit a business continuity plan to assess its effectiveness? • Whether procedures are included in the change management program to comply with the business continuity plan.	1

and incorrect execution) is identical to the *isomorphism problem* of the two systems under *some mapping*. In order to clarify the *necessary and sufficient conditions* for the system *isomorphism* as well as to determine the *isomorphism mapping qualitative and quantitative parameters*, a *similarity theory of the mathematical apparatus* was developed. The key similarity theory points were formulated by *A. A. Gukhman. (1949), Kirpichev M. V. (1953), Venikov V. A. (1966), Sedov L. I. (1977), Kovalev V. V. (1985), and Petrenko S. A. (1995)*. Initially, the theory provisions were developed in relation to the modeling of mechanical, electrical processes and heat transfer processes. However, in the late *1980s*, the results were applied in the field of modeling, applying the universal digital computers and then transferred to solve a much wider spectrum of problems, including cybersecurity and ensuring the required cyber resilience of the critical information infrastructure.

The most detailed provisions of the similarity theory were developed concerning the processes, described by the homogeneous power polynomial systems. There are three main theorems in the similarity theory: the direct, inverse, and *π*-**theorem** [202].

Let us consider two processes of p_1 and p_2, which complete equations have the following form:

$$\sum_{i=1}^{q} \varphi_{ui} = 0, u = 1, 2,\ldots,r; \tag{2.1}$$

$$\sum_{i=1}^{q} \Phi_{ui} = 0, u = 1, 2, \ldots, r; \tag{2.2}$$

where $\varphi_u = \prod_{j=1}^{n} x_j^{\alpha_{u1}}$ and $\Phi_u = \prod_{j=1}^{n} X_j^{\alpha_{u1}}$ are homogeneous functions of their parameters.

The ***direct similarity theorem*** states that if the processes are homogeneously similar, then the following system takes place:

$$\frac{\varphi_{ui}}{\varphi_{uq}} = \frac{\Phi_{ui}}{\Phi_{uq}}, \tag{2.3}$$

$$u = 1, 2, \ldots, r; s = 1, 2, \ldots, (q - 1).$$

Expressions

$$\pi_{us} = \frac{\varphi_{ui}}{\varphi_{uq}}, u = 1, 2, \ldots, r; s = 1, 2, \ldots, (q - 1) \tag{2.4}$$

are called *criteria or similarity invariants* and, as a theorem deduction, are numerically equal to all processes belonging to the same subclass of mutually similar processes.

Thus, the *direct theorem* formulates the necessary conditions for the correlation of the analyzed process with one of the subclasses. Sufficient conditions for the homogeneous similarity of two processes are given in the **inverse similarity theorem**: if it is possible to reduce the complete processes equations to an isostructural relative form with the numerically equal *similarity invariants*, then such processes are homogeneously similar.

The **similarity theorem**, known as "*π-theorem*," allows identifying the functional relationship between variable processes in relative form. The deductions form the **direct theorem** and the "*π-theorem*" of similarity allowed formulating invariant informative features for the correct behavior of some critical information infrastructure software.

Mathematical problem formulation:
Imagine the *computational process (CP)* in the following form:

$$CP = <T, X, Y, Z, F, \Phi>. \tag{2.5}$$

T – the set of points in time t at which the computational process is observed;
X, Y – sets of input and output parameters of the computational process;

Z – the set of computational process states. Every state Z_{kj} ($j = l;m$) of the computational process is characterized at each $t\ 2\ T$ moment in time by a sequence of performing arithmetic operations at the selected control point k.

F – the set of transition operators fi, reflecting the mechanism of changing the states of the computational process during its execution, including the arithmetic operations being performed;

Φ – a set of the output operators ϕ_i, describing the result formation mechanism during the computations.

We introduce the following notation:

λ – violation mapping of an arithmetic operation at a specific time t_i for given input parameters;

ψ – mapping of the computational process regular invariants formation;

μ – comparative mapping of standard and reference invariants of the computational process;

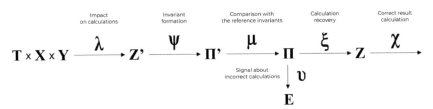

Figure 2.53   The mapping diagram of the computation correctness recovery.

υ – mapping of the signal generation about incorrect computations;

ξ – mapping of the arithmetic operations recovery based on reference similarity invariants;

χ – performed computation correctness mapping based on the recovered arithmetic operations.

In order to exclude the possibility of discreet modification, made by the computations, it is necessary to perform a dynamic control of the executed computational process (Figure 2.53). Under the dynamic control of the computations correctness, we will understand the correctness control of the performed arithmetic operations semantics during their actual execution. Data for dynamic control must first be obtained as a program passport, resulting from its additional static analysis [203].

The system is defined as an impact on computation, computations invariants formation, comparison with the reference invariants, signal about incorrect computations, computations recovery, correct result computations.

In order to form the passport program, the following actions are required:

1. solving the observative problem (the computational process simulation by an oriented program control graph);
2. solving the problem of presenting computations by similarity equations on linear graph parts, i.e., to transform the arithmetic operations of the form:

$$z_i(x_1, x_2, ..., x_m) = \sum_{j=1}^{p} z_{ij} (x_1, x_2, ..., x_m). \qquad (2.6)$$

To dimensionless form:

$$[z_{ij} (x_1, x_2, ..., x_m)] = [z_{il} (x_1, x_2, ..., x_m), j,l = \overline{1,p}. \qquad (2.7)$$

In order to ensure the computations correctness, the following actions are required:

1. Solving the problem of managing the computational process by comparing the semantic invariants with the program passport which means that it is necessary to find the maps:

$$\psi: Z' \rightarrow \Pi'$$
$$\mu: \Pi' \rightarrow \Pi$$
$$\xi: \Pi \rightarrow Z \quad\quad\quad\quad\quad (2.8)$$

Limitations and assumptions:

1. Considered set of arithmetic operations $\{+, -, *, /, =\}$
2. $t_i < t_{max}$, where t_i is computation time recovery and t_{max} is maximum allowable time to recover the computations correctness.

Solving these problems allowed developing a new method to control the computational program semantic correctness, which complemented the known method capabilities to ensure the required cyber resilience of the secured critical information infrastructure.

Program control graph:
In order to control the software correctness, it was necessary to construct a program control graph.

Let us imagine some computational process in the form of a program control graph:

$$G(B.D) \quad\quad\quad\quad\quad (2.9)$$

where $B = \{B_i\}$ — set of vertices (linear program part),
$D = \{B \times B\}$ — set of arcs (control connections) between them.

Here, each linear graph part $B_i \in B$ has its own arithmetic operator sequence, i.e.,

$$B_i = (b_{i1}, b_{i2}, ..., b_{il}). \qu\quad\quad\quad (2.10)$$

An ordered vertex sequence corresponds to each elementary (without cycles) route of the graph input vertex to output vertex

$$B^k = (B_1^k, B_2^k, ..., B_t^k), \qu\quad\quad\quad (2.11)$$

where $B^k \subseteq B$ and $B_i^k = (b_{i1}^k, b_{i2}^k, ..., b_{il}^k)$ $B_i^k = (b_{i1}^k, b_{i2}^k, ..., b_{il}^k)$ $\forall i = \overline{1, p}$ form a sequence of the executed arithmetic operators, called a program implementation or a computational process. The arithmetic expression sequence data is the potentially dangerous program fragments.

The computational process algorithm was reduced to the graph representation form to derive the arithmetic expression operators from the control operators (*conditional transitions, branching, cycles*, etc.). As a result, in the control graph, all arithmetic expression operators were grouped on a set of linear program parts – the graph vertices, into which *checkpoints (CP)* were entered. Here, checkpoints were needed to determine the route context within which the computations take place. Moreover, the special systems of defining relations were constructed in the form of similarity equations at each checkpoint for arithmetic operators. The equation system solution allowed to form the matrices of similarity invariants to control the computational process semantics [34, 56, 89].

A similarity equations system development:
The studies have shown that the most effective way to control the computation semantics is to test relations, based on theoretically based relations and computation features. Here the key relation in the approach for detecting the parameters of the incorrect computational process functioning is some invariant, which is understood as the auto modeling (constant) presentation of program execution in the actual operating secured infrastructure conditions. The invariant generation problems, from the different program representations, are non-trivial and poorly formalized. In the program execution dynamics, only semantic invariants remain fully computable (reproducible) (since they do not depend on the specific values of the program variables).

Let us imagine the implementation of B^k of the program control graph as an ordered primary relation sequence, corresponding to arithmetic operators:

$$\begin{cases} y_1 = f_1^k(x_1, x_2, \ldots, x_N), \\ y_2 = f_2^k(x_1, x_2, \ldots, x_N, y_1), \\ \ldots \\ y_M = f_M^k(x_1, x_2, \ldots, x_N, y_1, y_2, \ldots, y_{M-1}). \end{cases} \tag{2.12}$$

Having performed the superposition $\{y_i\}$ on X on the right relation sides, we obtain a relation invariant system according to the displacement:

$$\begin{cases} y_1 = z_1^k(x_1, x_2, \ldots, x_N), \\ y_2 = z_2^k(x_1, x_2, \ldots, x_N), \\ \ldots \\ y_M = z_M^k(x_1, x_2, \ldots, x_N). \end{cases} \tag{2.13}$$

The relation $y_i = z_i^k(x_1, x_2, ..., x_N)$ can be presented as:

$$y_i = \sum_{i=1}^{p_i} z_{ij}(x_1, x_2, ..., x_N) \qquad (2.14)$$

where $z_{ij}(x_1, x_2, ..., x_N)$ is a power monomial.

In accordance with the **Fourier rule**, the summands (2.14) should be homogeneous in dimensions, i.e.,

$$[y_i] = [z_{ij}(x_1, x_2, ..., x_N)], j = \overline{1, p_i} \text{ or}$$

$$[z_{ij}(x_1, x_2, ..., x_N)] = [z_{il}(x_1, x_2, ..., x_N)], j = \overline{1, p_i}. \qquad (2.15)$$

System (2.15) is a defining relations system or a similarity equation system.

Using the function $\rho = X \rightarrow [X]$, we associate each $x_j \in X$ with some abstract dimension $[x_j] \in [X]$. Then the summand dimensions (2.14) will be expressed as

$$[z_{ij}(x_1, x_2, ..., x_n)] = \prod_{n-1}^{N} [x_n]^{\lambda_{jn}}, j = \overline{1, p_i}. \qquad (2.16)$$

Using (2.15) and (2.16), we develop a system of defining relations

$$\prod_{n-1}^{N} [x_n]^{\lambda_{jn}} = \prod_{n-1}^{N} [x_n]^{\lambda_{1n}}, j = \overline{1, p_i}. \qquad (2.17)$$

which is transformed to the following form:

$$\prod_{n-1}^{N} [x_n]^{\lambda_{jn} - \lambda_{1n}} = 1, j = \overline{1, p_i}. \qquad (2.18)$$

Using the *logarithm method*, as it is usually done, when analyzing the similarity relations, we obtain a homogeneous system of linear equations from the system (2.18)

$$\sum_{n-1}^{N} (\lambda_{jn} - \lambda_{1n}) \ln[x_n] = 0, j = \overline{1, p_i}. \qquad (2.19)$$

Expression (2.19) is a criterion for semantic correctness.

Having performed a similar development for $\forall B_i^k \in B^k$, we obtain a system of homogeneous linear equations for κ-implementation:

$$A^\kappa \omega = 0. \qquad (2.20)$$

Generally, we can assume that the function $\rho = X \rightarrow [X]$ is *surjective* and, therefore, the B^k implementation is represented by a matrix $A^k = \|a_{ij}\|$ of size $m_k \times n_k$, whose number of columns is not less than the number of rows, i.e., $n_k \geq m_k$.

We say that the implementation of B^k is representative if it corresponds to the matrix A^k with $m_k \geq 1$, i.e., the implementation allows developing at least one similarity criterion.

Usually, a program corresponds to a separate functional module or consists of an interconnected group of those and describes the general solution of a certain task. Each of the implementations $B^k \in B$ describes a particular solution of the same problem, corresponding to the certain X component values. Since $B^k \cap B^l \neq \emptyset$, $\forall B^k, B^l \in B$ then the mathematical dependencies structure should be preserved during the transition from one implementation to another, i.e., similarity criteria should be common. Then the matrices $\{A^k\}$, corresponding to the implementations $\{B^k\}$, can be combined into one system.

Let the program have q *implementations*. Denote by A the union of the matrices $\{A^k\}$ corresponding to the implementations $\{B^k\}$, i.e.,

$$A = \begin{pmatrix} A_1 \\ \ldots \\ A_q \end{pmatrix} \tag{2.21}$$

The A development can be carried out using selective vertices covering the implementations.

Thus, the matrices A union are part of the program passport and are a database of semantic standards $\{A^k\}$ for the linear program $\{B^k\}$ sections [34].

The similarity equation example:

Let us consider an assignment operator:

$$p = a*b + c/(d - e). \tag{2.22}$$

Here, the correct expression must be generated by some selected grammar, which depends on both the possible term meanings and the chosen operations set. For a *context-free grammar*, each expression can be matched to an output tree in a unique way. Thus, an output tree can be used as an alternative expression representation.

When constructing a tree by the expression, the computation order plays its role. Obviously, the vertex descendant values are computed earlier than the ancestor vertex value. Therefore, the operation last performed will take place at the tree top. In order to construct a tree unambiguously, it is necessary to determine the operation calculation order in the expression,

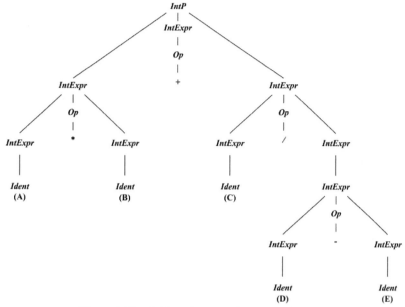

Figure 2.54   Arithmetic expression generation tree.

taking into account their priorities and the operation order with the same priority, including the case when computing the same operation (associativity property). Usually, such expressions are computed from left to right.

The constructed tree will definitely correspond to the specified expression considering the computation order. The above expression [2.22] will correspond to the tree presented in Figure 2.54.

We formalize the arithmetic expressions:

Let *Op* {+, −, *, /} be an arithmetic operations set under consideration.

"Terms" is a set of terms consisting of possible objects that can be operation arguments.

Expr is a set of all possible expressions, and *Terms* ⊂ *Expr*.
elem(o,e) ∈ *Expr* − many other elements, and *o* ∈ *Op*, *e* ∈ *Expr*.

Thus, an arithmetic expression is either a term or an operation connecting several expressions.

The expression [10] with the set of terms *Terms* = {*p, a, b, c, d, e*} and the binary operations set *Op* {, −, *, /} will be represented as:

elem: (=,p,(,(*,a,b),(/,c,(-,d,e))))).

The arithmetic operator execution correctness can be assessed using the appropriate semantic function. When applied to expressions, the

Table 2.23   The operations on the program variables dimensions.

Operator	Denotation	Correctness condition	Linear equations	Similarity criterion
Addition	$R = L + P$	$[L] = [P]$	$[R]^0[L]^1[P]^{-1} = 1$	0 1 −1
Subtraction	$R = L - P$	$[L] = [P]$	$[R]^0[L]^1[P]^{-1} = 1$	0 1 −1
Multiplication	$R = L * P$	$[R] = [L][P]$	$[R]^1[L]^{-1}[P]^{-1} = 1$	1 −1 −1
Division	$R = L/P$	$[R] = [L][P]^{-1}$	$[R]^1[L]^{-1}[P]^1 = 1$	1 −1 1
Exponentiation	$R = L^s$	$[R] = [L]^s$	$[R]^1[L]^{-s}[P]^0 = 1$	1 −s 0
Assignment	$L = P$	$[L] = [P]$	$[R]^0[L]^1[P]^{-1} = 1$	0 1 −1

semantic function $T : a \rightarrow [a]$ assigns to each argument some abstract entity or dimension [a]. Thus, the arithmetic operations, performed on program variables during program execution, are, in fact, operations on physical dimensions, and the semantics reflections, performed at runtime, are linear mappings. The axiomatic of extended semantic algebra, which defines operations on the variable dimensions, is presented in Table 2.23.

where R is the operation result; L,R are left and right operands; $[\ ]$ is *dimension*.

For a correctly running program in the context of this operator, the following relations between the physical dimensions of the terms $\{p,a,b,c,d,e\}$ should be fulfilled:

$$[p] = [a * b] = [a][b],$$
$$[d] = [e],$$
$$[p] = [c / (d - e)] = [c][d]^{-1} = [c][e]^{-1} \qquad (2.23)$$

where $[X]$ is a physical object X *dimension*.

A computation model in memory can be represented using the context-free grammars. It allows describing the computation process structure as a whole. Context-free grammar has the following form:

$$G = (\Sigma, N, R, S), \qquad (2.24)$$

where

$\Sigma = \{identifier, constant, address ... register\}$ is a set of assembler terminal symbols (Table 2.24);

$N = \{Addition, Subtraction, Multiplication, Division, Appropriate\}$ is a non-terminal character set;

$R = \{AddCommand, SubCommand\ MulCommand, ..., DivCommand\}$ is an output rule set;

$S \in \Sigma$ is a starting symbol.

Table 2.24   Sets of non-terminal symbols.

Non-terminal symbols N	Generalizing feature	Terminal symbols Σ
Addition	Addition commands	fiadd | fadd | faddp | ...
Subtraction	Subtraction commands	fisub | fsub | fsubr | ...
Multiplication	Multiplication commands	fimul | fmul | fmulp | ...
Division	Division commands	fidiv | fdiv | fdivr | ...
Appropriate	Data transfer commands	fist | fst | fstp | ...

The terminal symbols include arithmetic coprocessor command lexical tokens, including addition, subtraction, multiplication, division, assignment (data transfer) commands, etc. A non-terminal symbol set is a set of lexical tokens, united by a generalizing feature, as well as their combinations, using products. An example of non-terminal symbols is given in Table 2.24.

The output rule represented by expression (1.42) determines the use of the "fadd" command. Thus, we will present all possible inference rules in assembly language.

AddCommand → Addition_Register, Address
| Addition_Register, Register
| Addition_Register, Register faddp st(1), st
| ...   (2.25)

where

Addition – a non-terminal set of coprocessor addition commands;
Register – a non-terminal set of coprocessor stack registers;
Address – a memory identifier set or actual memory addresses.

Each output in a context-free grammar, starting with a non-terminal symbol, is uniquely associated with a directed graph, which is a tree and is called an output (parse) tree. An output tree example related to the disassembled expression code (1.39), as well as its representation as the similarity equations in terms of the dimension theory, is shown in Figure 2.55.

The solution to this equation system is a similarity coefficient matrix, constructed as follows:

$$[ebp+p] = [ebp+a][ebp+b]$$
$$[ebp+d] = [ebp+e]$$
$$[ebp+p] = [ebp+c][ebp+d]^{-1}$$
$$= [ebp+c][ebp+e]^{-1}$$

$$[ebp+p]^1[ebp+a]^{-1}[ebp+b]^{-1}[ebp+c]^0[ebp+d]^0[ebp+e]^0 = 1$$
$$[ebp+p]^0[ebp+a]^0[ebp+b]^0[ebp+c]^0[ebp+d]^1[ebp+e]^{-1} = 1$$
$$[ebp+p]^0[ebp+a]^0[ebp+b]^0[ebp+c]^{-1}[ebp+d]^1[ebp+e]^0 = 1$$

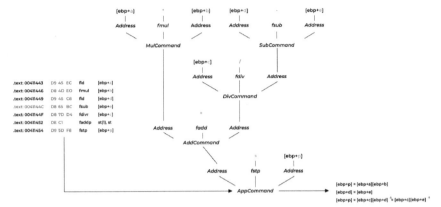

Figure 2.55   Computations representation by similarity equations.

By taking a logarithm, we obtain a homogeneous linear equation system with a coefficients matrix:

$$A^1 = \begin{pmatrix} 1 & -1 & -1 & 0 & 0 & 0 \\ 0 & 0 & 0 & 0 & 1 & -1 \\ 0 & 0 & 0 & -1 & 1 & 0 \end{pmatrix} \qquad (2.26)$$

In order to organize the similarity relations development, it is necessary to construct a translation grammar for assignment operators of the arithmetic type. The translational (attribute) grammar in addition to the syntax allows describing the action characters, which are implemented as functions, procedures, and algorithms. According to dimensions, these functions should implement algorithmic computations and the similarity relation development, power monomials, equations, and solutions.

Thus, the observation problem solution (control graph) and the computations representation (similarity equation) made it possible to form the image of a system for monitoring destructive software actions on the secured infrastructure, and infrastructure and restoring computation processes based on similarity invariants.

The possible destructive action control:
The plan of destructive software impacts control and the computational processes recovery includes preparatory and main stages (Figure 2.56). The preparatory stage includes the program passport formation in similarity invariants; the main ones are the stages of:

- similarity invariants formation under exposure;

1. Program passport formation in the similarity invariants;

2. Similarity invariant formation under exposure;

3. Similarity invariants database formation at the checkpoints of the program control graph;

4. Validation of the semantic correctness criteria of computational processes;

5. Signal generation of the computation semantics violation and a partial calculations recovery according to the program passport;

Figure 2.56   Distortion control and computation process recovery scheme.

Figure 2.57   The correct computation scheme.

- similarity invariants database formation at the checkpoints of the program control graph;
- validation of the semantic correctness criteria of computational processes;
- signal generation of the computation semantics violation;
- partial computations recovery according to the program passport.

A general representation of the information infrastructure that implements correct computations under the hidden intruder program actions is reflected in Figure 2.57. We will reveal the stages of the destructive software impacts control and the computation processes recovery in more detail.

Stage 1. The program passport formation in similarity invariants:
In order to implement a dynamic control, it is necessary to use the static verification results in the form of a program passport.

At the stage of a static verification using the disassembled correct computation code (Figure 2.57), the program control graph is constructed.

At each checkpoint for each arithmetic operator, a production tree of an arithmetic expression is generated to develop a linear homogeneous equation system in the dimension terms. The result of solving the equation systems for each linear program part is a similarity invariant matrix.

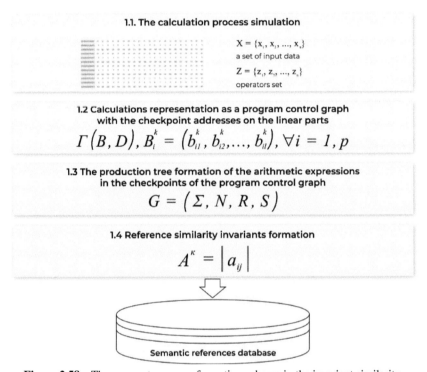

Figure 2.58   The passport program formation scheme in the invariant similarity.

The semantic standard database is made up of reference matrices of similarity invariants for each checkpoint (Figure 2.58).

Stage 2. The similarity invariants formation under exposure:
The similarity invariants formation of the computational process, which is subjected to the hidden arithmetic operations impacts, runs according to the same algorithm as the computational process reference invariant formation.

For a given program, a set of *checkpoints (CT)* is formed, which are embedded in the studied program. The initial program model is the control graph of the computation process in terms of linear program sections. The similarity equations are analyzed and a coefficient matrix is developed in embedded *CT* for each linear program section, where the computations take place (Figure 2.59).

Incorrect computations will differ in the state set of the *computational process Z*, i.e., in arithmetic operator sequence. The incorrect computations scheme is presented in Figure 2.60.

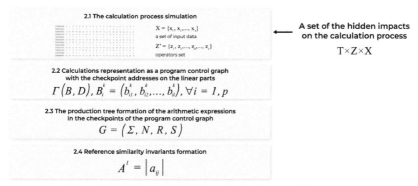

Figure 2.59   The similarity invariants scheme under exposure.

Figure 2.60   The incorrect computations scheme.

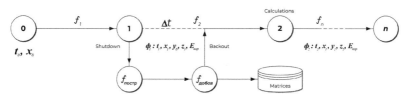

Figure 2.61   The similarity invariants database formation scheme.

Stage 3. The similarity invariants database formation at the checkpoints of the program control graph:
At this stage, the similarity invariant matrices constructed for each check-point form a similarity invariants database. The scheme of adding matrices to the database is presented in Figure 2.61.

Stage 4. The validation of the semantic correctness criteria of the computational processes:
In order to control the semantic correctness of the performed computations, it is necessary to check the semantic correctness criterion by the formula (1.36) applying the reference and standard invariants matrix (Figure 2.62). A necessary criterion for semantic computations correctness is a solution existence to a system in which none of the variables ($ln[x_j]$) are turned to 0.

If the validation of this checkpoint has been completed, then proceed to check the criteria in the next CT until the program ends.

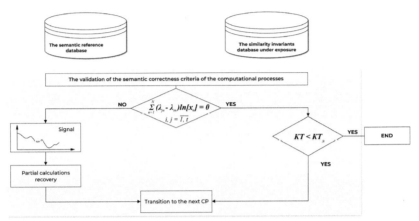

Figure 2.62   The computational processes validation scheme.

Stage 5. The signal generation of the computation semantics violation and the partial computations recovery according to the program passport:

If the semantic correctness violation of the program execution is detected, that is, if for a given checkpoint $\lambda_{jn} - \lambda_{in} \neq 0$, then a signal is formed and an attempt is made to recover the computations from the inverse transformation of the reference matrix invariants (Figure 2.62).

This approach allows not only to determine the fact of the computation semantics violation but also to indicate the specific impact location on the program, using the mechanism for introducing checkpoints.

Thus, the dimensions and similarity theory application allowed synthesizing new informative features – the so-called similarity invariants for controlling the computational processes correctness. The similarity invariants use made it possible to bring the monitoring system of destructive program actions and the computation processes recovery closer to the controlled computational process semantics. The obtained results allowed presenting a controlled computational process as a corresponding equations system of dimensions and similarity invariants, and its solution was to analyze the computations semantics under the destructive program impacts on the secured critical information infrastructure.

2.11.2   Intellectual Cyber Resilience Orchestration

Let us consider a possible method of intellectual administration, the so-called intellectual orchestration of cyber resilience, based on multilayer similarity invariants. Here, the observed computations is represented by defining relations or similarity equations. The solution of these similarity equations allows synthesizing the invariant informative features that together form the so-called *"passport"* of computation programs (some standard of the regular behavior of the protected infrastructure). These standards are formed in the course of computations and are compared with a predetermined passport of computation programs. The technical implementation of this new approach was brought to the beta version of the special supervisor, the corresponding technical device, and software and hardware complex for managing the cyber resilience.

The listed developments allow making a correlation analysis of the detected inconsistencies and to timely detect and resolve the problem situations that arise in real time. It is significant that the proposed approach allows us to control the *computation semantics* of the protected information infrastructure in the conditions of previously unknown heterogeneous mass cyber-attacks by intruders.

The task of the computational semantics control:
The following classes of computational tasks are distinguished: *measurement, information, computational,* and *information-computational.* In practice, it is especially important to control the implementation of computation and information-computationn tasks since a minor modification of one program operator can lead to an error accumulation, as a result of which the incorrect computer computations will be obtained. In addition, these results may be in a specific confidence interval; so an error without additional controls will not be detectable. Due to the high construction complexity and the potential danger of undeclared functioning of hardware and system-wide software, critically important information infrastructures become extremely vulnerable to covert impacts on the process of calculating software and hardware bookmarks (*"logical or digital bombs"*) and malware.

Let us consider the structure and characteristics of the vulnerabilities of information-computing tasks. Table 2.25 presents the possible ways to influence the computer computations at different execution levels of computational programs in some typical operating environment of critically important information infrastructure.

Typical risks of malfunctioning and unacceptable lowering of cyber-resilient indicators of critically important information infrastructure include:

- distortion of the machine data, algorithms, and computer computations;
- block or violation of the information exchange between the key components of the information infrastructure;

Table 2.25   Ways to modify computations.

Operating environment levels	Ways to modify computations
Level 7. Tasks *Level 6.* Programs	• Masking program execution • Difficulties associated with program analysis at the application level
Level 5. Program components *Level 4.* System calls and interrupts *Level 3.* Command system	• Using the system library replacement • Intercepting the access to system functions • Changing the process import table • Substituting the export table • Substituting the interrupt handler
Level 2. "Processor–memory" interaction processes *Level 1.* Register commands	• Making changes to the machine code commands

- violation of the access rights to information infrastructure components;
- partial or complete disruption of the timing of computer computations;
- partial or complete block of the execution of emergency control algorithms in emergency situations;
- transfer to the irreversible catastrophic state of the information infrastructure;
- physical destruction of information infrastructure.

As a rule, the cyber-attacks by intruders are aimed at disorganizing the computations' algorithms, changing the order of actions performed, and calculating properties' distortion.

The main approach ideas:
A new model of the executable computer programs was proposed in order to organize the control of computer computation semantics. At the same time, it was taken into account that typical settlement software systems are characterized by a certain hierarchical multilevel structure. This structure includes system software, integration buses (the basis of data exchange protocols), as well as the special application software (a set of information and computational tasks). This stratification determines the typical form of computer computations and allows constructing the desired semantic standard of the correct behavior of computer programs in the form of a multilayer similarity invariant.

In turn, each computer program loaded into the operational memory of a specific processor is characterized by a unique internal multilayer structure that reflects the knowledge of the total number of subprograms, procedures and functions, program blocks, and atomic operations (Figure 2.63). This

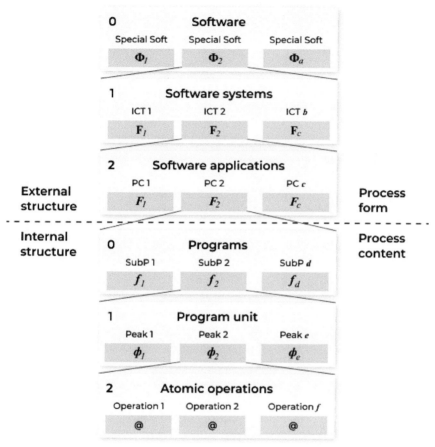

Figure 2.63 Multilevel representation of computer computations.

knowledge allowed the systematical research some computer computation content and forms the desired semantic standard for the correct behavior of the protected information infrastructure under the growth of security threats based on the mathematical apparatus of the theory of dimensionality and similarity.

Let us consider the structure of some typical computer computations as a complex static system with a finite number of elements, when the stratum number is 3. All system elements, in this case, are divided into three types: elements of the zero (upper) stratum, elements of the first (middle) stratum, and elements of the third (lower) stratum. The top-level structure is defined using a binary relation on the base set of this stratum:

$$\Phi = \{\Phi_1, \Phi_2, ...,\Phi_a\}, r_s = \langle \Phi, \Phi; R_s \rangle = \langle \Phi^2; R_s \rangle. \qquad (2.27)$$

The elements Φ_1 are subsets, which can be represented as

$$\Phi_i = \langle F_j, r_{si} \rangle, i = \overline{1,a}, j = \overline{1,b} \qquad (2.28)$$

where

$$F_{ij} = \left\{ F_{ij1}, F_{ij2}, \ldots, F_{ijb} \right\}, r_{si} = \left\langle F_{ij}, F_{ij}; R_{si} \right\rangle = \left\langle F_{ij}^{\,2}; R_{sj} \right\rangle$$

Finally, the elements F_{ij} can be represented as

$$F_{ij} = \left\langle \tau_k, r_{sjk} \right\rangle, i = \overline{1,c}, j = \overline{1,b}, k = \overline{1,c} \qquad (2.29)$$

where

$$\tau_k = \left\{ \tau_{ij1}, \tau_{ij2}, \ldots, \tau_{ijk} \right\}, r_{sij} = \left\langle \tau_{ij}, \tau_{ij}; R_{sij} \right\rangle = \left\langle \tau_{ij}^{\,2}; R_{sij} \right\rangle$$

Here, Φ is software, F is a set of information and computational tasks, τ is a set of software complexes, and $r \subseteq R$ is a relation characterizing the internal structural and quantitative characteristics of a certain software package.

We introduce the equivalence relation on the set of structural components of the calculation programs; it uniquely determines the partition of the base set into disjoint subsets (equivalence classes):

$$A = A_1 \cup A_2 \cup \ldots \cup A_v, A_i \cap A_j \cap A_k = \varnothing \text{ when } i \neq j \neq k \qquad (2.30)$$

where A_i are the equivalence classes of the structural program set elements. The top structure level is represented by an aggregated graph:

$$\Phi_i \approx c_i, C = \left(c_j \right)_1^k, R_{aep} = \left\{ \left\langle c_i, c_j, c_k \right\rangle \middle| \exists \left\langle \Phi_\alpha, \Phi_\beta, \Phi_\gamma \right\rangle \right\} \subseteq F^2 \qquad (2.31)$$

and the following ones are graphs of the corresponding aggregates.

Imagine the internal structure of the computer computations [56] in the form of the control program graph:

$$G(B,D) \qquad (2.32)$$

where $B = \{B_i\}$ are many vertices (linear program sections) and $D = \{B \times B\}$ is a set of arcs (control connections) between them.

The path in the control graph is determined by the sequence of vertices

$$R^B (B_1, B_2, \ldots, B_n) \qquad (2.33)$$

or a sequence of arcs

$$R^D (d_1, d_2, ..., d_{n-1}).$$

Here, each arc d connects the vertices of the oriented graph B_i and B_k.

Each elementary (without cycles) path R of the graph corresponds to an ordered sequence of vertices

$$R^k = (B_1^k, B_2^k, ..., B_i^k),$$

where $B^k \subseteq B$, and $B_i^k = (b_{i1}^k, b_{i2}^k, ..., b_{il}^k)$, $\forall i = \overline{1, p}$ $B_i^k = (b_{i1}^k, b_{i2}^k, ..., b_{il}^k)$ form a sequence of arithmetic operators on each linear part of the graph, i.e.,

$$B_i = (b_{i1}, b_{i2}, ..., b_{il}) \tag{2.34}$$

is called a program implementation or computer process.

Let us consider the control subroutine graph of the information and computation task (Figure 2.64).

As a result, the control flow graph of the computation program is transformed into a form with all operators of arithmetic expressions are grouped in a set of linear program sections – the graph vertices (Figure 2.65) into which the control points (CP) are embedded. Here, control points are necessary to determine the path context to make computations.

For the most critical computational routes, a set of control points (CP) are formed, which are embedded in studying the subroutine. The initial subroutine model is the control flow graph (the computation route under study) in terms of linear sections. In embedded CP for each linear subroutine section, where critical computations are performed, the similarity relations are analyzed and a coefficient matrix is constructed.

Thus, combining structural and semantic invariants, a multilayer program invariant is formed, which is a new model of knowledge about computer computations and allows controlling the implementation of computational programs along the most probable ways of its implementation depending on the distribution of input data.

Forming the standards:
The multilevel representation of a typical computational process can be displayed in the form of a certain tree structure (F). Figure 2.66 is trees with nodes; each node, excluding the root and leaves, can contain subtrees (from one to m). We will say that the «tree» root is the higher hierarchy level (zero level) and is a special software, a set of root nodes, forms the first hierarchy level and represent a set of information and computation tasks, a set of

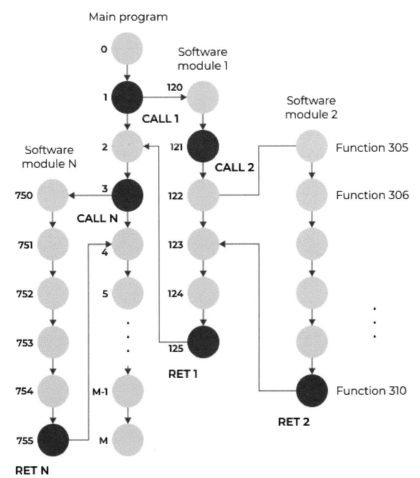

Figure 2.64   Control flow graph of the information and computation task.

nodes included in the nodes of the first hierarchy level characterize its second level, representing a variety of software packages, etc. The leaves form the last, lowest hierarchy level and are atomic computational operations.

It should be noted that the top-level elements of the internal graph structure are the control flow graphs of the subprograms f_i, which are decomposed into program blocks φ_i in terms of the linear portions of the control flow graph. The representation form of the computational process in the form of an ordered graph allows us to derive the atomic operators of arithmetic expressions @ from program blocks (conditional transitions, forks, cycles, etc.).

The modeling of a certain computational process by a control flow graph is caused by the need to analyze (and research) the program functionality, considering the domain structure and certain properties of its

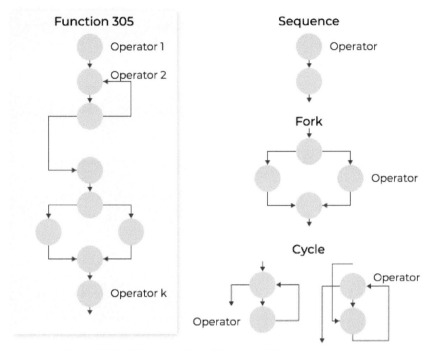

Figure 2.65   Decomposition of the control flow program graph.

variables. In addition, this representation of the internal program structure
allows creating a semantic invariant to control its integrity.

The existing possibilities of special tools for disassembling and studying the program structure, for example, *IDA Pro or IRIDA*, allow the computational process of the executable code of some computation program to
be represented by a control flow graph. Next, describe the call graph of subroutines as well as classify the transfer of control into subroutines (short-
range calls, register-based calls, calls via the import table, long-distance
calls, calls with or without returning to the calling subroutine, unclassified
calls, which include non-disassembled IDA Pro parts of the code and parts
of the code that are not related to one of the subroutines, etc.). In addition,
the IRIDA toolkit has a mechanism for setting *control points (CP)* along
the path of the computation process (Figure 2.67); further, the semantic
standards will be formed in these *CP*.

Thus, the data obtained using these tools (control flow program graph
with embedded *CP* on the linear parts of the computational process) are
the input data for creating a multilayer program invariant.

At each control point for arithmetic operators, it is necessary to develop
systems of constitutive relations in the similarity equations' view. The

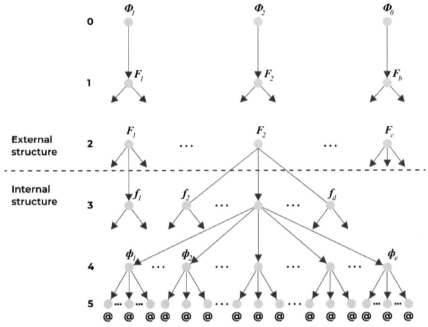

Figure 2.66   Graph representation of the computation structure.

solution of these systems allows us to form invariant matrices, which, in turn, allow us to control the semantics of the computer processes.

Imagine the implementation (program block) B^k of the control graph in the view of an ordered sequence of primary relations corresponding to arithmetic operators:

$$\begin{cases} y_1 = f_1^k(x_1,x_2,\ldots,x_N), \\ y_2 = f_2^k(x_1,x_2,\ldots,x_N,y_1), \\ \ldots \\ y_M = f_M^k(x_1,x_2,\ldots,x_N,y_1,y_2,\ldots,y_{M-1}). \end{cases} \tag{2.35}$$

Having performed the superposition $\{y_i\}$ on X in the right-hand relations' sides, we obtain a system of relations invariant referred to the displacement:

$$\begin{cases} y_1 = z_1^k(x_1,x_2,\ldots,x_N), \\ y_2 = z_2^k(x_1,x_2,\ldots,x_N), \\ \ldots \\ y_M = z_M^k(x_1,x_2,\ldots,x_N). \end{cases} \tag{2.36}$$

Grammar Description	Program control flow graph presentation	Execution paths (protocols)

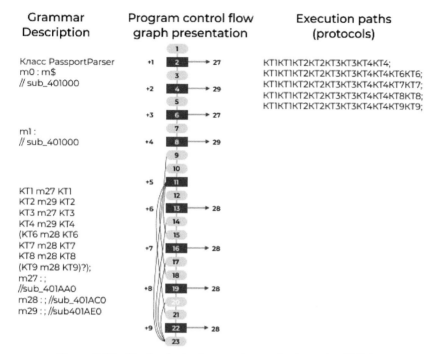

Figure 2.67   Checkpoint implementation mechanism using IRIDA.

The ratio can be represented as

$$y_i = \sum_{i=1}^{p_i} z_{ij}(x_1, x_2, \dots, x_N) \tag{2.37}$$

where $z_{ij}(x_1, x_2, \dots, x_N)$ is a power monomial.

In accordance with the Fourier rule, the members of sum (2.37) must be uniform in dimensions, i.e.,

$$[y_i] = [z_{ij}(x_1, x_2, \dots, x_N)], j = \overline{1, p_i}$$

or

$$[z_{ij}(x_1, x_2, \dots, x_N)] = \\ = [z_{ij}(x_1, x_2, \dots, x_N)], j = \overline{1, p_i} \tag{2.38}$$

System (2.38) is a system of defining relations or a system of the similarity equations.

Using the function $\rho = X \rightarrow [X]$, we associate each $x_j \in X$ with some abstract dimension $[x_j] \in [X]$. Then the dimensions of the members' sum (2.38) will be expressed as

$$\left[z_{ij}(x_1, x_2, \dots, x_n) \right] = \prod_{n=1}^{N} [x_n]^{\lambda_{jn}} \; j = \overline{1, p_i} \tag{2.39}$$

Applying Equations (2.38) and (2.39), we construct a system of defining relations

$$\prod_{n=1}^{N}\left[x_n\right]^{\lambda_{jn}} = \prod_{n=1}^{N}\left[x_n\right]^{\lambda_{ln}} \; j = \overline{1,p_i} \; .$$

We transform it to the form

$$\prod_{n-1}^{N}\left[x_n\right]^{\lambda_{jn}-\lambda_{ln}} = 1, \; j, \; l = \overline{1,p_i} \qquad (2.40)$$

Applying the logarithm technique, as is usually done when analyzing the similarity relations, from the system (2.40), we obtain a homogeneous system of linear equations

$$\sum_{n=1}^{N}\left(\lambda_{jn} - \lambda_{ln}\right)\ln\left[x_n\right] = 0, \; j, \; l = \overline{1,p_i} \; , \qquad (2.41)$$

Expression (2.41) is a criterion for semantic correctness.

Having performed a similar construction for $\forall B_i^{\,k} \in B^k$, we obtain for kth implementation, a system of homogeneous linear equations:

$$A^k \omega = 0. \qquad (2.42)$$

In the general case, we can assume that the function $\rho = X \rightarrow [X]$ is surjective and, therefore, the realization of B^k is represented by a matrix $A^k = \|a_{ij}\|$ of $m_k \times n_k$ size, whose number of columns is not less than the number of rows, i.e., $n_k \geq m_k$.

We say that the realization B^k is representative if it corresponds to the matrix A_k with $m_k \geq 1$, i.e., implementation allows creating at least one similarity criterion.

Usually, a program corresponds to a separate functional module or consists of an interconnected group of those and describes the general solution of a certain task. Each of the implementations $B^k \in B$ describes a particular solution of the same problem, corresponding to certain values of the X components. Since $B^k \cap B^l \neq \emptyset$, $\forall B^k, B^l \in B^k$, the structure of the mathematical dependencies should be saved during the transition from one implementation to another, i.e., similarity criteria should be common. Then the matrices $\{A^k\}$, corresponding to the realizations $\{B^k\}$, can be combined into one system.

Let the subroutine have q implementations. Denote by A the union of the matrices $\{A^k\}$ corresponding to the realizations $\{B^k\}$, i.e.,

$$A = \begin{pmatrix} A_1 \\ \dots \\ A_q \end{pmatrix} \qquad (2.43)$$

The construction of A can be made using selective implementations, which provide covering the vertices. The non-trivial compatibility of the matrix system A, according to this method, is a criterion for controlling the semantic process correctness.

When developing computation programs in a certain procedural programming language for call points of procedures and subroutines, the question of matching their formal parameters arises. In this case, square permutation matrices T_e are formed, reflecting the correspondence between the formal procedure parameters and the main process variables. As a result, the system (2.43) is converted to the form:

$$
A = \begin{Vmatrix}
A_1 \\
A_2 \cdot T_1 \\
A_3 \\
\cdots \\
A_{q-2} \cdot T_{e-1} \\
A_{q-1} \cdot T_e \\
A_q
\end{Vmatrix}
\tag{2.44}
$$

The direct method of calculating the modified criterion is creating (based on the matrix A) the equations' system coefficients of dimension (2.44) of the matrix R, which has a special form:

$$
R = \begin{Vmatrix}
1 & 0 & \cdots & 0 & c_{1,1} & \cdots & c_{1,n-k} \\
0 & 1 & \cdots & 0 & c_{2,1} & \cdots & c_{2,n-k} \\
\cdots & \cdots & \cdots & \cdots & \cdots & \cdots & \cdots \\
0 & 0 & \cdots & 1 & c_{k,1} & \cdots & c_{k,n-k}
\end{Vmatrix}
$$

Imagine the matrix R in this form of

$$
R_{k \times n} = E_{k \times k} \big| c_{k \times (n-k)}
$$

where E is the identity matrix and k and n are the number of rows and columns of the original matrix A, respectively.

To create the matrix R, it is sufficient to use three types of operations:

1. addition of an arbitrary matrix row with a linear combination of other rows;
2. row permutation;
3. column permutation.

As applied to the solution of the dimension constraint system, the matrix R is identical to the matrix A, with the exception of possibly made column permutations, i.e., there is an equivalent

$$(S \cdot X = 0) \Leftrightarrow (R \cdot T \cdot X = 0)$$

where T is the square permutation matrix of dimension $n \times n$ corresponding to the column permutations of A made at the stage of creating R.

This result is due to the nature of the transformations performed on the matrix A in the process of creating the matrix R.

Formula (2.44) allows us to use matrix R when calculating the modified semantic correctness criterion.

Thus, the semantic invariant characterizing the internal program structure is a database of semantic standards $\{A^k\}$ for linear sections of the program $\{B^k\}$, and, in the general case, the union of matrices A into matrix R forms the database of semantic standards $\{R_i\}$ for subroutines f_i.

Let us note that for the semantic invariant formation, program operands were used as variable equations. To form a structural invariant, we will use the names of subroutines, complexes, and tasks as variable equation systems.

We define the additive operation "+" and the multiplicative operation "*" in the above external structure of calculations. We assume that if two structure elements do not interact with each other, then they are interconnected by the additive operation "+." Otherwise, they are connected by the multiplicative operation "*." Thus, we have the opportunity to describe the resulting structure in the form of equation systems.

Imagine the multiplication operation as multiplication of the structure polynomials

$$x^0(\Omega_1(x)) \text{ and } x^0(\Omega_2(y)):$$
$$x^0(\Omega_1(x)) * x^0(\Omega_2(y)). \tag{2.45}$$

Here, the symbol $x^0 = 1$ means the root of the structure tree; instead of this one, later, some symbols of the structure will be written, and, thus, the tree will be transformed into some new subtree.

The multiplication of polynomials looks like the following:

$$x^0(\Omega_1(x)) * x^0(\Omega_2(y)) = x^0(\Omega_1(x)) * (\Omega_2(y)). \tag{2.46}$$

Here, the number of factors characterizes the structure hierarchy level number, and the structure of each factor characterizes the structure of the corresponding hierarchy level. The expression on the right-hand side

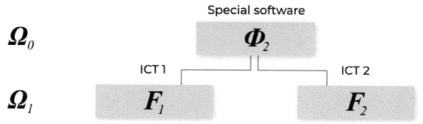

Figure 2.68   Block diagram of software and application systems.

reflects the actual structural complexity of each element at the most elementary level of the structure hierarchy.

Imagine the operation of addition as the addition of structural polynomials

$$x^0(\Omega_1(x)) \text{ и } y^0(\Omega_2(y)):$$
$$x^0(\Omega_1(x)) + y^0(\Omega_2(y)). \quad (2.47)$$

Polynomial is a structure with the maximum complexity of subordination dependencies in the modules, its components $(\Omega_1(x)$ и $\Omega_2(y))$.

Consider the following block diagram (Figure 2.68):

We describe the scheme in Figure 2.69 by the equation of the form:

$$\Omega_1 = \omega(\Omega_0) = \Phi_2 * (F_1(\Phi_2) + F_2(\Phi_2)). \quad (2.48)$$

Using the substitution method, all these polynomials can be expanded by expressing the polynomial Ω_i in Ω_1. Further expansion of the structure (Figure 2.69) results in a polynomial of the form:

$$\Omega_2 = \omega(\Omega_1) = (\Phi_2 * (F_1(\Phi_2) + F_2(\Phi_2))) (F_1((\Phi_2 * (F_1(\Phi_2) +$$
$$+ F_2(\Phi_2)))) + (F_2((\Phi_2 * (F_1(\Phi_2) + F_2(\Phi_2))))))). \quad (2.49)$$

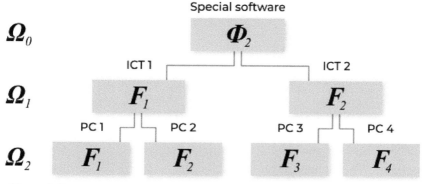

Figure 2.69   Block diagram of software levels, application systems, and complexes.

Now let us consider the convolution operation of structural polynomials. As a result of the convolution, the structural polynomial is transformed in such a way that its structure will be identical to the original polynomial. The convolution process is that the structure is transformed to a simpler form.

Let us consider the structural polynomials of the form

$$\begin{cases} \Omega_3 = \omega(\Omega_2) = F_1 * (f_1(F_1) + f_2(F_1)); \\ \Omega_2 = \omega(\Omega_1) = F_1 * (F_1(F_1) + F_2(F_1)); \\ \Omega_1 = \omega(\Omega_0) = \Phi_2 * (\Gamma_1(\Phi_2) + \Gamma_2(\Phi_2)). \end{cases}$$

Here each subsequent polynomial is a convolution of the previous one, i.e., each previous polynomial in such structures plays the role of an elementary member of the structure (basic element).

Let us consider an equation system describing the structure of performing calculations of a software complex of an information and computation task using special software containing three subprograms:

$$\begin{cases} F = F_1 * (f_1 + f_2 + f_3); \\ F = F_1 * F; \\ \Phi = \Phi_2 * F. \end{cases} \tag{2.50}$$

In terms of dimensions, the system (2.50) can be represented as

$$\begin{cases} [f_1]^1 = [f_2]^1 \\ [f_1]^1 = [f_3]^1 \\ [F]^1 = [F_1]^1 [f_1]^1 \\ [F]^1 = [F_1]^1 [F]^1 \\ [F]^1 = [F_2]^1 [F]^1. \end{cases} \tag{2.51}$$

From system (2.51), we obtain a matrix of coefficients by logarithm:

$$S = \begin{pmatrix} 1 & -1 & 0 & 0 & 0 & 0 & 0 & 0 & 0 \\ 1 & 0 & -1 & 0 & 0 & 0 & 0 & 0 & 0 \\ -1 & 0 & 0 & -1 & 1 & 0 & 0 & 0 & 0 \\ 0 & 0 & 0 & 0 & -1 & -1 & 1 & 0 & 0 \\ 0 & 0 & 0 & 0 & 0 & 0 & 0-1 & -1 & 1 \end{pmatrix} \tag{2.52}$$

The solution of the system (2.52) allows us to form a matrix of coefficients describing the desired structural invariant of a certain computation program. Thus, a multilayer similarity invariant (Figure 2.70) can be represented as some multidimensional matrix (Figure 2.70), *which allows controlling the semantics of the observed computational process.

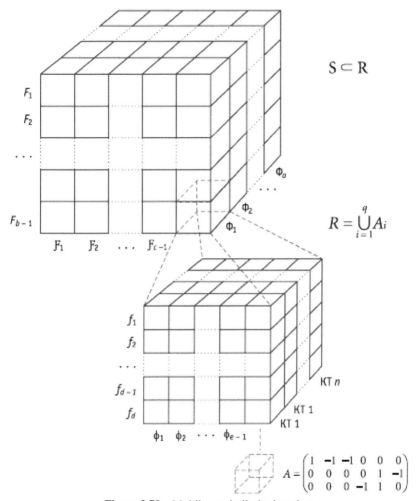

Figure 2.70   Multilayer similarity invariant.

Here, the matrix A is an invariant of a program block, the union of which forms the matrix R is an invariant of the subprogram. The invariant of the structure S includes the matrix R and forms the desired multilayer similarity invariant to control the semantics of computer computations in the protected information infrastructure.

Cyber resilience orchestration:
The semantic correctness control stage of computer computations in the protected information infrastructure includes the following sub-steps:

- forming the observable similarity invariants under the impacts;

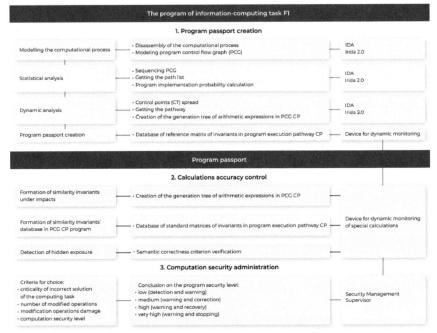

Figure 2.71   Cyber resilience management system.

- forming a similarity invariants' database in control points of the control graph of the computation program;
- detecting the exposure as a result of verification of the computations' semantic correctness criterion using previously prepared "*passport*" of computation programs.

At the stage of administration, the developed supervisor of cyber resilience monitoring of the information infrastructure analyzes the detected modifications of the computation paths and decides to handle critical situations.

A general view of the information infrastructure that implements correct computations under the hidden actions of intruders is shown in Figure 2.71.

Let us note that the above transformations of the representations of the computational process to control the computer computations' semantics require a significant amount of computing resources of the information infrastructure. For optimal resource use, taking into account the due dates for the execution of design tasks, a utility model was developed that allows real-time task execution to control the computer computations' semantics

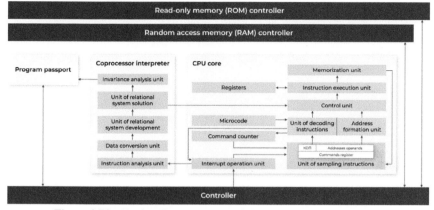

Figure 2.72 Device diagram to control the computations' semantics.

with a minimum delay time. The said device (Figure 2.72) is a separate chip (the so-called "*memory key*"), containing:

- block analysis of the machine instructions from the processor (math coprocessor);
- data processing unit containing a programmable logic integrated circuit for high-performance data processing and allowing the semantic function use:

$$T : a \rightarrow [a] \tag{2.53}$$

In order to assign to each argument a somewhat abstract essence or dimension [a]:

- creation block of the defining relations in terms of dimensions in accordance with the processed machine (assembly) instruction;
- block of solving the system of defining relations, the result of which is the matrix of similarity invariants;
- unit for analyzing and comparing the resulting invariant matrix with reference matrices;
- database is stored in the form of a program passport in a permanent storage device.

At the same time, the coprocessor board provides tasks' parallelization for controlling semantics and managing computer computations in actual operating conditions of the protected information infrastructure. The device and the main processor of some key components of the critically important information infrastructure exchange information via serial interface channels (Figure 2.73).

Figure 2.73   Circuit device interaction with the central processor.

The device works as follows.

The set of commands for the executable program (assembler commands) is divided into three subsets:

- K_A – additive commands (addition, subtraction, comparison, etc.);
- K_m – multiplicative commands (multiplication, division, exponentiation, etc.);
- K_N – not interpreted commands.

The device interprets the processor instructions (math coprocessor) as follows:

- If the processor executes a command $k_i \in K_A$, the coprocessor performs a comparison of the dimensions of its operands.
- If the processor executes the command $k_i \in K_M$, the coprocessor performs manipulations with the dimensions of the operands (addition or subtraction).
- If the processor executes the command $k_i \in K_N$, then the coprocessor is idle.

The main result is that the utility model allows representing the computational process in the form of the corresponding system of dimensional equations. The equation system solution allows us to study the semantics of the computations made by the processor (mathematical coprocessor). Comparison of the obtained results with the reference ones makes it possible to draw a conclusion about the semantic correctness of the implemented computer computations and the absence of covert modifications of arithmetic operations.

A prototype of the *program apparatus complex (PAC)* of intelligent administration of cyber resilience of the protected critically important information infrastructure was also developed (Figure 2.74).

The mentioned complex allows us to control the most critical ways of executing computation programs in the computation flow, perform a correlation analysis of the detected modifications using the supervisor, and highlight the most critical events that require immediate response.

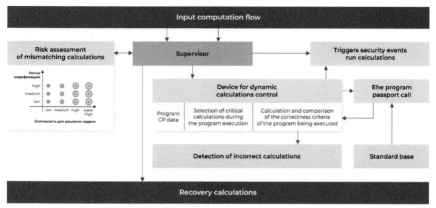

Figure 2.74 Architecture of the cyber resilience management PAC.

Here, the main functions of cyber resilience management are performed by a *specialized supervisor*, which is a device that, for each interruption received from the computation execution controller, calculates the risk assessment of the computation semantics violation based on the hierarchy analysis method.

Let us note that the *analytic hierarchy process (AHP)* [56] makes it possible to optimize a decision-making and contains a procedure for synthesizing priorities calculated on the basis of information received from a device for dynamic computations control. The *supervisor* performs mathematical computations and processes the incoming information, performs a quantitative assessment of alternative solutions, and, based on the data obtained, a decision is made to prevent or restore computations on the *program passport* (Figures 2.75 and 2.76).

Here, to determine the correctness of the computational process, it was necessary to compile an appropriate matrix and express pair judgments.

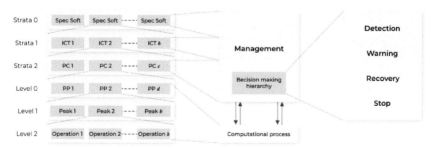

Figure 2.75 Example of a decision levels' hierarchy for cyber resilience management.

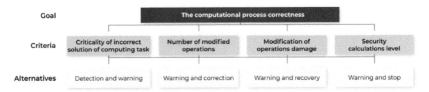

Figure 2.76   Supervisor decision criteria to ensure the required cyber resilience.

Due to the heterogeneity of the evaluation, criteria were formed on a scale of relative importance (Table 2.26). For example, when comparing the relative weights of criteria A weighing W_A and B weighing W_B, the ratio W_A/W_B was entered into the matrix as the ratio of criterion A to criterion B. And the return value W_A/W_B was entered into the matrix as the ratio of criterion B to criterion A.

For each subsequent hierarchy level, additional matrices were created. Table 2.26 shows a criteria comparison example by importance.

The alternative choice was determined based on the matrices and subjective pair judgments. In this case, a set of local priorities was formed

Table 2.26   Scale of relative importance.

Relative intensity	Definition	Description
1	Equal importance	Equal contribution of two criteria to a goal
3	Modest superiority of one over the other	Experience and judgment give a slight superiority to one criterion over another
5	Substantial or strong superiority	Experience and judgment give a strong superiority to one criterion over another
7	Significant superiority	One criterion gives significant superiority over another
9	Very strong superiority	Evidence. The superiority of one criterion over another is confirmed most strongly
2, 4, 6, 8	Intermediate decisions between two adjacent judgments	Apply in a compromise case
Reciprocals of the above numbers	If, when comparing the criteria, one of the above numbers (3) is obtained, then when comparing the second criterion with the first, we get the reciprocal (1/3)	

Table 2.27   Getting priorities vector.

Matrix					Evaluation of the eigenvector components by rows	Normalization of the result to obtain a priorities' vector assessment
	A1	**A2**	**A3**	**A4**		
A1	$\dfrac{w_1}{w_1}$	$\dfrac{w_1}{w_2}$	$\dfrac{w_1}{w_3}$	$\dfrac{w_1}{w_4}$	$\sqrt[4]{\dfrac{w_1}{w_1} \times \dfrac{w_1}{w_2} \times \dfrac{w_1}{w_3} \times \dfrac{w_1}{w_4}} = a$	$\dfrac{a}{\text{sum}} = x_1$
A2	$\dfrac{w_2}{w_1}$	$\dfrac{w_2}{w_2}$	$\dfrac{w_2}{w_3}$	$\dfrac{w_2}{w_4}$	$\sqrt[4]{\dfrac{w_2}{w_1} \times \dfrac{w_2}{w_2} \times \dfrac{w_2}{w_3} \times \dfrac{w_2}{w_4}} = b$	$\dfrac{b}{\text{sum}} = x_2$
A3	$\dfrac{w_3}{w_1}$	$\dfrac{w_3}{w_2}$	$\dfrac{w_3}{w_3}$	$\dfrac{w_3}{w_4}$	$\sqrt[4]{\dfrac{w_3}{w_1} \times \dfrac{w_3}{w_2} \times \dfrac{w_3}{w_3} \times \dfrac{w_3}{w_4}} = c$	$\dfrac{c}{\text{sum}} = x_3$
A4	$\dfrac{w_4}{w_1}$	$\dfrac{w_4}{w_2}$	$\dfrac{w_4}{w_3}$	$\dfrac{w_4}{w_4}$	$\sqrt[4]{\dfrac{w_4}{w_1} \times \dfrac{w_4}{w_2} \times \dfrac{w_4}{w_3} \times \dfrac{w_4}{w_4}} = d$	$\dfrac{d}{\text{sum}} = x_4$

from the group of matrices of pairwise comparisons. Then the set of eigenvectors for each matrix was determined, and the result was normalized to unity, which allowed us to obtain the desired vector of priorities (Table 2.27).

The matrix multiplication by the priorities' vector was made as follows (2.54):

$$A_{4\times4} \times \begin{pmatrix} x_1 \\ x_2 \\ x_3 \\ x_4 \end{pmatrix} = \begin{pmatrix} Y_1 \\ Y_2 \\ Y_3 \\ Y_4 \end{pmatrix} \tag{2.54}$$

Here the priorities were synthesized, starting from the second level down. Local priorities were multiplied by the priority of the corresponding criterion at the higher level and were summed over each element in accordance with the criteria. As a result, the composite (or global) element priority that was used to weight the local priorities of the elements was determined. The procedure continued until reaching the lower level.

Thus, the use of models and methods of the similarity theory and the theory of dimensions made it possible to synthesize the new informative features (represented by multilayer similarity invariants) to control the semantic correctness of the computer computations in the protected information infrastructure. The results obtained allow us to create an equation system of dimensions and similarity invariants of the above computations, their solution allows us to research the semantics of the computations under the hidden modifications, and the proposed supervisor model allows us to determine the modification place in the program.

The developed prototype of program apparatus complex of the intelligent cyber resilience management of the protected critical information infrastructure has confirmed the effectiveness of the proposed approach and the corresponding cyber resilience metric.

3

BC Project Management

Currently, the so-called *"Best practice"* has been formed to create the enterprise continuity programs, ECP. This is primarily the practice of such recognized consulting companies as *Accenture, E&Y, KPMG, PWC, Deloitte* [204, 205, 206, 207, 208, 209, 210, 211, 212, 213, 214, 215, 216, 217], as well as leading manufacturers of hardware and software *IBM, HP, EMC, Microsoft, Symantec, etc.* [218, 219, 220, 221, 222, 223, 224, 225, 226, 227, 228, 229, 230, 231, 232]. Let us take a look at this ECP development practice in more detail.

3.1 Accenture Practice

Accenture's practice was formed during the implementation of a number of consulting projects in the field of business continuity management (BCM) (Table 3.1).

The main Accenture consulting services in the field of BCM include the following:

- risks analysis (RA) of interrupting the company's business processes;
- business impact assessment (BIA);
- classification of business processes and IT services by the degree of business criticality;
- development of methods for calculating quantitative indicators of business continuity and cyber stability;
- defining a business continuity strategy;
- develop business continuity program (BCP) and DRP plans, as well as relevant BC policies, standards, regulations, instructions, etc.

In Accenture, business continuity management refers to a special method of organizing work that will manage the residual risks of business interruption to minimize possible losses to the organization. Here, RA and BIA will develop effective preventive measures to prevent and neutralize potential threats, to develop adequate ways to respond to catastrophic and other emergencies.

Table 3.1   The practice of Accenture in the field of BCM.

| Client | Industry | Category | | | | | | Platforms |
		Assessment and analysis	Strategy development	Plan development	Training	Testing	Integration	
ACCO	Manufacturing	X	X	X	X	X	X	Mainframe, AS/400, C/S
Adobe	Document Management	X	X					Business; SAP
Ahold	Foot Retailer	X	X	X	X	X	X	Mainframe (3 data centers)
American Lung Assoc.	Not for profit	X						Business
ANSES (Argentina)	Government Agency	X	X					Mainframe
Apple Computer	Manufacturing	X	X	X	X	X	X	Mainframe, HP, Sun, Stratus
Ardent Health Services	Health Care	X	X					Systems, business
Aristech Chemical	Chemical Processing	X	X	X	X	X	X	DEC, Wintel
Astra Zeneca	Pharmaceuticals	X	X	X	X	X	X	Business
Astra-Merck	Pharmaceuticals	X	X	X	X	X	X	HP, Sun, C/S; Business
Banco Union (Mexico)	Fin. Services - Banking	X	X					Mainframe
Bank One	Fin. Services - Banking	X	X	X	X	X	X	Mainframe, AS/400, C/S

Bankers Trust	Fin. Services - Banking	X	X	X	X	X	Mainframe, HP, Sun, C/S
Bell South Mobility	Wireless Telecommunications	X					Business
Best Buy	Consumer Electronics Retail	X	X	X	X	X	Mainframe, HP, AIX, Wintel
Bond Brewing (Australia)	Brewing	X	X	X	X	X	Mainframe
Central Illinois Light	Utility	X					Enterprise systems
Chicago Transit Authority	City government	X	X	X	X	X	Emergency response
Dade County, Florida	Local government	X	X	X	X	X	Systems, emergency response
Dial	Consumer products	X	X	X			Systems, business
First Bank	Financial services – banking	X	X	X	X	X	Mainframe, AS/400, C/S
First Tennessee Bank	Financial services – banking	X		X			Mainframe, HP, Sun, C/S
Fortis (U.S. and Belgium)	Financial services – corp banking	X	X	X	X	X	HP, Wintel, business
Georgia-Pacific	Paper and wood products	X	X	X	X	X	HP, SUN
H. B. Fuller	Chemical processing	X	X	X	X	X	Mainframe, AS/400, C/S
Hamilton Medical Cntr	Health care	X	X	X	X	X	DEC

Continued

Table 3.1   Continued

Client	Industry	Category							Platforms
		Assessment and analysis	Strategy development	Plan development	Training	Testing	Integration	
HDFSI	Financial services	X	X	X	X	X	X	HP, Sun, C/S
Home Savings	Financial services – banking	X	X	X				Mainframe
Landis & Gyr Powers	Manufacturing	X	X	X	X	X	X	Mainframe, IBM RS/6000, HP
Milliken Industries	Manufacturing	X	X					Mainframe, business
Milwaukee County	County government	X	X					Mainframe, enterprise systems
Nationwide Insurance	Fin. Services - Insurance	X	X					Mainframe, HP, Sun, C/S
Northern Trust	Financial services – banking	X	X	X	X	X	X	Mainframe, HP, Sun, C/S
Northwestern Mutual	Financial services – insurance	X	X		X	X		Mainframe, HP, Sun, C/S
NYNEX	Telecommunications	X	X					Mainframe, HP, Sun, C/S
Pacific Gas & Electric	Electric and gas utility	X	X	X	X	X	X	Business, IT
Paragon Trade Brands	Manufacturing	X	X					HP (SAP)

							Business
PPL	Electric utility	X	X	X	X	X	
Progressive Insurance	Financial services – insurance	X	X	X	X	X	Mainframe, HP, C/S
Sara Lee (Hanes Knit)	Manufacturing	X	X	X	X	X	Mainframe, enterprise systems
Sears	Consumer products	X	X	X	X	X	Sun (PeopleSoft)
Sony (Canada)	Consumer electronics	X	X	X	X		Mainframe, enterprise systems
T. Rowe Price	Financial services – mutual fund	X	X	X	X	X	Mainframe, trading systems
Texas Instruments	Manufacturing	X	X	X	X	X	Mainframe, HP, Sun, C/S
Victoria's Secret Catalog	Consumer products	X	X	X	X		Mainframe, customer service
Visa	Financial services – credit card	X	X	X	X	X	Mainframe, business
Westinghouse	Waste management	X	X	X	X	X	DEC, business
West One Bank	Financial services – banking	X	X	X	X		Mainframe, enterprise systems
Whitney National Bank	Financial services – banking	X	X	X	X	X	Mainframe, enterprise systems
YPF (Argentina)	Energy (oil)	X	X	X	X		Mainframe, business

Accenture's approach is in line with *The Good Practice Guidelines (GPG) 2018 Edition*) of the British Business Continuity Institute (BCI)[69] (*The Professional Practices for Business Continuity Management 2017 Edition*) of the *Disaster Recovery Institute International (DRI)*[70], (*ISO 22301:2019 – Security and resilience – Business continuity management systems – Requirements, ISO 22300:2018 – Security and resilience – Vocabulary, ISO 22313:2020 – Security and resilience – Business continuity management systems –Guidance on the use of ISO 22301, ISO 31000* "Risk management," *ISO/IEC 20000-1* "*Information technology – Service management," ISO/IEC 27001:2013* "*Information security management systems," ISO 28000* "*Specification for security management systems for the supply chain,*" as well as *national standards ASIS ORM.1-2017, NIST SP800-34, NFPA 1600:2019, and libraries COBIT 2019, RESILIA 2015, ITIL V4,* and *MOF 4.0* in terms of business continuity management.

Accenture consultants suggest clearly defining the three components of the company's process management phases in order to build a successful ECP program:

- goal setting system – the main strategic and tactical goals and objectives of the business;
- the system of management business processes – the methods to achieve the goals;
- a process environment (environment or conditions) that allows the business process management system to effectively achieve its goals.

Then the project area is defined and fixed, and the expected effect of the project is calculated. In the expanded form, Accenture's approach to creating an enterprise ECP program is shown in Figure 3.1. Here, the first four stages are aimed at the development and proper implementation of the ECP program, and the fifth and sixth stages are aimed at improving this program. It is assumed that the results of the sixth stage, after appropriate analysis and processing, will serve as the initial data for the first stage, thus initiating a new BCM cycle. According to Accenture, this continuous and the cyclical process can maximize the involvement of all the company's main business units in the maintenance and development of the ECP program and increase the overall awareness of the company's employees on business continuity issues. In the future, this will allow the BCM process

---

[69] www.thebci.org
[70] www.drii.org

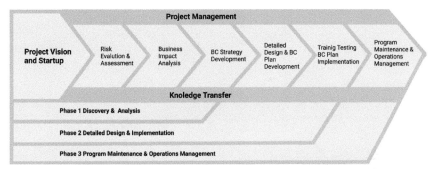

Figure 3.1 Accenture's approach to creating a successful ECP program.

to become part of the organization's culture and, ultimately, become an effective tool for achieving the company's goals.

The approximate plan of the Accenture project for creating and implementing the enterprise continuity program (ECP) program is presented in Table 3.2.

Table 3.2 Sample plan for creating and implementing ECP.

Step description	Step duration, days	Duration, days
Project preparation – phase 0	**5**	
Defining the project framework, goals, and objectives		1
Clarifying project goals and objectives		1
Development of the project charter		1
Approval of the work plan		2
Analysis of IT risks of business continuity violations – phase 1	**29**	
Preparing an overview of existing BCM approaches		2
Analysis of the organizational structure and development of materials for working meetings		2
Building a map of it services		3
Building a catalog of threats and assessment of damage		3
Building a map of infrastructure elements		2
Building a map of existing business processes, for example, based on the eTOM model		5
Creating a report that analyzes the maturity level of the BCM implementation		2

Continued

Table 3.2   Continued

Step description	Step duration, days	Duration, days
Development of business requirements for the functional stability of technological processes		3
Creating a technical task for the development and implementation of a continuity management strategy in the company		3
Creating a work plan for the next stage		2
Approval of the results by the company's management		2
Business continuity strategy development – phase 2	36	
Developing a business continuity policy		12
Development of the business continuity management CRI methodology		6
The development of indicators of stability of critical business processes		6
Preparation methodologies for analyzing the business continuity		4
Development of procedures for supporting and supporting the BCM strategy		4
Evaluation of the received effect		2
Approval of the results by the company's management		2
Implementation of the business continuity strategy – phase 3	60	
Adaptation of the HCM strategy – phase 4	16	
Development of a plan for the implementation of the BCM strategy		2
Adapting the HCM strategy		3
Adapting the KPI		2
Adaptation of the system of indicators of stability		2
Development of standard solutions for the organization of the company's IT infrastructure		3
Development of a feasibility study for the implementation of standard solutions		2
Approval of the results by the company's management	12	

Continued

Table 3.2   Continued

Step description	Step duration, days	Duration, days
Development of a technical specification for an automated system for functional stability analysis		3
Building a pilot system for functional stability analysis		4
A system demonstration		2
Preparation of a report on market analysis of systems for the organization of the business continuity process		3
The organization of the pilot – phase 5	4	
Developing a test plan		2
Approval of the results by the company's management		2
BCM training – phase 6	10	
Development of training materials		2
Conducting business continuity team training		1
Conducting staff training		2
Conducting management training		1
Analysis of the training conducted		2
Approval of the results by the company's management		2
BCM testing – phase 7	15	
Run tests		3
Analysis of test		2
Development of standard solutions for the organization of the company's IT infrastructure		3
Analysis of the effectiveness of implementing standard solutions		2
Development of a presentation on the results of the implementation of BCM		3
Approval of the results by the company's management		2
The completion of the project – phase 8	3	
Development of the final report		2
Presentation of the main results to the company's management		

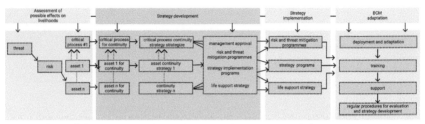

Figure 3.2 An illustration of Accenture's approach to ECP development.

Let us take a look at some stages of the plan for creating and implementing an ECP program (Figure 3.2) using the example of a national mobile operator in more detail.

3.1.1 RA and BIA

It defines the procedure for collecting information and prepares the necessary materials (checklists and questionnaires) for conducting interviews and working meetings.

A map of IT services that provide critical business process is compiled, which reflects, for example, the following:

- service definition;
- assessment of the impact (criticality) of this service on the company's business in general:
  - high – this service is a key part of the company's service package and its operation is highly dependent on the company's revenue, stability, and prestige;
  - average – this service is a popular service, but its availability has little impact on the company's business;
  - low – this service is not popular among users; its unavailability may go unnoticed by most of the company's clients;
- defining the functional elements of the infrastructure that provide these services;
- assessment of the degree of dependence of the service on this functional element of the system.

The collected data is presented in a table, a simplified version of which is shown in the example of Table 3.3.

Then potential threats and vulnerabilities are identified, possible risks of interruption of technological processes are calculated, and proposals for organizational and technical measures for risk management are prepared. At the same time, an assessment of previously taken risk management

Table 3.3 Analysis of the dependence of services provided on infrastructure elements.

Functional element Service	Criticality	BTS	BSC	MSC/VLR	HLR/AC	PrePaid platform	OAM&P	Foris
Network service								
Local telephone service mobile2mobile	High	Average	High	High	High	High	Low	
Local telephone connection between a mobile subscriber and a PSTN or another local mobile operator	High	Average	High	High	High	High	Low	
Inbound roaming (service of the subscribers of other operators in its network)	Average	Average	High	High	High	–	Low	
Outgoing roaming (support for servicing its subscribers when they work in the networks of other operators)	High	–	–		High	Low	Low	
Enterprise information system								
Connecting a new subscriber to the communication network service	Average	–	–	–	–	High	High	
Formation of detailed invoices for company subscribers	High	–	–	–	–	–	–	High

Table 3.4   Assessing the stability of the information infrastructure.

Feature / element	Implemented fault tolerance	Geographical distribution components	Personnel reserve	Presence of procedures to ensure fault tolerance	Fault tolerance test procedure
BTS		Not applicable	Yes	No	Yes
BSC	High	Not implemented	Yes	Yes	No
MSC/VLR	High	Not implemented	Yes	Yes	Yes
PrePaid	Average	Implemented	Yes	Yes	Yes
HLR/AC	High	Not implemented	Yes	Yes	No

measures is made. For example, the stability of an organization's information infrastructure can be assessed using the following scheme (Table 3.4).

Then the analysis of the organization's initiatives and projects aimed at improving the level of business continuity is performed. The current state of the BCM enterprise program is evaluated, including the adequacy of the existing BCM enterprise strategy and DRP and BCP plans. These studies relate to a certain business process model, such as eTOM (Figure 3.3).

Accenture consultants use the recommendations of the COSO and NFPA 1600:2019 standards for assessing cyber risks; in particular, they determine the so-called residual (acceptable) risk of business interruption (Figure 3.4).

Possible results of damage assessment are presented in Table3.5.

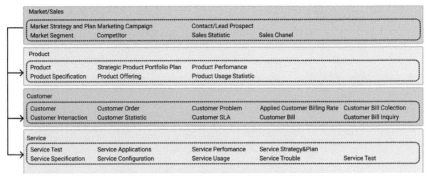

Figure 3.3   *e*TOM model for the telecom operator.

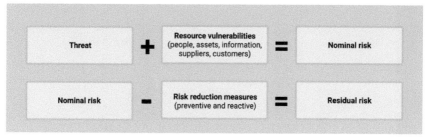

Figure 3.4 Definitions of the cyber risk structure.

The maturity analysis of the ECP program is based on the recommendations of *ISO 22301:2019, ISO/IEC 27001:2013 (A. 17)*, and *ISO/IEC 27031:2011, COBIT 2019*, and *RESILIA 2015, and* contains a maturity assessment:

- organization of business continuity management processes:
  - organizing the business continuity planning process:
    - residual risk assessment;
    - developing a BCM strategy;
    - defining areas of responsibility and roles for employees;
  - development and implementation of DRP and BCP plans:
    - key service providers;
    - availability of backup duplicate business chains;
  - support of DRP and BCP strategy and plans:
    - BCP monitoring and verification;
    - BCP update process;
    - BCP testing process;
    - documenting the BCP;
- organizations of reserve resources (suppliers, personnel, services, etc.):
  - crisis management programs:
    - culture (strategy, periodic monitoring and review, awareness and support of management, etc.);
    - crisis management team (composition, locations, roles, functions, etc.);
    - internal and external communications strategy;
- enterprise system policies (enterprise information resources, communications, data transmission systems, etc.):
  - identification of risks and key IT processes;
  - key service providers (partners, contractors, sub-contractors, etc.);
  - fault tolerance of systems (infrastructure, data storage, information security, etc.);

Table 3.5   Analysis of possible damage.

Resource / Threat	MSC	PrePaid platform	CRM-system
Power system failure	Probability: 1%	Probability: 1%	Probability: 1%
Damage Loss of profit: • loss of customers • delay in receiving revenue • one-time profit losses	• The outflow of 0.01% of the customers • X $ • X million minutes per hour of traffic	• The outflow of 0.01% of the customers • X $ • X million minutes per hour of traffic	• The outflow of 0.01% of the customers • X $ • X million minutes per hour of traffic
The cost of repairing: • the cost of repairing • cost of attracting additional resources	X $	X $	X $
Personnel error that caused the system stop	Probability: 0.1%	Probability: 2%	Probability: 1%
Damage Loss of profit: • loss of customers • delay in receiving revenue • one-time profit losses	• The outflow of 0.01% of the customers • X $ • X million minutes per hour of traffic	• The outflow of 0.01% of the customers • X $ • X million minutes per hour of traffic	• The outflow of 0.01% of the customers • X $ • X million minutes per hour of traffic
The cost of repairing: • the cost of repairing • cost of attracting additional resources	• X $	• X $	• X $

○ procedures for evaluating methodologies and updating them;
○ testing methods;
• policies for building enterprise platforms:
  ○ site planning (power supply, water supply and sanitation, physical safety, evacuation routes, emergency services, periodic operation checks, etc.);
• training and awareness programs:
  ○ planning and informing staff;
  ○ training and knowledge control.

In the course of determining business requirements for business continuity and cyber stability of the information infrastructure, materials are developed for meetings with the company's management to identify critical technological processes and develop the required business metrics for the functional stability of the services provided. These materials include a list of critical processes (services) and their corresponding set of key performance indicators. The set of identified risks and parameters of impact on the company's business is coordinated.

At the development stage of the technical specification for ECP development, the following are defined:

• ECP goals;
• areas of responsibility and roles of the organization's employees;
• application;
• business continuity and cyber stability indicators system and methods of their assessment;
• business continuity and cyber resilience policies;
• a set of measures to protect the system's infrastructure from destructive factors;
• the processes of resource allocation;
• solutions and suppliers that need to be involved in the process;
• a set of necessary regulatory documents, lists of resources, materials, and actions for implementing the ECP program.

The list of possible results of the work performed at this stage is presented in Table 3.6.

3.1.2 Definition of BC Strategy

At this stage, the following regulatory documents are prepared:

• draft BC strategy;
• draft standard "Assessment methodology and system of criteria and indicators of business continuity and cyber stability," which defines,

Table 3.6   Recommendations on the ECP program improvement.

No.	Title	Description
1	Overview of the well-known BCM practices	Various approaches to ensuring the continuous operation of companies in the telecommunications sector are considered. Data on the organization of the continuity process, employee involvement in the overall management structure of the company, critical processes of companies, required operational characteristics, envisaged emergencies and measures, and generalized plans for overcoming them are provided.
2	Classification of business processes and IT services	The document provides a list of business processes and IT services, the decomposition of the network elements used, resources, and the degree of dependence of processes and services on the components of the information infrastructure.
3	Catalog of threats, countermeasures implemented, and potential residual damage	The document contains a list of threats of natural and technological nature, threats to cybersecurity for the continuous functioning of the organization, potential damage from the implementation of threats, and a list of existing measures to reduce and prevent these threats.
4	Map of infrastructure elements and their classification in terms of business continuity	The document describes the infrastructure elements and their classification in terms of business continuity.
5	A report assessing the level of ECP maturity in the company	The report contains an analysis of the ECP maturity level. The matrix of responsibilities as well as the implemented opportunities for the development of the BCM process, existing policies, standards, regulations and instructions, BCP/DRP plans, etc., are considered.
6	Business requirements of the organization for business continuity and cyber stability	Critical technological processes and business metrics of their functioning are defined.
7	Technical specification for the ECP development and implementation	The document describes the requirements for the development and implementation of ECP: goals and objectives, responsibilities, and roles of employees of the organization, a system of criteria and indicators for business continuity and cyber stability, documents, and draft solutions.
9	Priority action plan for improving ECP	The plan contains a list of priority actions to improve ECP, expected results of work.

among other things, the normative values of quantitative indicators of business continuity and cyber stability;

- draft regulations, plans, and instructions for implementing the strategy and managing the continuity of critical technological processes (the set of regulatory documents is agreed upon developing technical design assignment (TDA).

As part of the TDA, the main provisions of the BC strategy formulated at the first stage are filled in:

- goals of BC strategy;
- areas of responsibility and roles of the organization's employees;
- application;
- business continuity and cyber stability indicators system and methods of their assessment;
- set of measures to protect the system's infrastructure from destructive factors;
- processes of resource allocation;
- solutions and suppliers that need to be involved in the process;
- defining a set of necessary regulatory documents, lists of resources, materials, and actions to implement a business continuity strategy.

The key task of developing this strategy is to build and integrate the BCM process into the business environment of the organization in accordance with the requirements of regulators and the specifics of the organization's business.

The following points are defined:

- rules for classifying existing technological processes by the degree of criticality for business and the map of technological processes;
- rules for classification of functional elements of IS involved in ensuring the functioning of the technological process and built dependencies for them;
- developed requirements for cyber stability of individual functional elements in accordance with their criticality parameters.

Accenture's approach to developing optimal strategies and plans for managing the states of complex systems, including financial and technological risks, is based on an analysis of the states of system components, their mutual influence, and consideration of dynamic changes in external factors. The main ideas of this approach are to consider in detail the essential elements of the analyzed system, their relationship, as well as to identify (based on available statistical data) external and internal factors of influence. In general, the problem is posed as a dynamic analysis of the state of structures similar to mathematical objects-mixed graphs, where each node (element of the analyzed system) at different levels of generalization can be considered as an infrastructure component, service, or key performance indicator (KPI). In this case, the edges of the graph model the type (directional or non-directional) and the degree ("strong"–"weak" using weight coefficients) of the connection between nodes (elements of the system).

This approach will "play out" various scenarios of external impact on the system at any selected time horizon, take into account the internal state and relationships between elements, dynamically introduce control actions (simulating a set of measures to support KPIs in a given range), and develop strategies for the behavior and development of the system in accordance with the specified target functions.

It is noted that the proposed approach cannot claim to be complete, without considering financial factors. Therefore, we consider not only technological indicators, business indicators, and related risks but also full factors registration that affect the cost characteristics. The ability to consider the costs of maintaining the system and its components in a given state (or to consider the required dynamics of development) will link the technological, operational, and financial dimensions of the company. The use of various methods of cost accounting and allocation (for example, *activity based costing (ABC)* analysis) provides a key to understanding how current or planned costs affect the achievement of specified KPIs and the continuity of critical technological and business processes (Figure 3.5).

A sequence of steps for creating a model is proposed.

Model A:
There are three main levels (the number of levels is not important and is given for illustrative purposes) [ISO 22301: 2019], containing various groups of elements, services, and indicators, which, at a certain level of generalization, represent a complete description of the business model.

Level of infrastructure:
For example, for a telecom operator, this level is intended to consider the relationship between elements of the network infrastructure and elements of the IT infrastructure used directly or indirectly to provide services to end users. For example, such elements may include channel – forming equipment, communication equipment, radio subsystem elements, billing systems, switches, gateways, authentication and authorization systems, communication quality control devices and systems, etc. The mutual

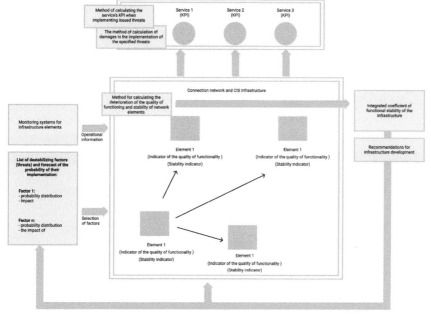

Figure 3.5 Data processing diagram.

influence of elements, as well as consideration of external factors that affect the ability of the infrastructure to perform its functional tasks, is based on existing operational experience and statistical data provided by the customer.

Level of service:
The service level is a higher-order generalization. In this case, services are everything that can be directly used by the subscriber or used by the customer's services, as an internal service necessary to provide a higher-level service. Examples of such services can be GPRS, MMS, SMS, WAP, or mobile communication services (voice, roaming, etc.). In addition, such important internal services as payment acceptance, financial reporting, planning, etc., should be also considered. It is obvious that a well-founded definition of the relationship between the elements of the infrastructure level and the corresponding services lays the foundation for the model applicability.

Level of key performance indicators (KPIs):
The KPI level is the next level of generalization above the service level. Here we assume the use of the following important business parameters: the growth in the number of subscribers of a particular service, the

company's revenue from providing various types of services, the overall availability of mobile services, and so on.

In many ways, the adequacy of the model application will be determined by the KPI function settings, the degree, and nature of their dependence on the elements of the service level.

Model B:

Each element that corresponds to one of the above levels is assigned a set of relationships (in general, incoming, and outgoing) that determine the interdependence between elements at this level. For example, for the infrastructure level, the simplest analog of connections can be considered a section of the network (channel, connecting line, etc.) between two devices. At the service level, the connection between mobile Internet and GPRS services could be determined. The KPI level assumes that there are links between indicators such as the company's revenue from the sale of the service and the number of subscribers of this service. Note that interlink relationships between elements are not always a necessary condition. It is acceptable to have isolated elements (or groups of elements) at any of the levels under consideration. The presence of these groups does not have a direct impact on the elements of its level, but it can have a connection with a hierarchically higher level and be considered when calculating the corresponding integral parameters.

In addition to the relationships that determine the mutual influence between elements, there is a need to define sets of parameters that characterize the functioning and /or state of elements. Time acts as an independent parameter that characterizes the dynamic process of changes in the state of elements and systems.

External factors (destructive or positive, such as preventive measures to reduce the risk of reducing the availability of services) are defined as unplanned or planned changes to the parameters of elements or their groups. It uses the apparatus of probability theory methods, mathematical statistics, and available statistical data or expert assessments of the behavior of analysis objects.

Model C:

There is a reasonable and clear relationship between the levels of infrastructure, services, and KPIs. At each level, the groups of elements with successful functioning (maintenance) positively affecting elements (or groups of elements) at a higher level could be specified. The degree of dependence between different levels and elements could vary (for example, using weight coefficients), taking into account the total contribution

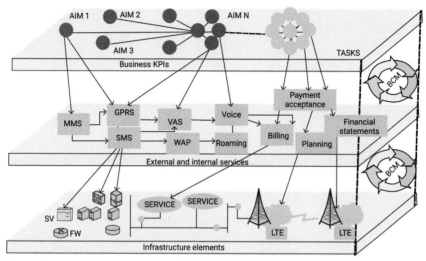

Figure 3.6 Three-level model of the organization and its business continuity management system, BCMS.

of a particular parameter or integral indicator to the resulting function (Figure 3.6).

This defines the model components, parameters, and factors that characterize their functioning or state, the relationships between elements, and the hierarchy of interaction between the three levels of the model.

For each of the levels, integral indicators (target functions) of the state are set – indicative functions of finding the system in the normal state (or a state close to the set one).

Accenture consultants draw attention to the fact that the system model may contain some contradictions between individual target functions – for example, excessive reliability of the system (infrastructure level) in the first approximation may have a negative impact on financial indicators (KPI level), or the allocation of excess resources to a new type of service (service level) may cause the quality of other services to deteriorate. This situation is very common in system models where the optimal solution (or behavior strategy) is clearly missing or difficult to determine.

It is assumed that it is acceptable to use the concept of balance between those state functions that are given priority at this stage of the company's development. The balance of state at various levels, considering the limitations, caused by the realities of the business, will determine and justify the company's strategy for a given period of time.

Figure 3.6 provides an illustrative action plan for applying the Accenture approach in practice.

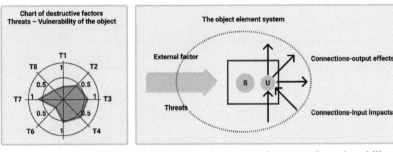

Figure 3.7   An illustration of a dynamic model system element and a vulnerability diagram for this object.

A system of indicators (KPIs) for the stability of critical technological processes and a methodology for analyzing the stability of the customer's infrastructure are being developed. The purpose of this methodology is to identify critical infrastructure elements and develop recommendations for improving its stability, which would be able to provide the specified performance indicators for any weighted implementation of a risk combination.

The model elements corresponding to the levels described earlier (infrastructure, services, and KPIs) are highlighted.

Threats of business interruption are identified: destructive factors that affect infrastructure elements and their degree of influence. Threats are ranked by the degree of influence on the element's state. In some cases, an integral threat level indicator is defined – a vulnerability indicator that is a weighted function of a set of significant threats (Figures 3.7–3.9).

Connections between elements and the degree of their mutual influence are revealed. The degree of influence is ranked and a preliminary assessment of the indicator of the importance of an element in the system is made.

The relationships between critical processes and infrastructure elements (assets) are determined from the perspective of continuity.

The analysis of statistical characteristics of destructive factors and processes (risks) leading to interruptions in the provision of services and a decrease in KPI is performed. A set of measures to counteract the decline in the availability of services and KPIs is defined. The functions corresponding to these measures are set (if necessary, a statistical approach or consideration of the direct impact on the elements of the system is used).

Based on statistical data and expert evaluations, the probability functions of state changes for the key elements of the model are set.

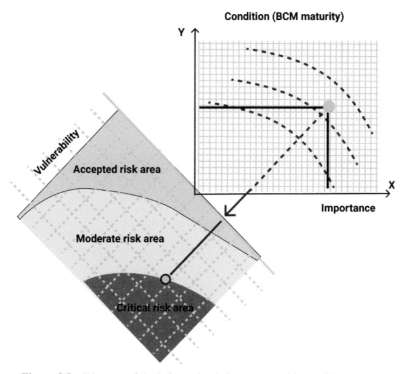

Figure 3.8 Diagram of the information infrastructure object's risk zones.

Simulation of the system behavior is performed at the selected time horizon under the specified conditions and the state of the system elements.

Based on the simulation results and the obtained integral indicators, a conclusion is made about the state of maturity of the system and its elements. For elements or their groups, defined by their degree of importance (both expert assessments and results of direct simulation of the degree of influence of elements on the integral indicators of the system can be used) and the level of vulnerability, a risk zone diagram is constructed (Figure 3.9).

Based on the obtained risk zone estimates, a set of measures and a program for managing the company's BCM are developed for groups of elements. At the same time, we are able to assess the impact of the developed measures during the repeated run of the model and direct modeling of measures to counteract negative changes in KPIs. If there is a sufficient set of statistical and expert data, the model can provide reasonable and consistent computations of financial indicators.

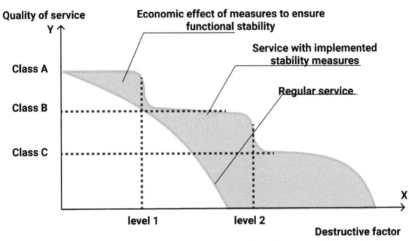

Figure 3.9   A step change in the level of service, with an increase of destructive factors.

This approach will continuously monitor the current and forecast the expected level of business continuity and cyber stability of the information infrastructure. In addition, this method helps to identify the critical elements of the system, which builds an infrastructure that will resist increasing destructive factors and maintain high-quality performance of business processes and IT services .

As a result of modeling the effects of destructive factors, key performance indicators of processes (services) are calculated, as well as an indication of critical infrastructure elements and the development of generalized recommendations for their development.

By performing iterative computations using this method with various sets of destructive factors in combination with measures that reduce the vulnerability of the system and prevent threats to the continuity of business processes, it becomes possible to optimize capital and operational investments in maintaining and developing the system in terms of ensuring business continuity and cyber stability of the information infrastructure.

As part of the BC strategy definition, the following are developed:

• regulations for adaptation and implementation of the developed BC strategy;
• high-level plans and instructions for implementing the BC strategy.

The main results of the BC strategy definition stage are presented in Table 3.7.

Table 3.7   Project documentation for the stage.

No.	Title	Description
1	Target business continuity strategy of the organization	The document provides a detailed description of the business continuity strategy for the organization: the goals of the BC strategy and its application areas, the developed system of indicators of continuity and cyber stability, and a set of measures to ensure the BC strategy.
2	Example of classification of business processes and IT services	This document provides rules for classifying processes and IT services, functional elements of information infrastructure, and requirements for business continuity and cyber stability of critical components.
3	Recommended system of business continuity and cyber stability indicators	The document describes a system of indicators of business continuity and cyber stability of the organization's information infrastructure.
4	Recommended method for analyzing the state of business continuity and cyber stability	The document describes the methodology for analyzing the state of business continuity and cyber stability of the organization – identifying infrastructure elements, building relationships, identifying external and internal factors, determining the algorithm for processing the built model, and calculating targets.
5	Plan of priority actions for implementing and testing the BC strategy of the organization	This document contains steps for implementing and testing the organization's BC strategy.

3.1.3   Improving the BC Strategy

At this stage, the following steps are performed:

- identification of key technological processes;
- threat identification and analysis;
- key relation definition (partners, contractors, sub-contractors, etc.);
- defining requirements for system fault tolerance;
- analysis of the level of maturity;
- analysis of methods for ensuring continuous operation;
- modification and refinement of the BCM strategy in the pilot zone.

The standard solutions for the organization of IT infrastructure of companies are developing based on the results of the analysis of the previous stages, testing results, existing solutions, and initiatives to ensure continuity, ensuring the realization of established business, the required values

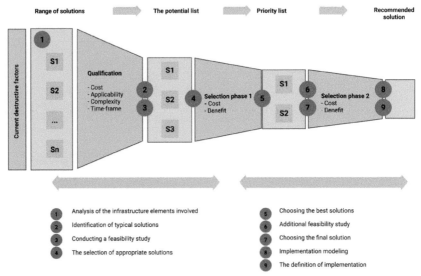

Figure 3.10 Building a typical BC solution for an organization.

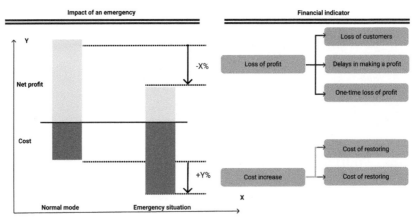

Figure 3.11 Relationship between the emergency situation and the organization's financial indicators.

of indicators of business continuity, and cyber resilience of information infrastructure (Figure 3.10).

The potential damage analysis and classification of losses from the implementation of threats identified in the previous stages are performed to conduct a feasibility study (Figure 3.11).

The cost of implementing measures to ensure the functional stability of the processes of the calculated potential damage is correlated, as a result of which the positive effect of implementing business continuity tools and cyber stability of the information infrastructure is calculated (Figure 3.12).

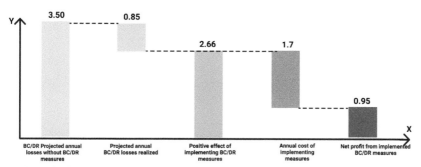

Figure 3.12 Calculating the positive effect of implementing standard BC solutions.

It specifies system tasks, input data types and interfaces, computation algorithms, control, and display tools, and analyzes system performance parameters. Within this stage, it is planned to build a prototype of a dynamic model for calculating business continuity parameters and cyber stability of the organization's information infrastructure using automated modeling tools (no more than five technological processes, taking into account no more than 100 elements for each technological process). It is important to note the need to provide statistical data and expert assessments on changes in the state and behavior of system elements and the degree of influence of external and internal factors on the relevant business processes and IT services selected for modeling. Here, the adequacy of mathematical models depends significantly on the representativeness of statistical samples (for a period of two years or more). Expert assessments and heuristics can help to find acceptable solutions for managing business continuity and ensuring the required cyber stability of the organization's information infrastructure.

Here we recommend paying attention to the functionality of the known automation tools for managing business continuity and ensuring the required cyber stability. Critical analysis of the known automation solutions (Figure 3.13 and Tables 3.8 and 3.9) allows choosing the appropriate cloud platform or cluster solution. For example, *Fusion Risk Management, Clear View, Recovery Planner, Assurance Software, Continuity Logic, Dell Technologies (RSA)*, and others[71]. However, the BCM balanced scorecard may be applied in order to select the preferred automation system.

The main results of the stage are presented in Table 3.9.

---

[71] https://www.gartner.com/en/documents/3957353/magic-quadrant-for-business-continuity-management-progra, https://www.cirmagazine.com/cir/reports/CIR-Business-Continuity-Software-Report-2018-19.pdf

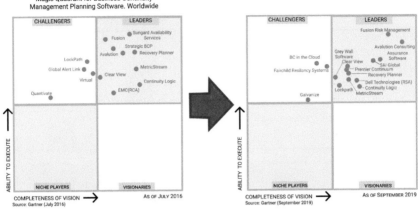

Figure 3.13 Magic quadrant for business continuity management planning software, Worldwide 2016 and Worldwide 2019, Gartner.

3.2 Ernst & Young (E&Y) Experience

Ernst & Young (E&Y) provides the following services in the field of business continuity:

- assessment of the existing business continuity management system (BCMS);
- building and implementing complex BCMS systems;
- development and implementation of the enterprise continuity programs (ECP0;
- certification audit for compliance with the requirements of *ISO 22301:2019 – Security and resilience – Business continuity management systems – Requirements*, etc.

The company has over 100 professionals (CISA, CISSP, CISM, MBCI, CPA, CIA, ACCA, etc.) in the field of information risk management, information security, and business continuity.

During this time, E&Y consultants have completed a number of BCM projects (Table 3.10).

The methodological basis of the E&Y approach is a set of mutually complementary methodologies and standards for business continuity management, risk management, information security management, IT governance, and enterprise governance (Table 3.11).

In addition, in practice, E&Y consultants, depending on the specifics of the project, can additionally use other practices, such as:

- eTOM (KPIs) is a methodology for identifying and using key business indicators for service operators.

Table 3.8 Comparative analysis of BCM software[72].

Features	Alive-IT	BC in the cloud	Business continuity Plan Template	Catalyst	Clear view	Continuity Logic v5	Crisis control	Fusion Framework System	Inoni Pro	Mataco	MIMS	ORBIT4BC	Parasolution	PdrWeb	RealBCP	ResilienceOne	RPX	Shadow planner	Shield
Plan navigator	+		+	+	+	+	+	+	+	+	+	+	+	+	+	+	+	+	+
Dependency mapping	+	+		+	+	+		+	+	+		+	+	+	+	+	+	+	+
Graphical call list				+	+	+		+	+			+	+	+	+	+	+	+	+
Location resource manager	+			+	+	+		+	+	+		+	+	+	+	+	+		+
Recovery site layout planning				+	+	+		+	+			+	+	+	+	+	+	+	
Reports – preformatted	+	+	+	+	+	+	+	+	+	+	+	+	+	+	+	+	+	+	+
Reports – own build	+				+	+	+		+	+		+	+			+	+	+	
Process modeling capabilities	+				+	+			+	+		+	+	+	+	+	+	+	
Technology modeling	+			+	+				+	+		+	+	+		+	+	+	
"What-if" analysis	+			+	+	+			+	+		+	+	+	+	+	+	+	+

Continued

[72]https://www.cirmagazine.com/cir/reports/BCSoftwareReport2019-20.pdf

Table 3.8 Continued

Features	Alive-IT	BC in the cloud	Business continuity Plan Template	Catalyst	Clear view	Continuity Logic v5	Crisis control	Fusion framework System	Inoni Pro	Mataco	MIMS	ORBIT4BC	Parasolution	PdrWeb	RealBCP	ResilienceOne	RPX	Shadow planner	Shield
Data collector	+	+	+	+	+	+	+	+	+	+		+	+	+	+	+	+	+	+
Automatic analysis	+			+	+	+		+	+	+		+	+	+	+	+	+	+	+
Simulation capability	+	+		+	+	+	+	+	+	+	+	+	+	+	+	+	+		
Dynamic updating from database	+				+	+	+	+	+	+		+	+	+		+	+	+	+
Education and training	+	+		+	+	+	+	+	+	+	+	+	+	+	+	+	+	+	+
Test and exercise	+			+	+	+	+	+	+	+	+	+	+	+	+	+	+	+	+
Test scripting	+	+		+	+	+	+	+	+	+	+	+	+	+	+	+	+	+	
Dynamic incident management	+			+	+	+	+	+	+	+	+	+	+	+	+	+	+	+	
Dynamic question setting/reviews				+	+	+		+	+			+	+	+			+		+
RTO/RPO desired/actual analysis				+	+	+	+	+				+	+	+	+	+	+	+	+
Standards compliance	+		+	+	+	+	+	+	+	+		+		+	+	+	+	+	+
Integrates with GIS mapping	+			+	+	+		+				+		+		+	+		
Workflow management with email alerts and reporting	+	+		+	+	+	+	+	+	+		+	+	+	+	+	+	+	+

	Multilanguage capability – interface	Multilanguage capability – user data	User roles and groups	Document update management	Comprehensive audit trails	Mobile device support	Templates available	Change control and tracking	Screen customization	Help	24/7 live support	Internal search engine	Charts, reports, and graphs	Filters	Personal filter	Drag and drop	Mobile app for offline viewing	Integrates with EMN software	Published APIs for data interface	Remote hosting	SaaS option
1	+	+	+	+	+	+		+	+		+					+	+	+	+	+	+
2	+	+	+	+	+	+	+	+	+	+	+	+	+	+	+	+	+	+	+	+	+
3	+	+	+	+	+	+	+	+	+	+	+	+		+		+	+	+	+	+	+
4	+	+	+	+	+	+	+	+	+	+	+	+		+		+	+	+	+	+	+
5			+	+	+	+	+	+		+	+	+		+	+	+		+	+	+	+
6	+	+	+	+	+	+	+	+	+	+	+	+	+	+		+	+	+	+	+	+
7	+	+	+	+	+	+	+	+	+	+	+		+	+	+	+	+	+	+	+	+
8	+	+	+	+	+	+	+	+	+	+	+	+	+	+	+	+	+	+	+	+	+
9	+	+	+	+	+	+	+	+	+	+	+	+						+			
10	+	+	+	+	+	+	+	+	+	+	+	+	+	+					+		+
11	+	+	+	+	+	+	+	+	+	+	+	+	+	+	+	+			+		+
12	+	+	+	+	+	+	+	+	+	+	+	+	+	+	+	+	+	+	+	+	+
13	+	+	+	+	+	+	+	+	+	+	+	+	+	+	+			+		+	+
14	+	+	+	+	+	+	+	+	+	+	+	+	+	+	+	+	+	+	+	+	+
15	+	+	+	+	+	+	+	+	+	+	+	+	+	+	+	+	+	+	+		+
16			+		+					+				+		+	+	+		+	+
17							+		+				+	+		+	+			+	+
18	+		+	+	+	+	+	+	+	+	+	+	+	+	+	+	+	+	+	+	+

Table 3.9   Project stage documentation.

No.	Title	Description
1	Recommended program for implementing a business continuity management strategy.	The document contains an action plan for implementing the business continuity management strategy in the macroregion.
2	Adapted strategy for ensuring business continuity in the selected region.	The document contains a description of the adapted Strategy of VSM. Adaptation consists in clarifying the parameters of the system of indicators of functional stability of technological processes, policies, and a set of measures to ensure them, as well as the list of IT risks and potential damage.
3	Adapted catalog of key performance indicators for critical business processes.	This document describes an adapted catalog of key performance indicators for critical business processes and their achievable values.
4	Adapted system of indicators of stability of critical technological processes.	The document updates the stability indicators of infrastructure elements and the method for calculating integrated stability indicators for critical technological processes.
5	Typical solutions for the organization of the company's IT infrastructure.	The document contains a description of typical solutions for organizing elements of the company's IT infrastructure in accordance with the approved key indicators of functional stability of critical technological processes.
6	Feasibility study of the implementation the developed standard solutions.	The document contains a feasibility study for the implementation of standard solutions developed.
7	Terms of reference for building a dynamic model of the system for analyzing the functional stability of critical technological processes.	This document describes the technical requirements for building a dynamic model of a system for analyzing the functional stability of critical processes. It specifies the system's tasks, describes interaction interfaces and types of input and output data, and controls, displays, and defines system performance parameters. In addition, during this stage, a prototype of a dynamic model will be built and the results of business continuity modeling will be presented, taking into account the risks and vulnerabilities of the system at a given time interval (from 1 to 5 years).
8	Report on the analysis of the market of systems for the organization of the business continuity process.	This document contains an analysis of the market for business continuity management systems in terms of their applicability.
9	Report with analysis of residual risks and reached maturity level.	The document contains an analysis of the remaining risks and the company's achieved maturity level for BCM.
10	Recommendations for further development of the BCM process in the company.	The document contains recommendations for further development of the BCM process in the company.

Table 3.10   Characteristics of some E&Y projects in the BCM area.

Customer	Project
Azercell	Risk assessment and analysis of the impact of process interruptions on the company's business; Development of a business continuity and disaster recovery plans.
The Head Office of "VimpelCom"	Assessment of the development level of the business continuity management and IT recovery program (BCP/IT DRP) in accordance with the PAS 56 business continuity management standard.
Knauf	Comprehensive project to ensure business continuity of the company; Risk assessment and analysis of the impact of process interruptions on the company's business; Development of a business continuity strategy; Development of a business continuity and disaster recovery plans.
Raiffeisen Bank Austria	Comprehensive business continuity planning project (risk assessment, analysis of the impact of process interruptions on the Bank's business, development of a business continuity policy and plan, development of recovery plans for key business units); The project to develop plans to restore IT after a crash; BCP/IT DRP projects for regions.
International Moscow bank	Assessment of risks affecting the continuity of critical business processes of the Bank and analysis of the impact of process interruptions on the Bank's business; Develop business continuity and disaster recovery plans.
MTN Group	Assessment of business continuity management and IT disaster recovery (BCP/IT DRP). Development of a business continuity strategy.
Colombia Telecomunicaciones S.A. E.S.P. (Telefonica)	Assessment of business continuity management and IT disaster recovery (BCP/IT DRP) identification, assessment, and analysis of IT risks; Developing a continuity strategy for the SAP R/3 platform; Development of a business continuity and disaster recovery plans.
BT	Assessment of business continuity management and IT disaster recovery (BCP/IT DRP); Development of a business continuity strategy.

Table 3.11   The main approaches adopted by E&Y.

Standard/ methodology	Description
ISO 22301:2019	ISO 22301:2019 contains requirements for evaluating and improving the existing business continuity management system (BCMS). It is possible to conduct a certification audit on the compliance of the business continuity management system with the requirements of ISO 22301:2019.
GPG 2018 (Good Practice Guidelines)	A set of norms and principles developed by members of the Business Continuity Institute (BCI) to ensure the practical implementation of business continuity management principles by providing a standardized approach and best practices.
COBIT 2019	An internationally recognized IT management and control methodology that also includes business continuity management. Implementation of the controls proposed by the methodology should guarantee efficient and reliable use of IT in the organization.
ISO 27001: 2013 (A.17)	A key model that provides a structured view of best practices for creating, implementing, operating, monitoring, reviewing, maintaining, and improving an information security management system (ISMS), which also includes business continuity management issues.
Capability maturity model integration (CMMI)	Methodology for improving processes in organizations. CMMI serves as a tool for forming criteria for evaluating the quality of processes, methods for improving them, and provides fragments of effective processes.
Enhanced telecom operations map (eTOM)	The key model is a structured view of the main representatives of the telecommunications industry on their activities from the point of view of business processes. The model is presented as a map of business processes with decomposition up to level 3, grouped by key business and operational areas.
Benchmarking metrics framework (BMF)	A model of performance targets which is specialized for the telecommunications industry, based on the balanced scorecard methodology. The model will compare similar indicators with the industry and other operators.
Ernst & Young BCM methodology	A comprehensive methodology developed by Ernst & Young for building a business continuity management system and implementing projects in this area. It also contains a complete set of work programs for all key stages of the BCM/BCP project, forms, and report projects.

- IT BSC – E&Y balanced scorecard for information technology.
- TAM – a model that describes the relationship between business processes in the eTOM model and types of information systems. This model will classify company applications in terms of business processes.
- ISO 22300 – family of business continuity management standards: *ISO 22301:2019 – Security and resilience – Business continuity management systems – Requirements, ISO 22300:2018 – Security and resilience – Vocabulary, ISO 22313:2020 – Security and resilience – Business continuity management systems – Guidance on the use of ISO 22301*, etc.
- ISO 31000 is a family of risk management standards for organizations.
- ISO/IEC 20000 (BS 15000) is an international standard for managing company IT services based on the well-known ITIL V4 methodology.
- ISO/IEC 27001 (A. 17) is an international standard for information security management in terms of business continuity.
- COSO ERM – development of the internal control – Integrated framework in terms of enterprise risk management. The model defines the main components of risk management and their contribution to achieving the set business goals.
- *ASIS ORM. 1-2017* – national standard for business continuity management.
- *NFPA 1600:2019* – national business continuity management standard.
- AS/NZ 4360 is an Australian standard for risk management. The standard defines a general risk management model. The following stages are highlighted and discussed in detail: defining the goals and objectives of risk management, identifying and calculating risks, analyzing risks, developing countermeasures for risk management, monitoring the risk management process, organizing interaction on risk management issues, and so on.
- NIST SP 800-34 – guide to IT continuity planning. This guide provides instructions and recommendations for managing IT continuity.
- E&Y Metrics DB – E&Y metrics for developing indicators of information security processes.
- NIST SP 800-55 – guide to selecting and implementing security measures. The guide defines the approach to selecting and implementing security measures.
- NIST SP 800-80 – guide to the development and implementation of IS metrics. The proposed processes and methods will link the process

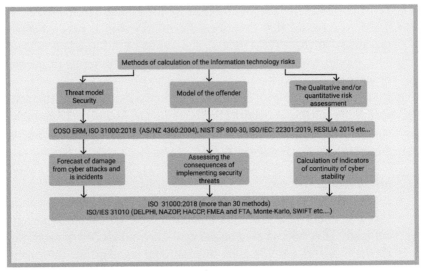

Figure 3.14   E&Y methodology for ECP development.

of ensuring information security with the goals and objectives of the organization.
- NIST SP 800-160 – guide to cybersecurity and resilience of critical information infrastructure, etc.

An example of the E&Y methodology for developing an ECP for a certain national telecom operator is shown in Figure 3.14, and the corresponding stages of ECP development are described in Table 3.12.

In the expanded form, the specified stages of ECP development look like the following.

Phase 1: Project planning in the BCM:
- forming a working team and coordinating committee for the project;
- conducting an introductory seminar;
- development of the work schedule.

Stage 2: Threat analysis and business interruption risk assessment, RA:
- identification of assets that are significant for ensuring business continuity;
- identification of threats and risks related to business continuity;
- classification and characteristics of failures/interrupts.

Table 3.12   Characteristics of ECP development stages.

Project stages	Tasks	Results
Planning	The definition of the project objectives and creating the project structure	The work schedule
Threat and risk analysis	Analysis of threats and vulnerabilities and identification of risks and controls	Identify key risks and threats
Evaluating the impact of failures and interruptions on business	Analysis of business divisions and IT infrastructure	The definition of critical processes and requirements of business units
Development of a business continuity strategy	Analysis of failure and interruption scenarios and recovery approaches	Business continuity strategy
Development of a business continuity strategy	Planning, purchasing, and implementing business continuity solutions	System of business continuity measures and resources
Development of a set of regulatory documents	Develop a structure for disaster recovery plans	Recovery plans for key divisions
	The structure of the business continuity plan	Business continuity plan
	Development of the structure of the anti-crisis management program	Plan of the anti-crisis committee
Testing, support, and modification	The approach definition	Plan support regulations; plan testing procedures

Stage 3: Business impact assessment, BIA:

- analysis of the impact of failures/interruptions on the company's operations and a report preparation;
- development of a list of resources required for rehabilitation activities;
- classification and ranking of processes, services, and resources;
- the formation of recovery requirements.

Stage 4: Development of a business continuity strategy;

- develop strategies for continuity of critical business processes;
- creating requirements for a backup data center;
- presentation for the company's management on business continuity strategy.

Stage 5: Implementation 5: Implementing a business continuity strategy:

- creating a package of solutions to ensure business continuity;
- the selection of suppliers of solutions and services;
- implementation of a package of solutions to ensuring business continuity.

Stage 6: Development of organizational and administrative documentation:

- development of the structure for disaster recovery plans;
- development of the plan of the business continuity plan;
- development of an anti-crisis management plan;
- development of the individual plan for disaster recovery plans.

Stage 7: Support and initial testing of BCP and DRP plans:

- developing a plan support policy;
- development of testing procedures for plans;
- development of a report on testing plans.

Let us take a look at some of these steps in more detail.

3.2.1 ECP Program Maturity Assessment

The goal of the stage is to assess the current enterprise business continuity management program, ECP, and develop recommendations for its improvement.

Project execution planning:
At this step, the project framework is defined, and the project plan and the charter are developed and approved. Then a working team and a project coordination committee are formed.

Before starting work on the project, an introductory seminar is held for employees of the organization, which addresses the following issues:

- content and sequence of work;
- project execution methodology;
- expected result;
- roles and responsibilities of project participants, etc.

Analysis of known work experience:
At this step, a sample analysis of similar BCM projects in international and domestic companies is performed. A variety of information sources, such as our own database of E&Y projects, databases of analytical companies Gartner, Forrester Research, materials from specialized organizations Business Continuity Institute, British Standards Institute, Information

Systems Audit and Control Association, International Information Security Systems Certification Consortium, and so on are applied to do this. As a result, the target state of the ECP program is formed and a set of measures necessary for proper business continuity is determined.

Assessment of the ECP maturity level:
The current state of the ECP program is determined. We first develop a method of expert assessment of the ECP maturity level, taking into account the recommendations of *ISO 22301:2019, ISO 27001:2013 (A. 17), CobiT 2019, RESILIA 2015, ITIL V4, MOF 4.0,* and *CMMI* (in terms of process maturity assessment). What is important here is that the methodology has appropriate metrics and measures that will observe, compare, control, and optimize the ECP program (Figure 3.15).

This is followed by a series of interviews with representatives of the company's business units, including IT and IS services. The following components of the ECP program are evaluated by experts:

- enterprise policies and standards for business continuity management;
- methods for assessing information risks and the results of such assessments;
- methods for assessing the impact on business and the results of assessments (in particular, residual risks of business interruption);
- strategy of business continuity management;
- business continuity management plan;

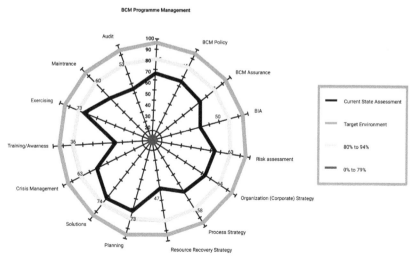

Figure 3.15 Sample of an ECP maturity level assessment representation.

- disaster recovery plans;
- crisis management plans;
- issues of organization and ensuring the required availability of reserved areas and resources in general;
- procedures for supporting and testing business continuity plans;
- organizational and technical measures to ensure business continuity;
- training and awareness programs on business continuity, etc.

Then the results of expert evaluation of the ECP program are compared with known practices and summarized. The final conclusions and recommendations are submitted to the company's management for approval (usually as a presentation). Based on the results obtained, a detailed plan of priority actions to bring the ECP program to a certain target state is developed.

Business interruption risk assessment, RA:
At this step, the risks of interrupting the organization's business and significant risks are identified (Table 3.13). The focus is on the components of the ECP program that require increased attention. Understanding the risks of business interruption will optimally increase the level of maturity of business continuity management processes. At the same time, a visual representation of significant risks of business interruption, as well as a plan of priority actions to minimize them will ensure that timely and reasonable risk management decisions are made.

First, an appropriate methodology that addresses the following issues for risk assessment is developed:

- identifying and ranking threats to business continuity;
- assessment of the probability of implementation of the mentioned threats;
- assessing the organization's vulnerability to threats;
- the definition and computation of risks of business interruption;
- identification and ranking of residual risks;
- business impact assessment;
- determining the necessary preventive and corrective controls for each risk;
- determining the necessary preventive and corrective controls for each risk.

Then the risks of business interruption are directly assessed. In this case, attention is drawn to possible discrepancies between the target and observed values of continuity indicators, in particular, the Recovery Time objects, and the amount of data lost during Recovery Point objects. Further, to eliminate the identified inconsistencies and minimize risks, a corrective action plan is developed. At the same time, a certain weight (priority) is determined for each corrective action, depending on the degree of influence of the identified risk on the business. As a result, the focus is on measures to minimize the risks that are most significant for the business (Table 3.13). Attention is also drawn to the risks that counter business interruptions with lower costs for the organization (temporary, labor, financial, etc.).

At the end of this step, the results of the business interruption risk analysis and the plan of corrective actions to minimize risks (usually in the form of a presentation) are submitted to the organization's management for approval.

Business impact assessment, BIA:
The relevance of this step is explained by the fact that the results of the BIA will select and use, in practice, the necessary measures to ensure business continuity.

First, the appropriate methodology, which determines the order of solving the following tasks, is developed for BIA:

- identification and ranking of critical business processes and supporting assets;
- identify existing relationships between business processes;
- decomposition of business processes into supporting resources;
- determining the degree of dependence of critical business processes on suppliers and business partners;
- definition and ranking of IT services, providing business processes;
- development of an IT process map linked to a business process map with a level of detail that ensures the solution of BCM tasks;
- quantitative and qualitative assessment of the business impact of various business interruptions (depending on the time of occurrence and downtime);
- defining requirements for maintaining the minimum acceptable level of functionality of business and technological processes.
- determine the maximum allowed downtime (consider the situation when there is a need to restore the minimum level of functionality of the business processes);
- evaluate and organize the resources needed to restore critical business processes and systems.

Then the business requirements for the level of availability of IT services that support the business (information systems and their components) are

Table 3.13 Example of presenting the results of risk analysis.

ID	Risk area	Threat	Issue	Key conclusions
IT03	IT failures	Disclosure of critical information.	Policies, procedures, and standards in the field of IT. Logical access control. Separation of powers. Monitoring.	• There is an information security policy. • Access to information systems is provided on the basis of formal requests. • There is no internal IT audit service. • There is no regular audit of access rights.
IT04	IT failures	Unauthorized modification of critical information.	Policies, procedures, and standards in the field of IT. Logical access control. Separation of powers. Monitoring. Backup copying.	• There is an information security policy. • Access to information systems is provided on the basis of formal requests. • Firewall systems are used. • Powers divided between administrators. • Backup is performed.
IT05	IT failures	Infrastructure component failures (server, storage management data, etc.).	Change control. Incident management. Performance and bandwidth management. Hardware redundancy. Maintenance agreements. Backup and restore procedure.	• Implemented a change management procedure. • Performance monitoring is performed. • There are backup servers for the Tycoon and performance and domain controller. • There are framework agreements with equipment suppliers. • Annual maintenance work is carried out. • There is no crash recovery plan.
IT06	IT failures	OS crashes		

Possibility	Vulnerability	Risk	Control maturity	Residual risk	Priority	Description of measures
5	4	23	3	2	3	Develop and implement a rights audit procedure user access. Consider creating a position "Information security officer," one of the duties of which will be to conduct internal IT audit
5	3	20	4	0	No	
5	4	23	3	2	3	Implement incident management procedures if possible and implement reserved server configurations (such as blade systems) with locations in different buildings.
5	4	23	3	2	3	To implement the procedure.

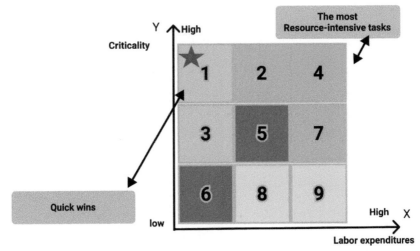

Figure 3.16   The prioritization of risk management business interruption.

clarified. These requirements are checked for completeness and consistency, summarized, and used to develop appropriate business continuity criteria and indicators.

As a result, business impact assessment data is recorded on maps of IT services and businesses (Figure 3.17).

Exchange of experience:
The exchange of experience involves the organization of round tables and conferences, trips and working meetings, consultations with well-known specialists and experts in the field of BCM, etc. This lively exchange of experience with domestic and Western companies that have successfully implemented the ECP program will take a critical look at the results achieved. In addition, it becomes possible to get valuable advice on the practice of implementing such projects in a timely manner.

Development of the technical task:
The development of the terms of reference is aimed at defining and documenting business requirements for business continuity measures, including measures for disaster recovery of IT services, reducing the existing risks of business interruption, and increasing the level of ECP maturity in general.

Presentation of the project to the company's management:
Management's support for business continuity initiatives is one of the key factors for the project's success. For this reason, timely preparation and

- The target recovery time is defined as the maximum allowed service termination time that a business can sustain without financial loss.
- All kinds of services (from basic to complex)
- Compliance service-system, device, tools

Bus.Unit	LOB/Dept	Function	RTO
CCBU	Fin & SBS	Distributed Printers & Copiers	48
CCBU	Fin & SBS	Print Production & Creation	48
CCBU	Fin & SBS	Information Management	48
CCBU	Fin & SBS	Accounts Payable	72
CCBU	Fin & SBS	CCBU Financial Reporting & Analysis	72
CCBU	Group	Group Life Claims/ Commissions to Brokers	48
CCBU	Group	Group Disability Claims-Adjudication	72
CCBU	Group	Accounting/Billing/ Collection-ASO	72
CCBU	Group	Accounting/Billing/ Collection-Self-Admin	72
CCBU	RI	Customer Service Center (Ottawa)	8
CCBU	RI	Agent Service Centre (Ottawa)	48
CCBU	RI	Quebec Service Team	48
CCBU	RI	Billing & Remittance	48
CCBU	RI	Reinsurance	48
CCBU	RI	Life Claims	48

Figure 3.17   Example of an RTO definition for IT services.

presentation of the project to the company's management plays an important role for the further progress of the project.

Updating the work plan:
The main results of this stage include:

- detailed project plan;
- project charter;
- assessment of the ECP maturity level;
- methodology for risk assessment, RA;
- methodology of business impact analysis, BIA;
- report on the current state of the ECP.

The report may include sections containing:

- description of experience in implementing BCM in large foreign and domestic public companies;

- description of the company's current level of maturity in business continuity management;
- comparison of the company's current level of maturity with global practices;
- planned maturity level of the company as a result of ECP implementation;
- matrix with the results of IT risk assessment;
- action plan to reduce identified risks;
- business requirements for the functional stability of IT services with a description of the planned (project) maturity level of BCM after the development and implementation of ECP;
- results of evaluating the impact of process interruptions on the company's business;
- map of IT services linked to a business process map with a level of detail that ensures the solution of tasks in the field of BCM;
- technical specification;
- updated project implementation plan and expected results.

3.2.2 Developing a BCM Strategy

The goal of the stage is to develop and justify a long-term strategy for ensuring business continuity (Figure 3.18).

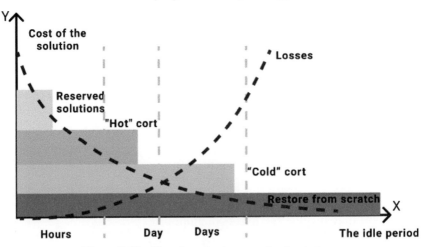

Figure 3.18 Choosing a business continuity strategy.

Development of organizational and administrative documentation:
The named step assumes the following.

First, the development of a business continuity policy, as follows:

- goals and objectives in the business continuity;
- roles and responsibilities;
- general rights and duties of officials;
- requirements for supporting and testing BCP and DRP plans;
- training requirements for company employees;
- links to detailed technical documents in the BC and DR area;
- measures to monitor compliance with the policy;
- mechanisms for reviewing and updating documentation.

Second, the development of the enterprise "Methodology for evaluating and system of indicators for the continuity of critical business processes" standard. This standard is necessary for unambiguous definition of uniform requirements for the ECP program, considering the heterogeneous organizational and technical measures taken at present. It is important to note that the requirements of the standard must be implemented in practice; so it is possible to include various alternative options for recommended measures and solutions in the standard. In addition to defining requirements, the standard should define appropriate controls and monitor key performance indicators of the process. For this reason, it is recommended to include in the mentioned standard:

- business requirements for RTO and RPO of IT services;
- acceptable options for technical implementation of solutions;
- requirements for the IS architecture;
- organizational requirements;
- staffing requirements;
- requirements for reserved areas;
- equipment requirements
- metrics and key performance indicators of the ECP program.

Third, the development of a set of documents, including a strategy for ensuring business process continuity, BCP and DRP regulations and plans, BC and DR procedures, metrics, and so on. For example, BCP and DRP plans should contain the following:

- order of the plan initiation;
- composition and structure of responsible groups (management, recovery, backup center, etc.);
- rights and obligations of officials involved in the BCM process;

- issues of general management in a crisis situation;
- procedures for reporting, monitoring, and documenting actions in emergency situations;
- actions of key recovery commands;
- procedure for personnel actions in the event of an emergency;
- service availability requirements (what, where, and when);
- resource requirements for ensuring continuity and restoring critical processes;
- procedures for restoring technological processes, systems, and services;
- procedures for activating a backup site;
- requirements for material and technical support in a crisis situation;
- the issues of staff security;
- internal and external interaction issues;
- requirements for storing the most important information on remote sites;
- requirements for physical security of objects;
- contacts of suppliers, service providers, and third parties.

Protection of business continuity management strategy:
The further progress of the project, and, in particular, the implementation of the ECP program , depends on the results of this protection.

Clarification of the project implementation plan:
This step is necessary to clarify the requirements and determine possible approaches to implementing the developed business continuity strategy.

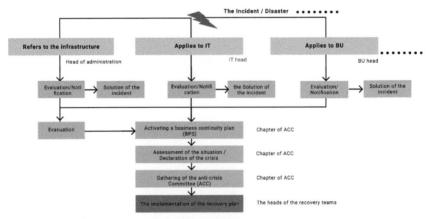

Figure 3.19   Presentation of an incident response strategy.

A set of regulatory and methodological documents is submitted to the company's management for approval, including:

- business continuity policy;
- enterprise "Methodology for assessing and system of indicators for the continuity of critical business processes" standard, which defines the normative values of critical BCM indicators for IT services, business requirements for the level of availability of IT services;
- strategy ensuring the continuity of business processes;
- regulations, standards, plans, and instructions for proper business continuity assurance;
- updated work plan.

3.2.3 Implementing a BCM Strategy

The goal of the stage is to implement and develop recommendations for further improvement of the business continuity strategy.

Development of recommendations:
This step defines the sequence of stages for implementing business continuity measures, the resources required for implementation, the time frame, the main risks, and so on. At the same time, recommendations of the best practices for implementing the ECP program in Western and domestic companies are considered.

The following recommendations are developed in practice:

- on the organizational structure of the coordinating committee;
- on the implementation of business continuity management processes;
- on the implementation of the types of solutions that will achieve the target recovery time.

Implementation of a business continuity strategy in a certain pilot zone. Feasibility study of solutions:
Typical business continuity solutions are developed in accordance with the requirements set out in the standard.

Testing a business continuity strategy:
Testing is recommended on a regular basis. The purpose of testing is to keep the continuity strategy up-to-date. The testing program can include various testing methods and techniques. We recommend using the test results to improve the implemented organizational and technical measures to ensure business continuity.

Support of implemented solutions:
The purpose of support and maintenance is to update and review the business continuity measures taken in a timely manner. For this purpose, the order and procedures for updating the developed plans, policies, regulations, and procedures for ensuring business continuity are defined.

Residual risk assessment:
Residual risks of business interruption are assessed in accordance with the previously developed risk analysis methodology. The purpose of this assessment is to determine the success of the project as well as to determine further steps to improve the ECP program.

Development of proposals for the development of the ECP program:
This step addresses the development of technologies, competencies, and existing practices that affect ECP.

Developing a BCM awareness program:
The goal of the step is to improve the enterprise culture on BCM issues. Using the recommendations and practices of the Business Continuity Institute, Information Systems Audit and Control Association, International Information Security Systems Certification Consortium, and others is recommended to develop an up-to-date awareness program.

Protection of project results:
It is recommended to conduct it in the form of a presentation, developed for the company's management. As a rule, the following are taken out for protection (Figure 3.20):

- ECP program and recommendations for its implementation;
- business continuity strategy;
- feasibility study of standard business continuity solutions;
- conclusion on the results of testing the business continuity strategy;
- proposals for the development of the ECP strategy and program, in general, with an estimate of the implementation costs;
- assessment of residual business interruption risks;
- assessment of the achieved ECP maturity level;
- program to raise awareness of the company's employees on BCM and other issues.

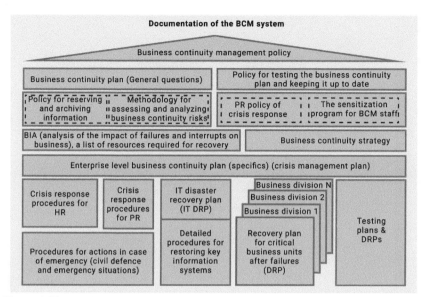

Figure 3.20   Possible composition of organizational and administrative documentation on BCM issues.

3.3   IBM Practice

The history of IBM's business continuity practice is as follows. In 1989, the company created a special division of *IBM Global Services – Business Continuity and Recovery Services (BCRS)*. Today, it is one of the international leaders in providing a wide range of business continuity services. The division employs approximately 2500 people who carry out projects in more than 100 countries around the world. In particular, *IBM BCRS* provides assistance in strategic planning and creating a *business continuity management system (BCMS)* (Figure 3.21), develops and implements the enterprise business continuity program (*ECP*), prepares the organization for certification for compliance with the *ISO 22301:2019* standard, and much more. Over *30* years of operation, *IBM BCRS* has completed approximately *20,000* contracts, including more than 10,000 successful disaster recoveries of its customers' IT infrastructure.

3.3.1   Methods of Work Performance

When developing ECP programs, IBM BCRS specialists use a number of general and special methods for performing work. The main common

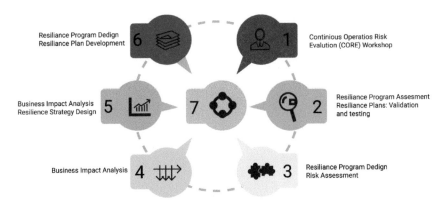

Figure 3.21   Typical IBM BCM services.

methods are: *Worldwide Project Management Method (WWPMM) and Global Services Method (GSM)*; for example, the *Method for Architectural Business Continuity Solutions (MABCS)*. Let us briefly take a look at the general methods and then focus in detail on the special one.

Worldwide Project Management Method:
In 1996, IBM identified its leadership in project management as one of its business development priorities. Currently, this goal has been achieved and all necessary measures are being taken to maintain the status of a leader in project management.

The IBM project management concept contains three key project characteristics: scope, depth, and frames of the project. In practice, WWPMM identifies three components of project management: areas of activity, action models, and work results.

Here, the areas describe in detail the main activities in the course of project management, which include the following:

- defining the project scope;
- management of information interaction;
- managing results;
- human resource management;
- quality management;
- risk management;
- sponsor management;

- supply management;
- managing the technical environment;
- managing the work plan;
- change control;
- management of special events;
- control of the project.

Action models describe the steps required to achieve the set project management goals. Action models are grouped into the following groups:

- defining project goals and objectives;
- work planning;
- project start;
- control of the project;
- actions in exceptional situations;
- actions to achieve results;
- project completion.

Work results define in detail the main and intermediate results of the project, the necessary work templates of documents, and draft reports. In this case, the work results are grouped by area.

IBM's project approach has been tested many times and provides the ability to manage a project or program (series of projects), regardless of their complexity. At the same time, the presence of highly qualified project managers as well as a set of software solutions (PM Tools Suite) available to the performer adds additional value to the IBM project approach.

GSM service delivery method:
It is used for the provision of services by IBM. Features of the method include:

- definition of detailed work plans for solving problems of project values;
- defining the format and structure of the content of working documents;
- development of technical manuals that describe in detail how to solve the project tasks;
- use of the special knowledge base that contains information on projects in the field of business continuity.

This method will accurately forecast the necessary costs to complete the project. At the same time, the experience and skills gained in the implementation of the similar projects significantly reduce the risks of project implementation and financial costs.

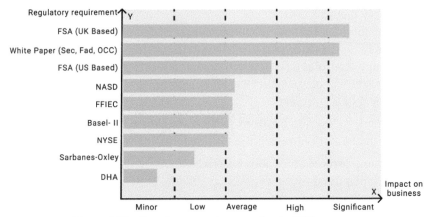

Figure 3.22   Regulatory requirements for the ECP program.

Table 3.14   Accounting for industry requirements.

1	Industry	• Governmental Accounting Standards Board (GASB) 1 • North American Electric Reliability Council (NERC) • NERC Security Guidelines for the Électriciry Sector Federal Energy Regulatory Commission • RUS 7 CFR Part 1730
2	Financial department	• Federal Financial Institutions Examination Council (FFIEC) Handbook, 2003–2004 (Chapter 10) • Basel II, Basel Committee on Banking, Supervision, Sound Practices for management and Supervision, 2003 • Expedited Funds Availability (EFA) Act, 1989
3	Healthcare	• HIPAA • Food and Drug Administration (FDA) Code of Federal Regulations (CFR), Title XXI, 1999
4	Services	• Sarbanes–Oxley • Solvency II • HIPAA for Health Insurers
5	Government	• FISMA • COOP and COG • NIST

Method for developing a business continuity architecture MABCS:
Designed to create a comprehensive business continuity architecture that best meets the requirements of business and regulators (Figure 3.22 and Table 3.14).

3.3.2   IBM BCRS Approach

The IBM BCRS approach is based on the well-known standards in the field of business continuity management *ISO 22301:2019 – Security and*

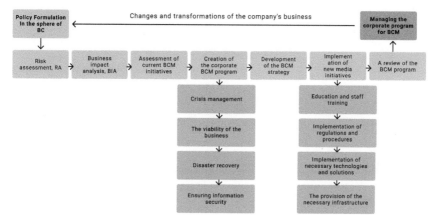

Figure 3.23   Characteristics of the IBM BCRS approach.

resilience – Business continuity management systems – Requirements, ISO 22313:2020 – Security and resilience – Business continuity management systems – Guidance on the use of ISO 22301, as well as the *ASIS ORM.1-2017, NIST SP800-34, NFPA 1600:2019*, and the Bible texts *COBIT 2019, RESILIA 2015, ITIL V4*, and *MOF 4.0*, etc. (Figure 3.23).

In practice, the implementation of the IBM BCRS approach begins with the decomposition of a complex business continuity process into components (subsystems, domains, processes, services, and objects). There is a clear description of the parameters to be achieved as a result of the work for them (Figure 3.24).

After the business continuity management process is decomposed, the business interruption risks are assessed.

Next, the effectiveness of business continuity countermeasures is evaluated, which begins with determining the current level of maturity of the ECP program. The following IBM BCRS developments can be used for this purpose (Figures 3.25–3.28):

IBM BCRS consultants recommend using GAP analysis to determine the target state of the ECP program. An example of this definition of the target state of the ECP program is shown in Figure 3.27.

IBM BCRS consultants offer the following recommendations to select the appropriate strategy for managing residual business interruption risks (Figure 3.28).

It is important to note that the IBM BCRS approach is used at all major stages of the business continuity management lifecycle:

• requirements analysis;
• planning;

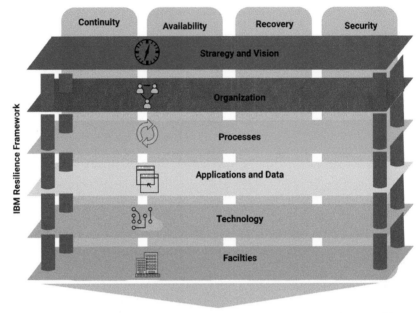

Figure 3.24   Example of business continuity management process decomposition.

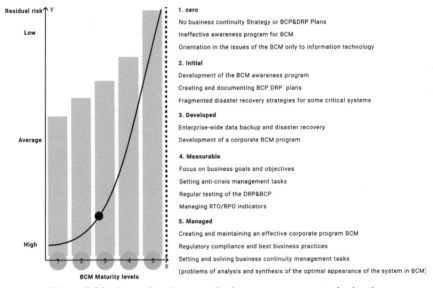

Figure 3.25   Map of business continuity management maturity levels.

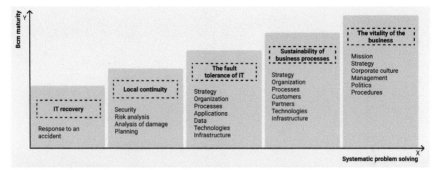

Figure 3.26 Main stages of development of the continuity concept.

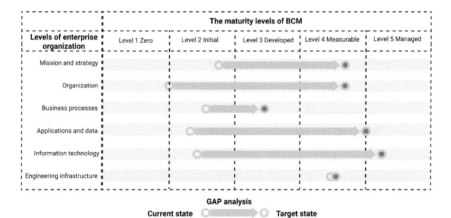

Figure 3.27 GAP analysis and setting goals for the development of the BCM process.

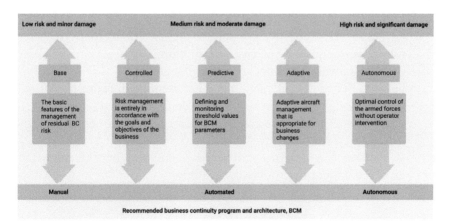

Figure 3.28 Ways to manage the residual risk of business interruption.

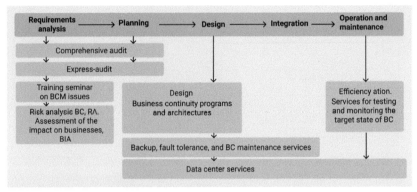

Figure 3.29   Typical IBM BCM services.

- design;
- integration;
- operation.

3.3.3   Services IBM BCRS

At each of these stages, IBM BCRS offers its own recommendations and related services (Figure 3.29).

   For example, to develop an ECP program, the following is recommended (Figure 3.30):

- develop requirements for the ECP program;
- define the main goals and objectives of the project;
- assess business interruption risks and perform business impact analysis, BIA;
- develop recommendations to prevent the most dangerous threats and minimize the risks of business interruption;
- develop a justified business continuity strategy;
- develop BCP and DRP plans and other working documentation;
- prepare proposals for the business continuity architecture;
- develop working documentation (policies, standards, regulations, and instructions) for emergency response;
- implement a program of awareness, training of the company's employees on business continuity issues.

Let us look at a number of IBM business continuity services in more detail.

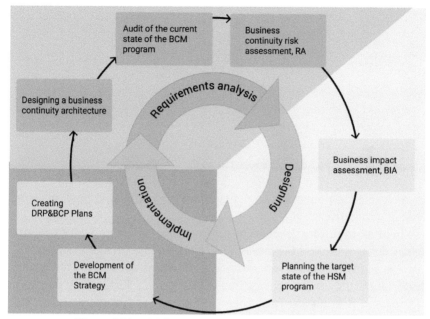

Figure 3.30 IBM's core services under the ECP program.

Risk assessment (RA) and business impact analysis (BIA) are performed.
Purpose of work:

- identifying the most critical business processes and assets, ensuring their work: staff, applications, infrastructure, etc.;
- qualitative and quantitative assessment of business interruption risks;
- determine the maximum allowed downtime (RTO) and information loss (RPO) for each process.

Process:

- study of project and technical documentation;
- interview with the owners of the business processes and organization;
- study of the company's work environment.

Results:

A result is a report with a detailed description of the existing risks of business interruption, an assessment of possible financial and other indirect losses, a list of critical business processes, and RTO/RPO parameters for each of them. Note that the IBM BCRS approach (Figure 3.31) is based

Figure 3.31   Characteristics of the IBM BCRS approach.

on a clear definition and documentation of the company's main goals and strategic initiatives.

In order to further develop the recommendations for adequate countermeasures against possible threats and minimize the risks of business interruption, a comprehensive analysis of the business continuity process at all levels of the enterprise organization (Figures 3.32–3.34) is required.

Indeed, it is almost impossible to develop and implement an effective business continuity program without applying a balanced approach to implementing practical measures to ensure the continuity of critical business processes. There is a need to study the parameters of the enterprise operating environment, decompose into critical processes and supporting components, and analyze the relationships between components to do this.

Audit and analyze current business continuity and disaster recovery capabilities: Purpose of work: evaluation of current indicators of recovery time and data loss prevention for critical business processes of the organization.

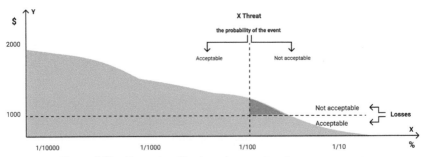

Figure 3.32   Example of business interruption threat assessment.

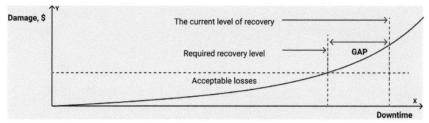

Figure 3.33 Example of possible damage assessment.

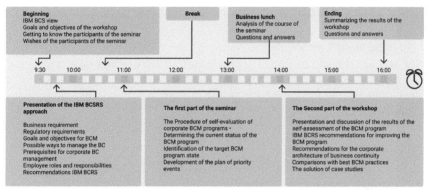

Figure 3.34 Example of an IBM BCM training workshop scenario.

Process:

- research on documentation;
- interviews with business process owners, key users, and IT staff;
- testing fault tolerance of IT systems, data recovery from backup media;
- exercises for working out a business recovery plan (staff actions, communications, backup office, etc.).

Results:
Report with a detailed description of existing disaster recovery capabilities and recommendations on how to bring them to the required indicators.

Conducting a training seminar on BCM issues:
Purpose of work:

- raising awareness of the company's employees on business continuity issues;

- express assessment of the current situation and development of a plan for further actions.

Process:

- overview presentations;
- discussion with key employees: heads of business units, CSOs, line managers, and heads of divisions;
- analysis of the obtained data using specialized tools;
- presentation of results and development of a plan for further actions.

Results:

- presentation materials;
- report on rapid assessment of the current state of business continuity.

Development of a business continuity strategy:
Purpose of work:

- defining the company's strategic development initiatives in terms of BCM;
- development of a high-level plan for achieving RTO/RPO targets for critical business processes.

Process:

- analysis of survey results;
- development of strategic initiatives;
- development of a portfolio of BCM projects aimed at implementing initiatives;
- project ranking and optimization of the BCM project portfolio.

Results:

- A report detailing strategic initiatives aimed at ensuring business continuity and implementing the relevant BCM project portfolio (Figures 3.35–3.37).

Design and implementation of business continuity architecture:
Purpose of work:
Creating a fault-tolerant IT infrastructure that meets the specified RTO/RPO requirements (Figures 3.38 and 3.39).

Features:

- Wide range of technologies will provide almost any level of fault tolerance up to constantly available systems with downtime of no more than a few minutes per year.

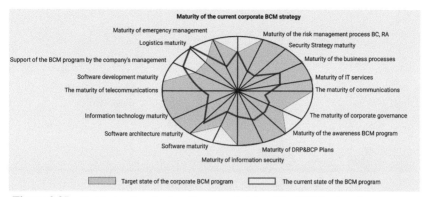

Figure 3.35 Example of evaluating the current state of the BCM enterprise program.

- Using optimal products and solutions from IBM and its partners to cover the majority of existing systems:
  - z/OS mainframes-IBM GDPS;
  - Power Systems c POWER7 – IBM I (i5/OS) Linux;
  - pSeries AIX – IBM HACMP & HACMP XD;
  - iSeries OS/400 – LakeView Mimix;
  - x86 Windows – LakeView Mimix, IBM GDOC, Microsoft Cluster Services.
- Using the IBM project approach and methods ensures high-quality and timely solution of project tasks.

Development of continuity and recovery plans, BCP and DRP
Purpose of work:
creating an organizational component of the enterprise continuity management program, ECP, in case of emergency.

Process:
- analysis of requirements for the continuity of key business processes;
- analysis of existing solutions for ensuring continuity and fault tolerance of the IT infrastructure;
- developing strategic and operational plans for disaster recovery and business renewal and related instructions;
- development of technological documentation;
- backup and restore guides;
- instructions for switching to alternative data processing tools;
- disaster recovery procedures for network and IP components;
- regulations for the company's employees' actions in emergency situations, etc.

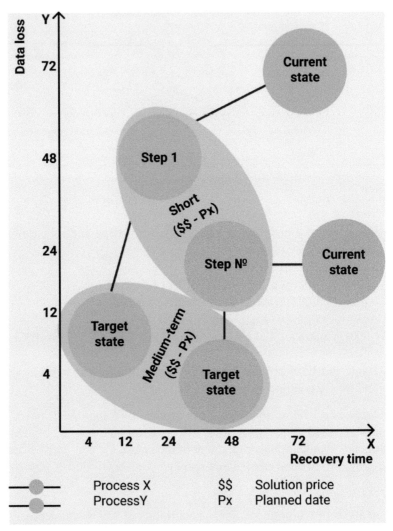

Figure 3.36   RTO evaluation sample.

Results:

- draft BCP and DRP plans as well as corresponding test plans;
- working documentation for disaster recovery and business renewal.

Outsourcing solutions in the field of business continuity
Goal:
The goal is providing the customer with reliable on-demand infrastructure without capital investment in business continuity solutions.

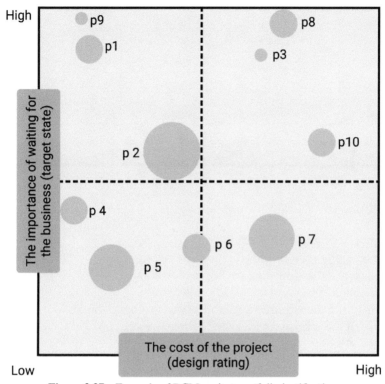

Figure 3.37   Example of BCM project portfolio justification.

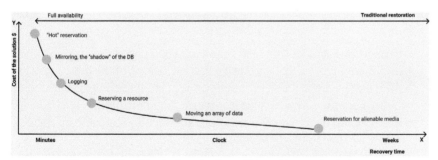

Figure 3.38   Criteria for selecting BCM architectural solutions.

Features:

- provision of equipment: servers, storage systems, communication equipment, system, and application software;
- placement of equipment in a professional data center, possible geographically distributed configuration;
- providing operation and management services;

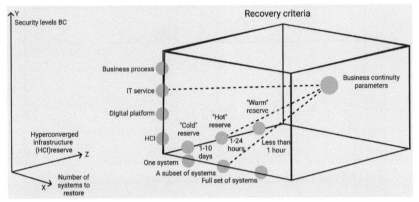

Figure 3.39   Example of a multicriteria evaluation of BCM countermeasures.

- monitoring in 24 × 7 mode;
- managing servers and storage systems;
- automated and manual backup and restore;
- manage apps up to the middle ware level;
- ensuring the required availability (according to the SLA);
- using the infrastructure as the primary or backup.

Implementing the lifecycle of electronic data management
Goal:
The goal is providing the customer with the technological capability to store and back up data without capital investment in the appropriate solutions.

Features:
- transporting and storing tape media on IBM premises;
- use of storage systems and tape libraries located in the data center for local use; payment for storage services is made depending on the amount of memory used as well as performance requirements and RAID level;
- data backup using specialized replication technologies from the customer's office to the IBM data center;
- organization of geographically distributed storage systems that use synchronous or asynchronous replication; the main storage system is located in the customer's office or the main IBM data center, and the backup system is located in the IBM backup data center.

Provision of backup office services
Goal:
The goal is providing the customer with a backup office to continue critical business processes in emergency situations.

Features:

- specialized rooms for storing backup equipment;
- providing an office equipped with the necessary infrastructure, including furniture, computer and office equipment, and communication facilities;
- fulfilling specific requirements, such as preparing traders' jobs, providing dedicated communication channels, and modern access control and surveillance systems;
- development and regular testing of plans for switching to reserved areas;
- keeping server hardware and workplace configurations up-to-date;
- support and maintenance of the backup office in 24 × 7 mode.

Features of the IBM BCRS approach:
The main distinctive feature of the IBM BCRS approach is the ability to create a comprehensive, cost-effective enterprise continuity program (ECP) that meets international standards and regulatory requirements.

The IBM BCRS approach accumulates its own experience as well as the world's best practices in implementing projects in the field of business continuity. The main practical recommendations of IBM BCRS include the following:

- There is a need to identify threats and risks of business interruption, as well as necessary and sufficient countermeasures to ensure business continuity.
- BCM enterprise program is based on well-developed business continuity goals and principles.
- With a step-by-step approach, all business units develop and implement all necessary business continuity measures that meet the requirements of the ECP program.
- As part of a business continuity policy, requirements may differ for individual business units of an enterprise and have a different impact on the distribution of roles within the company.
- Technical implementation of general business requirements and requirements in the field of business continuity leads to the construction of a rational continuity architecture and business continuity architecture.
- Requirements and recommendations for ensuring business continuity that cannot be met in these specific conditions require additional threat analysis and assessment of residual risks.

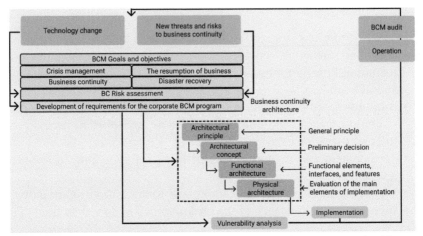

Figure 3.40   Impact of new threats and technologies on the BCM program and architecture.

- Any new threats and technologies affect the business continuity architecture and require additional research and revision of existing views on business continuity policy.

The critical factors of the business continuity strategy are the completeness and regular revision of the relevant working documentation (BCP and DRP plans, methods and policies, standards and regulations, procedures, and instructions).

The overall goal is to achieve an automatically renewable and evolving business continuity management process, which is the methodological, organizational, and technological basis for the continuous development and improvement of the ECP program.

The process of implementing countermeasures to ensure the required business continuity is cyclical and involves a number of sequential steps. The emergence of new threats and the development of business continuity technologies will have a direct impact on the enterprise program and business continuity architecture as well as cause the need for a new risk assessment and development of the business continuity process, as illustrated in Figure 3.40.

The development of the BCM program is an ongoing process and involves monitoring all the main stages of the business continuity management lifecycle.

3.3.4   Example of a Solution Selection

A distinctive feature of the IBM approach is making an informed choice of a suitable technical solution in the field of BC and DR, considering

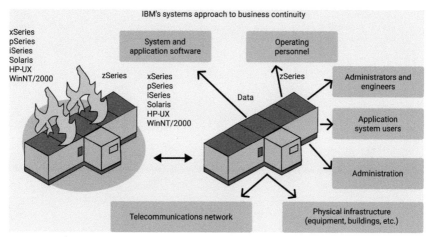

Figure 3.41 Example of BCM process decomposition.

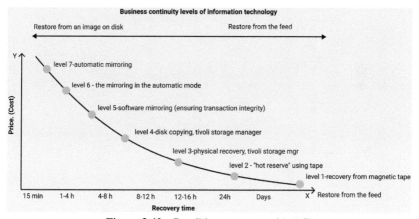

Figure 3.42 Possible ways to provide BC.

the needs of a particular organization. An acceptable recovery time (RTO) is determined for each critical business process and IT service, and then possible technical solutions are identified to ensure business continuity to do this. At the same time, with the advent of new technologies and solutions in the field of business continuity, new technical solutions are ranked according to the required RTO (Figures 3.41–3.43).

In practice, the appropriate solution in the IT business continuity can be chosen according to some level of business process recovery (Figure 3.44):

- Classification of business processes into groups based on their tolerance for possible interruptions in business: unacceptable, partially acceptable, and acceptable. Operational processes are also considered.

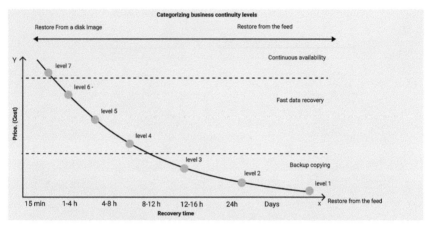

Figure 3.43 Example of ranking business processes by recovery level.

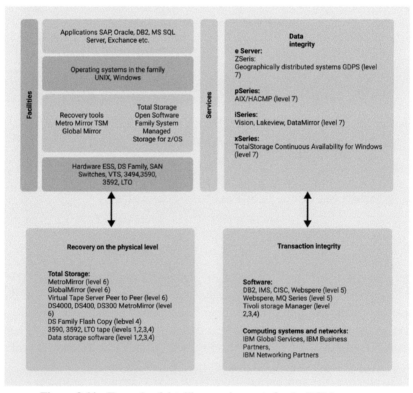

Figure 3.44 Example of detailing requirements for the BCM process.

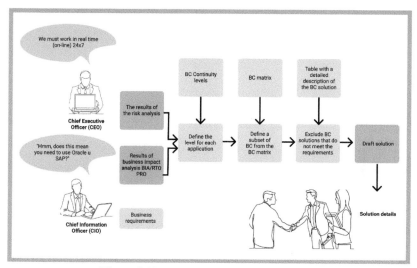

Figure 3.45 Typical BC software algorithm.

- Highlighting the main methods for ensuring IT business continuity within each group (of course, there is no need to specify all known methods here).
- Choosing the optimal set of IT business continuity methods and selecting appropriate technical solutions.

The process of detailing requirements for technical solutions in the IT business continuity can look like this (Figure 3.45).

Here, it is easy to determine which group a particular method and the corresponding technical solution belongs to. Now let us take a look at the method of choosing a technical solution for business continuity in more detail (Figures 3.45 and 3.46).

Note that this method only determines the business requirements for IT business continuity. For this reason, it does not contain detailed recommendations for the design and implementation of appropriate technical solutions.

Figuratively, the concept of choosing solutions for business continuity used in the methodology can be represented as an hourglass (Figure 3.47).

Possible questions and the algorithm for setting them are shown in Figure 3.48.

A fragment of the business continuity decision matrix is presented in Figures 3.49 and 3.50. In this example, the possible preliminary solutions are: z/OS Global Mirror, GDPS HyperSwap Mgr, eRCMF, and so on. Now that we have identified preliminary possible solutions for business

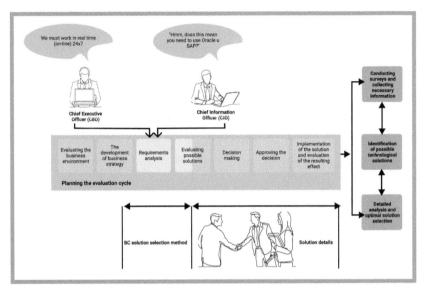

Figure 3.46   Role and place of BC decision selection methodology.

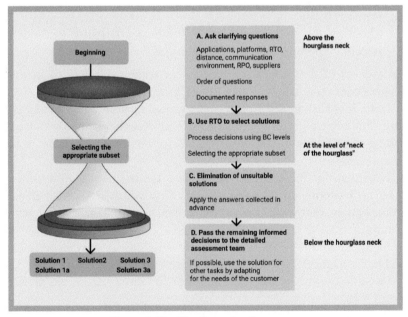

Figure 3.47   The "hourglass" concept.

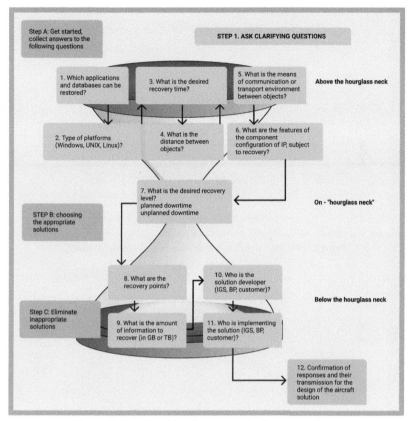

Figure 3.48   Algorithm for setting clarifying questions.

continuity, and we eliminate inappropriate solutions by applying the answers to the remaining questions collected in Step A.

For example, let us assume that the following responses were received in Step A:

- Which apps can be restored? – Different.
- What platform are they used on? – zSeries.
- What is the desired recovery time? – 3 hours.
- What is the distance between recovery sites (if there is more than one object)? – 35 km.
- What is the communications environment or environment to transport data to the recovery site? What is its bandwidth? – Fiber channel, DWDM, 50 MB/s.

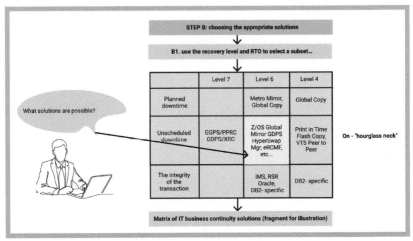

Figure 3.49   Example of a matrix of possible solutions BC.

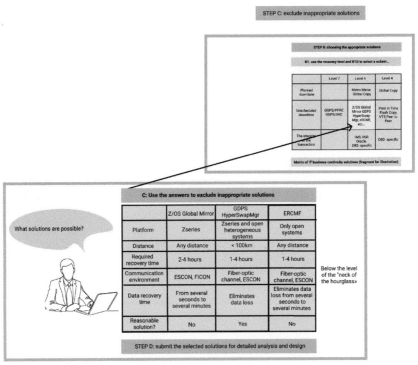

Figure 3.50   Excluding inappropriate BC solutions.

- What are the features of hardware and software configurations that need to be restored? – IBM ESS.
- What is the desired type of recovery? – No data loss.
- What is the amount of information to recover? – 4 TB.
- What is the desired level of recovery? (planned/unplanned/transaction integrity)? – Unplanned integrity.
- Who is the developer of the solution? – Requires clarification.
- Who will implement the solution? – Requires clarification.

After receiving this information, we proceed to Step B, where we are instructed to use RTO to determine a subset of possible solutions. A simplified version of the solution is presented in Table 3.15 (unplanned recovery, recovery time – 3 hours).

Here, the intersection of the "Level 6" column with the "Unplanned downtime" line will select preliminary solutions:

- z/OS Global Mirror;
- GDPS Hyper Swap Manager C;
- eRCMF.

Next, proceed to Step C: excluding inappropriate solutions (Table 3.16).

Thus, we see that after receiving answers to our questions and excluding inappropriate solutions, the following reasonable option remains: GDPS Hyper Swap Manager.

Step D: Transfer the solution for detailed analysis and design.
Comments to Table 3.16:
- Since the platform is IBM @Server zSeries(R), we can exclude zRCMF because it does not support zSeries (R).
- For a distance of 35 km, all the remaining solutions are suitable.
- From the point of view of the ESCON(R) communication environment, all the remaining solutions are suitable.
- From the vendor's hardware perspective, all remaining solutions are suitable for IBM ESS object 1.
- From the vendor's hardware perspective, all remaining solutions are suitable for IBM ESS object 2.
- From the point of view of zero RTO, only GDPS Hyper Swap Manager is suitable.

Table 3.15   Decision matrix for business continuity.

	7	6	5	4,3	2,1
RTO →	RTO up to 2 hours	RTO from 1 to 6 hours	RTO from 4 to 8 hours	For level 4 RTO: 6–12 hours; For level 3: from 12 to 24 hours	RTO more than 24 hours
Description	Automated recovery	Mirroring the server and storage devices	Ensuring transaction integrity	Hot reserve DiskPiT copy, Tivoli Storage Manager-DRM, fast tape	Software recovery from the tape
Planned downtime		Metro Mirror, Global Copy, Global Mirror, z/OS Global Mirror VTS Peer to Peer		FlashCopy/Global Copy VTS Peer to Peer, TSM tape	Tivoli Storage Manager
Unplanned downtime D/R	GDPS/PPRC, GDPS/XRC, AIX HACMP-XD & MetroMirror Total storage with continuous availability for Windows	z/OS Global Mirror, GDPS Hyper Swap Global Manager, eRCMF		Level 4: VTS Peer to Peer, Flash Copy, Flash Copy Migration Manager, Global Copy, eRCMF8 Global Copy	
Level 3: Flashcopy, TSM, tape	Tivoli Storage Manager, tape				
Transaction integrity	Clustering at the database and OS level		SAP, Oracle, DB2, SQL Server remote replication	Level 3: MS SQL Server Database cluster (tape recovery)	

Table 3.16 Example of choosing appropriate BC solutions.

Solution	z/0S Global Mirror	GDPS Hyper Swap Manager	eRCMF
Platform	zSeries	zSeries, heterogeneous, including zSeries	PSeries, LINUX, Sun, HP, Windows, heterogeneous (open)
Distance	Less than 40 km, 40–103 km	Less than 40 km, 40–103 km	Less than 40 km, 40–103 km
The communication environment	ESCON, FICON	ESCON, fiber-optic channel	ESCON, fiber-optic channel
Vendor (1)	IBM or Hitachi	PPRC – compatible devicestorage from the same supplier	IBM ESS, DS6000, DS8000
Vendor (2)	IBM or Hitachi		
RTO	From a few secondsup to a few minutes	Near-zero	Near-zero
Amount of data	Any	Any	Any

3.3.5 Example of Task Statement

According to IBM BCRS consultants, the main results of the project to create the ECP program include the following:

- Project management plan: The plan describes the scope of the project, defines its main goals and objectives, defines roles, and assigns responsibilities.
- Business impact analysis (BIA): It involves the identification and ranking of critical business processes and the decomposition of processes into components. The main threats to business interruption are evaluated, including a description of the threat sources, the intruder's model, and an assessment of the damage caused by the threat implementation.
- Business continuity risk analysis, risk analysis (RA). Audit of the current state of the ECP program based on the recommendations of *ISO 22301:2019*. The assessment of residual risks of business interruption is made considering the implemented business continuity countermeasures. The effectiveness of implemented continuity countermeasures is evaluated. A summary assessment of the maturity of the ECP program is based on a special IBM BCRS maturity model.

- A business continuity strategy that includes a number of elements:
  - target state model of the customer's ECP program:
    - general description of the target state of the ECP program, based on the results of business interruption risk analysis, legal norms and requirements, international standards and best practices, as well as the existing IT and security strategy;
    - possible options of the organizational structure to ensure business continuity;
    - possible options for the target state of the technological and architectural components of the BCM architecture;
    - description of the process for monitoring the effectiveness of the ECP program;
  - analysis of existing discrepancies between the current and target state (gap analysis) of the ECP program;
  - comparative analysis of options for eliminating identified inconsistencies;
  - key strategic initiatives on the way to implement the selected option to eliminate inconsistencies;
  - drafts of the main working documentation (BCP and DRP plans, BCM policies, standards and regulations, instructions, etc.).
- A portfolio of projects to implement the planned business continuity development strategy.

In practice IBM BCRS suggests creating a proper business continuity program ECP in a few steps:

phase 0 – initiation of the ECP development project;
phase 1 – business impact assessment, BIA;
phase 2 – business interruption risk assessment, RA;
phase 3 – developing a business continuity strategy;
phase 4 – offering a target portfolio of BCM projects;
phase 5 – presentation of the main results of the ECP project.
Let us take a look at the listed phases of the project in more detail.

Phase 0 – ECP project development initiation:
Goal:

- get a detailed description of the project and define:
  - the scope of the project;
  - expected result;
  - roles and responsibilities of project participants;
  - list of main works;
  - basic approaches to project management.

Source data:

- confirmation of the project start by the management;
- decisions made on the project goals and objectives;
- working materials on the organization of the company.

Main activity:

- create a description of the project (goals and objectives, description of the scope and organization of work, list of output materials, etc.);
- describe the project management procedure (escalation, changes, etc.).
- notify the company's divisions about the project and enlist their support;
- organize the process of creating a project team.

Results:

- project charter;
- working materials for the project.

The possible composition and structure of the project team is shown in Figure 3.51.

Phase 1 – business impact assessment, BIA:
Goal:

- identify and rank critical business processes of the company, including provision of IT services;
- assess the main threats of business interruption and possible damage to the organization in the event of an emergency.

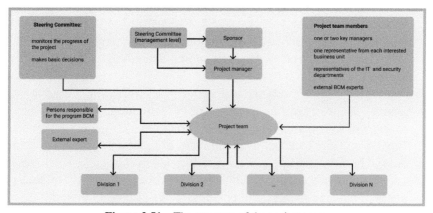

Figure 3.51 The structure of the project team.

Source data:

- business development strategy;
- business process map;
- map of IT services;
- IT strategy.

Main activity:

- preparing questionnaires and conducting a series of interviews with competent representatives of business units and IT specialists.

Results:

- a documented description of the list of threats to business interruption and an assessment of possible damage in the event of an emergency.

Phase 2 – business interruption risk assessment, RA:
Goal:

- get a documented assessment of business interruption risks taking into account the implemented business continuity countermeasures;
- get an assessment of the effectiveness of the mentioned countermeasures.

Source data:

- business development strategy;
- business process map;
- map of IT services;
- IT strategy.

Main activity:

- review current business continuity initiatives and related working documents;
- preparing questionnaires and conducting interviews with competent representatives of business units, including IT and security services;
- preparation of methods and instructions for conducting the survey;
- survey of the organization's physical infrastructure;
- survey of the organization's IT infrastructure;
- analysis of the received data and materials, obtaining an assessment of the basic and residual risks of business interruption;
- comparison of basic and residual risks, and assessment of the effectiveness of existing measures to counter threats of business interruption;

- evaluating the maturity of an existing BCM program and receiving confirmation;
- completion of preparation and presentation of the results of the project stage.

Results:

- documenting and evaluating identified vulnerabilities;
- assessing the likelihood of threats being implemented;
- assessment of the basic level of business interruption risks;
- assessment of residual risks of business interruption taking into account implemented countermeasures;
- evaluating the effectiveness of business continuity countermeasures;
- assessing the maturity of the current state of the ECP program.

Phase 3 – developing a business continuity strategy:
Goal:

- develop a draft business continuity strategy for approval by the company's management.

Source data:

- business development strategy;
- IT strategy;
- security strategy;
- requirements for the BC strategy;
- interim results of the project.

Main activity:

- developing a description of the target state of the business continuity strategy;
- identification of possible business continuity strategy development options;
- a reasonable choice of one of the possible options for developing a business continuity strategy based on GAP analysis;
- development of a business continuity strategy implementation plan;
- documenting key strategic initiatives.

Results:

- draft business continuity strategy approved by the project sponsor.

Phase 4 – offering a target portfolio of BCM projects:

Goal:

- Develop a draft of the target BCM project portfolio for approval by the company's management.

Source data:

- development of a business continuity strategy;
- list of ongoing projects in the field of BCM.

Main activity:

- defining the criteria and indicators for qualitative evaluation of target and current BCM projects;
- assessment of current project activities in the BCM area;
- creating a target portfolio of BCM projects;
- development of recommendations for the implementation of the most significant projects from the target portfolio of BCM projects.

Results:

- project of the target portfolio of projects in the BCM area, approved by the project sponsor.

Phase 5 – presentation of the main results of the ECP project:
Goal:

- approval of work results for previous phases.

Source data:

- working materials based on the results of previous work phases.

Main activity:

- preparation of materials for the mentioned presentation;
- approval of the presentation;
- conducting a presentation for management.

Results:

- project working materials reviewed and approved by the organization's management.

A possible schedule for the ECP development project is presented in Table 3.17.

Note that IBM BCRS consultants recommend using the IBM Worldwide Quality Assurance (QA) method to control the quality of the project to create

Table 3.17 Project execution schedule.

No.	Project stages	Beginning	End	Duration	2019
1)	Creating a corporate BCM program	01.08.2019	06.11.2019	...	
2)	Phase 0 – project planning	01.08.2019	07.08.2019	...	
3)	Phase 1 – business impact analysis (BIA)	08.08.2019	29.08.2019	...	
4)	Phase 2 – risk assessment	22.08.2019	12.09.2019	...	
5)	Phase 3 – BCM strategy development	05.09.2019	26.09.2019	...	
6)	Phase 4 – a target portfolio of BCM projects	27.09.2019	18.10.2019	...	
7)	Phase 5 – presentation of the project results	19.10.2019	06.11.2019	...	

ECP program. The method involves independent expertise of the project at the stage of its preparation. Verification is performed in terms of determining the technical feasibility of the project, economic feasibility, and the possibility of high-quality implementation of the project. The main task of QA is not so much to manage problems but to prevent them and make appropriate recommendations at the project implementation stage. IBM conducts a number of project reviews to prevent possible excess of planned costs:

- QA1 – solution assurance (SA) review: technical evaluation of the project;
- QA2 – business assurance review: an assessment of the economic feasibility of the project implementation;
- QA3 – project assurance review: assessment of project organization and contract preparation.

The most important component of the QA process is the technical evaluation of the project (SA). As a rule, shortcomings in the technical preparation of the project at an early stage negatively affect the quality of project implementation as a whole. SA is conducted by IBM BCRS specialists based on the experience of several thousand specialists working at IBM enterprises and laboratories. In this area, IBM has a well-developed and proven infrastructure, including more than 60 testing and integration centers located in 130 different countries in America, Europe, and Asia, employing more than 3600 technical specialists. Specialists conduct a comprehensive assessment of the proposed solution, including a detailed assessment of the possibility of interaction and compliance of the performance characteristics of individual technical components, work schedules, and project risks.

3.4   Hewlett-Packard Practice

The approach of Hewlett-Packard (HP) is characterized by the flexibility and technical sophistication of solutions offered by the company for creating comprehensive enterprise business continuity programs, ECP. HP has a number of interesting projects implemented around the world (Table 3.18).

The methodological basis for HP's BCM approach is the business continuity management lifecycle model presented in *ISO 22313: 2020* (Figures 3.52 and 3.53).

At the same time, the development and implementation of a comprehensive ECP program covers all the main areas of business continuity management in accordance with the recommendations of *ISO 22301:2019* (Figure 3.53).

In practice, HP distinguishes three main stages of ECP development and implementation projects: requirements analysis; design and implementation; operation and support. Here the distinctive feature is the use of number of well-developed business focused methodologies for the description of activity of the organization. For example, the methodology of the description of processes, CME Process Framework services, constructs some metamodel of processes of the organization which, along with the description of a current condition (model "as-is"), describes also a target condition (model "to-be") of the organization. For ECP, it means that it becomes possible to develop some optimal scenario of transition from the current (Figure 3.54) to the target (Figure 3.55) ECP state. Another process framework delivery methodology based on eTOM, NGOSS SID, and OSS/J reference models allows monitoring the transition states of an organization.

Using the example of a project to create and implement the ECP program, let us look at the specifics of the HP approach in more detail.

3.4.1   Evaluating the Current ECP State

At this stage, requirements are analyzed, project boundaries are defined, and specific project objectives are set.

Introductory seminar:
According to available materials of similar projects in domestic and foreign companies, an introductory seminar is held for employees of the

Table 3.18 A list of some of the projects HP BCM.

Company–customer	Project description
Asian operator of mobile communication	• Survey and analysis of the company's current operating processes. • Using the world's best practices for target design of operating processes. Development of the implementation plan. • Defining technical requirements and developing a long-term strategy for change. • Making the necessary organizational changes based on recommendations of the reference model of the eTOM model. • Standardizing processes and defining metrics for describing and measuring the target model of operational processes. • Using the HP CME Process Framework (now HP COSMOS), eTOM, and ITIL as reference models for ongoing analysis.

East-European operator of mobile communication	• Optimization of the problem management process.
	• Define, configure, monitor in real time, and build historical reports to determine:
	    ○ service levels: average troubleshooting time, cost of maintenance, and so on.
	    ○ client SLAs.
	• Define processes for managing the input flow of requests, processes for making changes, and fault distribution processes.
	• Research of current service assurance processes based on the eTOM model.
	• Designing the target state of incident management processes based on the data model.
	• Adaptation of fault management processes in accordance with best HP practices.
	• Creating and managing a consolidated service desk, covering various business units and systems of the company.

European operator mobile communication	• Defining and aligning business goals, business metrics, and operations – solving the company's tasks. Creating a business case. • Research of business processes, according to eTOM process areas. • Design of possible architectural solutions, including CRM, billing, business intelligence, business orchestration, etc. • Defining and standardizing metrics for service quality control of subscribers' lives: the time of servicing the subscriber's call, the cost services, etc. • Modeling the target state of processes based on HP COSMOS (CME Process Framework). • Consolidation and synchronization of reporting materials for creating a single enterprise architecture of the company.

Continued

Table 3.18 Continued

Company–customer	Project description
German fixed-line and mobile operator	• Project goals: improving the quality of customer service, reducing OPEX and CAPEX, reducing the time to market new products, etc. • The use of TOGAF, ITSA, and SCS methods for examination of OSS processes of the company. • Use of SPA and FRA methods for evaluating the current and modeling the target state of the company's business processes. • Development of a business case and protection of project results before the company's management. • Using NGOSS methodology to define and refine project goals, standardization of approaches used, and project implementation.
Algerian content providers	• Project goal: development of multimedia services for a content provider. • Implementation of the SPPA methodology for optimizing the time and money spent on project implementation. • Deep analysis of the current business situation, current business decisions, including analysis of the company's current service portfolio. • Analysis of segmentation of services and how they are presented to the end user. • Identify the shortcomings of the current model and implement a business strategy for the company's business development based on a comparison of business models' benchmarks with industry-leading company indicators. • The introduction of the HP convergent charging.
Hungarian mobile operator	• Project goal: identify and implement an effective service management program to consolidate trouble tickets management processes. • Standardization of internal interfaces and internal protocols of interactions between the company's divisions. • Unification and reduction to a single process for handling requests of clients (both internal and external). • Defining trouble ticketing for processes, activities, and responsible roles. • Defining metrics for measuring the performance of processes in accordance with SLA. • Implement HP OpenView quality manager (SQL) to support the target state of trouble tickets maintenance processes. • Defining and implementing additional measurement metrics according to the recommendations of the SMM model (business process maturity model of companies) responsible for optimizing processes.

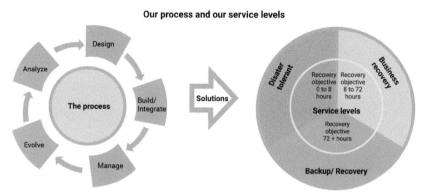

Figure 3.52   BCM lifecycle model.

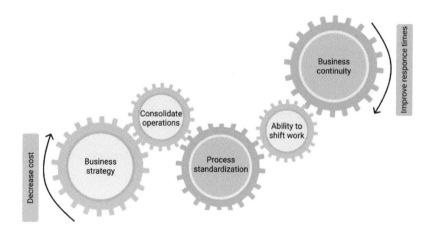

Figure 3.53   Role and place of ECP in the organization.

customer's company. This workshop discusses in detail all the main problems of developing and implementing the ECP program as well as possible solutions.

Modeling the organization's activities:
Based on a survey and interviews with representatives of the company's business units, information about the organization's activities, its main business processes, and IT services is collected automatically. The focus is on the so-called "end-to-end" processes (client-company-client) directly related to the provision of services and profit. For example, these processes may include:

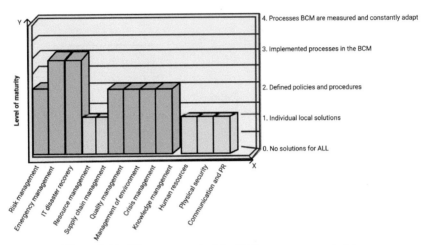

Figure 3.54   The assessment of the ECP program current state.

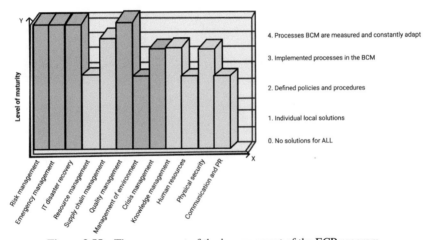

Figure 3.55   The assessment of the improvement of the ECP program.

- "From accepting an order to receiving payment for services";
- "From requesting a subscriber to providing information";
- "From receiving a change request to implementing a change";
- "From the discovery of the problem before it resolves," etc.

It also covers related operational processes and supporting IT services, including:

- managing problems with services;
- quality management;
- managing resource issues;

- managing resource performance;
- resource support and availability processes;
- service support and availability processes;
- collection and processing of data on the use of resources;
- managing supplier and partner issues;
- managing the performance of suppliers and partners;
- providing service capacities;
- provision of resource capacities;
- configuring and activating services;
- provision of resources, etc.

It is suggested to use special automation tools for quick collection and processing of information. For example, based on the Metastorm Provision software, Hewlett-Packard developed a package called *COSMOS* (previously called the *Hewlett-Packard CME Process Framework*). The mentioned package includes:

- business process map;
- SID model;
- TAM map;
- OSS/J and BMF models (Telemanagement Forum Business Metrics Framework);
- ITSM, COBIT, and TOGAF descriptions and templates.

Thus, the mentioned models and templates are considered as some primary material for the description and creation of adequate models of business processes of the organization.

Development of the technical task:
   At this step, a technical task is developed for creating and implementing an enterprise continuity program (ECP). This task reflects all the main requirements for ECP.

The coordination of the work results:
The results of the previous stage of work are clarified and approved (Tables 3.19–3.22 and Figure 3.56).

3.4.2 Developing a BCM Strategy

The goal of the stage is to develop justified solutions for business continuity management (Figures 3.57 and 3.58 and Table 3.23).

- *Developing a BCM strategy.*
- *Coordination of a BCM strategy.*

Table 3.19   Clarification of the results of the previous stage of work.

Report material	Description
Review of best practices in the field of BCM in domestic and foreign companies.	Presentation on typical problems of implementing similar projects in the field of BCM.
Workshop on best practices in the field of BCM.	Conducting a workshop for customer representatives on best practices in BCM.

Analysis of the subject area.	Preparation of a report on the results of the survey, including the following:
	• description of end-to-end business processes:
	  ○ name of the process;
	  ○ process areas that are part of this process and are linked to each other in threads;
	  ○ the owners of the business processes;
	  ○ organizational units that support the process;
	  ○ operational criticality of the process;
	  ○ strategic criticality of the process;
	  ○ business process metrics;
	  ○ link to business performance indicators (ARPU, OIBDA, MOU, etc.)
	• description of IT systems, including:
	  ○ name, type, class, and owners;
	  ○ eTOM process areas supported by the system;
	  ○ end-to-end process that the IT system supports;
	  ○ IT system risks (loss of availability scenarios for different time intervals);
	  ○ RPO, RTO, and other IT system indicators related to BCM;
	  ○ data level indicators (size, growth, redundancy, etc.);
	  ○ level of criticality of the IT system;
	  ○ description of the ECP program maturity level based on best practice recommendations BS25999, ISO27001, etc.

Terms of reference for the development and implementation of the corporate ECP program.	Development of a technical task containing, among other things:
	• general information about the customer's company:
	  ○ company name;
	  ○ planned start and end dates;
	• purpose and goals of the project;
	• characteristics of IT infrastructure and processes;
	• requirements for the business continuity strategy;
	• business requirements for the functional stability of technological processes;
	• requirements for the content of the ECP program;
	• content of the work on the creation and implementation of the ECR's strategic program;
	• procedure of control and acceptance of completed work;
	• requirements for the composition and content of works:
	  ○ expected actions on the part of the customer, if required for the implementation of the ECP;
	• set of regulatory documents (policies, regulations, standards, metrics, instructions, etc.), ensuring the implementation of the ECP;
	• the requirements for the documentation.

Agreed terms of reference for development and implementation of the ECP program.	Agreed terms of reference for the development and implementation of the ECP program.

Table 3.20   Example of a survey questionnaire for the subject area (processes).

Question group	Question	Answer
Respondent	Response date	27.09.2019
	Company	XXX
	Business unit	XXX
	Department	Department of network management
	Respondent	Ivanov I. I.
Process information	Process name	2-2 OPS Activation and configuration of services
	The operational criticality of the process	High
	Strategic criticality of the process	Average
	Process classifier	High/medium
	Process priority	3
Applications that support the process	Application 1	Service Activator
	Application 2	Microsoft Exchange
	Application 3	Other
	Application 4	
	Application 5	
	Application 6	
	Application 7	
	Application 8	
	Application 9	
	Application 10	

Project reporting materials are being approved (Table 3.24).

3.4.3   Implementing a BCM Strategy

The goal of the stage is to implement BCM solutions, assess residual risks, and develop recommendations and directives for further development of the ECP program.

Developing recommendations for implementing the BCM strategy:
A program for implementing a business continuity management strategy in the organization is being prepared.

Table 3.21 Information on application and data.

Question group	Question	Answer
Application data	Application ID	
	Name	Service activator application
	App description	Activation and configuration of services on hardware
	Application type	Transactional (read/write)
	Application class	Standard package
	Respondent	Sidorov I. T.
The scenario of unavailability of the application (what-if)		
... half-day (8 business hours) unavailability	The main type of business impact	Loss of revenue
	Description of the impact on business the	User cannot use the services-loss of average (ARPU)
	Level of influence	Very high
	App status	Critical
Recovery parameters	RTO	1–5 hours
	RPO	(0 hours) lossless
	Recovery plan	Exists
	Past issues, if any	

Implementing a BCM strategy:
The BCM strategy is being implemented and the provisions of the mentioned strategy are being checked (tested) in a certain "pilot" zone.

Development of standard solutions:
Development and feasibility study of standard solutions to ensure the required availability of IT services and business process continuity in general.

Residual risk assessment:
The residual risks of business interruption and the achieved (project) maturity level of the enterprise continuity program (ECP) are assessed.

Development of proposals for the development of the ECP program:
Proposals are developed for the elaboration of the enterprise continuity program (ECP) and the necessary costs are estimated.

Protection of the results obtained:
The ECP program and its development proposals are protected (Tables 3.25–3.27 and Figure 3.59).

Table 3.22   Example of a survey questionnaire for the subject area (map of processes).

Processing requests for services	Assessment of the possibility of providing services to the client
	Assessment of the client's creditworthiness
	Tracking the processing of requests for services
	Completion of work on the client's request
	Issuing a client request
	Reporting the process of processing requests for services
	The closing of the client application
Handling subscriber issues	Identification of a subscriber problem
	Notifications about subscriber problems
	Monitoring the resolution of a subscriber problem
	The closing of the subscription problems
	Creating a record of a subscriber problem
	Resolving a subscriber problem
Managing the quality/level of services for the client	Assessment of the quality/level of services for the client
	The violations control the quality/level of services for the client
	Reporting on the quality of services for the client
	Creating a record of service quality degradation for a client
	Tracking the resolution of service quality degradation for the client
	Closing a record of service quality degradation for a client
Customer retention and loyalty support	Establishing and ending relationships with the client
	Creating a client profile
	Analysis and management of client risks
	Personalization of the client's profile for retention and loyalty purposes
	Evaluation of customer satisfaction
Managing billing and payment collection	Managing customer requests for invoices
	Applying tariffs, discounts, and recomputations
	Printing and delivery of invoices
	Managing client billing
	Managing collection of payments
Activation and configuration of services	The design of the solution
	The selection of certain service parameters to services

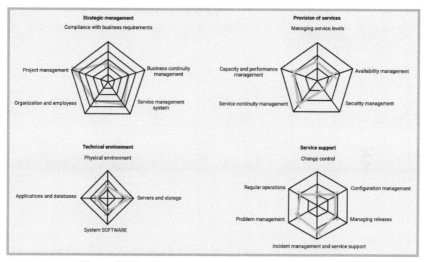

Figure 3.56 Examples of ECP maturity survey results.

No.	Actions	Expected time to complete	Total time spent
1)	Event (incident) occurred (date and time) and registered name (full name). The problem was not fixed. If necessary, go to step 6.	5	5
2)	The problem was addressed to the system administrator and was/was not resolved. If it was, go to step 6.	5	10
3)	Called the Process Manager, whose forces the problem was/was not fixed. If it was, go to step 6.	5	15
4)	Subsystem (DC) is set to ready state 1. Make sure that the following responsible persons are informed about the current state of the subsystem: Manager, Technical support; Manager, DC Management; Manager, Data Network. For current phone numbers, see Appendix <X. x>	15	30
5)	Determine the expected area of the problem (mark the appropriate areas). If there are several intended areas, go to the corresponding sections in parallel. Software – go to section 2 P. 1 (page x). Hardware – go to section 2 P. 2 (page x). Data network – go to section 2 P. 3 (page x). Service systems – go to section 2 P. 4 (page x). Undefined area – go to section 2 P. 5 (page x).	15	30
6)	Collect the data. Document the event. Close the incident.	30–120	60

Figure 3.57 Example of a fragment of a disaster recovery scenario.

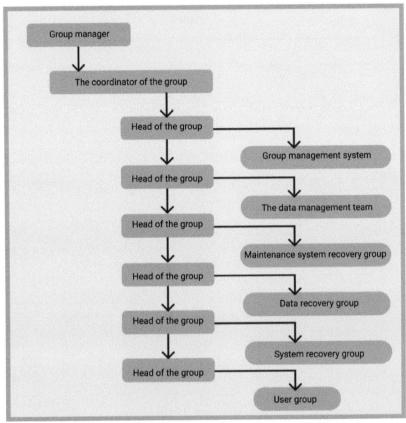

Figure 3.58   An example of the distribution of responsibility.

Table 3.23 Sample set of reporting materials for the stage.

Report material	Description
Strategy for managing the continuity of critical business processes.	Developing IT continuity policies for emergency services: • incident classification and order determination; • event registration; • general description of procedures for maintaining continuity; • business breakthroughs; • description and organization of responsible groups; • description of procedures and tools for alternative; • launch critical subsystems; • description of periodic procedures for each subsystem; • description of emergency procedures in case of failure and emergency situations. Creating an enterprise standard "assessment methodology and business continuity indicators system": • critical levels of end-to-end processes, including standard values of indicators for BCM for critical business processes and ensuring IT services; • target RPO, RTO, and other BCM indicators; • connection to business performance indicators; • development of organizational and administrative documentation (policies, regulations, plans, procedures, instructions, etc.).
Consistent business continuity management strategy.	Coordination of business continuity management strategy.

Table 3.24 Example of a reporting table and distribution of responsibility areas for groups.

The group restore the system data network	In the process	Completed
Mobilize a group		
Identify the source of problems in the transmission networkdata		
Make a list of damaged hardware		
Enable to configure the equipment		
To install to test hardware and/or software		
Install test hardware and/or software on a backup site		

Table 3.25 Example of reporting materials for a project stage.

Report material	Description
Approved program of priority measures for the implementation of the business continuity strategy.	Approval of the program of priority implementation measures of business continuity strategies.
Conclusion on the results of testing the strategy in the BCM.	Preparing a conclusion based on the results of testing the strategy in the BCM.
Standard solutions for disaster-resistant IT infrastructure.	Development of documents describing standard solutions, including: • description of the structure/architecture of typical information systems for five levels of criticality; • description of tools to support the survivability of typical subsystems (high availability, disaster recovery, etc.).
Feasibility study of a technical project for a proper IP architecture.	Feasibility study of the proposed solutions.
Technical specification for the creation of the BCP system.	Development of the technical specification for the BCP system.
Residual risks of business interruption.	Assessment of residual risks of business interruption and the achieved level maturity of the ECP.
Proposals for the development of the ECP program.	Development of proposals for the development of the ECP program.

Table 3.26 Example of classification of IT services.

Report material	Description
Classifier of IT systems by recovery priority	MC (mission critical)
The classifier of the IT systems at the recovery time	RC1 (very high speed)
Classifier of IT systems by type of failure handling	DT (disaster sustained)
Classifier of IT systems by rank	1U (single-link)
Classifier of IT systems by redundancy	FR (fully reserved)
Classifier of IT systems by functioning mode and support structure	S24×7+ (IT system supported by IT in 24 × 7 mode + level 3 support in 24 × 7 mode)
The classifier of IT systems by the method of failover	AR (automatic recovery)

Continued

Table 3.26 Continued

Report material	Description
Classifier of IT systems by stages (status) of the platform lifecycle and architecture	CUR (1 current/active)
Classifier of IT systems by hardware and system software manufacturers	FC (certified IT)
Classifier of IT systems by monitoring level	SLAM (corresponds to SLA)
Classifier of IT systems by user categories	MVS (employees of the Moscow region)
Classifier of IT systems by scaling type	AR (automatic recovery) VM (vertical zoom)

Table 3.27 System characteristics based on the example of the RC1 class (very high speed).

			Recovery in case of local system failure		Recovery in emergency cases	
Code	Title	Description	RTO	RPO	RTO	RPO
RC1	Very high speed	Systems whose unavailability has an immediate negative impact on the enterprise's ability to perform its functions	30 min	0–30 min	4 hours	0–30 min

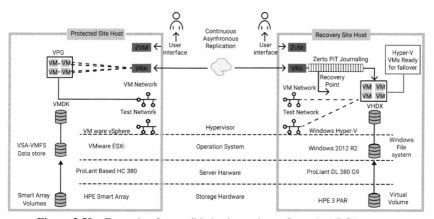

Figure 3.59 Example of a possible backup scheme for a class RC1 system.

3.5 EMC Practice

EMC has more than 20 years of experience in business continuity. During this time, EMC has completed more than 700 business continuity and disaster recovery projects around the world (Table 3.28 and Figure 3.60).

Table 3.28 ECP practice in the field of business continuity.

No.	Economic area	Customers
1	Telecommunications	• Mobile TeleSystems • Chunghwa Telecom • Data access • Philippine Long-Distance Telephone • SUNDAY • Version • STC
2	Financial sector	• Bank of East Asia • CheckFree • Depository Trust and Clearing Corporation • JB Hanauer • MidAmerica Bank • Omgeo LLC • NCB
3	Government departments	• Kaohsiung municipal Government authorities • New Zealand Ministry of social development • State of Michigan
4	Healthcare	• Austin Radiological Association • Montgomery Baptist Health • CHRISTUS health • El Camino Hospital • Guelph General Hospital • Norton Health
5	Extraction and processing of natural resources	• Lukoil-Inform • Eurasia, etc.
6	Other	• Chinese Estates Holdings Limited • Infineon Technologies • Information Resources, Inc. • Linklaters • LT of cement • Mayer Brown Rowe & Maw • New Horizons System Solutions • In the dream office • Paul, Hastings, Janofsky & Walker LLP • Station Casinos • George Washington University

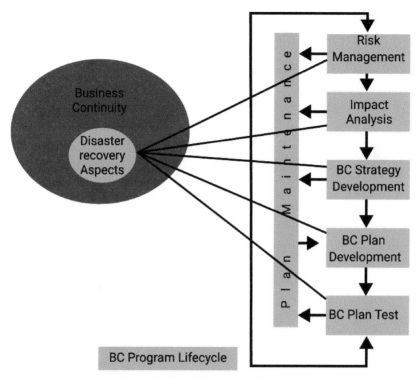

Figure 3.60 ECP program lifecycle.

3.5.1 Type of Work

EMC's core business continuity services include:

- assessment of the maturity level of the enterprise IT-dependent business continuity management program, BCM;
- implementation of effective methods for assessing and managing information technology risks (IT risks);
- creating organizational and methodological prerequisites for building an effective business continuity management system (business continuity management system (BCMS));
- development of a critical process continuity management strategy based on *ISO 22301:2019, ISO 31000 ISO/IEC 20000-1*, and *ISO/IEC 27001:2013 (A. 17)* business continuity standards;
- develop business continuity and disaster recovery plans (BCP and DRP);
- develop recommendations for creating up-to-date employee awareness programs on business continuity and disaster recovery (BCM and DRM);

- develop tests for business continuity and disaster recovery plans (BCP and DRP);
- building and implementing a sustainable information infrastructure for the organization.

EMC considers business continuity as one of the strategic development goals of any government or commercial organization (Figures 3.61 and 3.62). It is believed that the achievement of this goal will allow the move to a qualitatively new system level of ensuring cyber stability of the organization's critical information infrastructure.

Here, the business continuity refers to the strategic and tactical ability of an organization to actively counteract destructive impacts on critical business processes. A business continuity management is – a cyclical process of warning, timely response and recovery of critical business processes of the organization in the event of incidents, accidents and other emergencies (Figure 3.63).

To solve these tasks, EMC actively uses the well-known standards of the British and American institutes BCI/DRII (Table 3.29), supplemented by its own practical experience.

EMC consultants also adopted the best business continuity management practices *ISO 22301: 2019/ISO 22313: 2020/ISO 27001:2013*

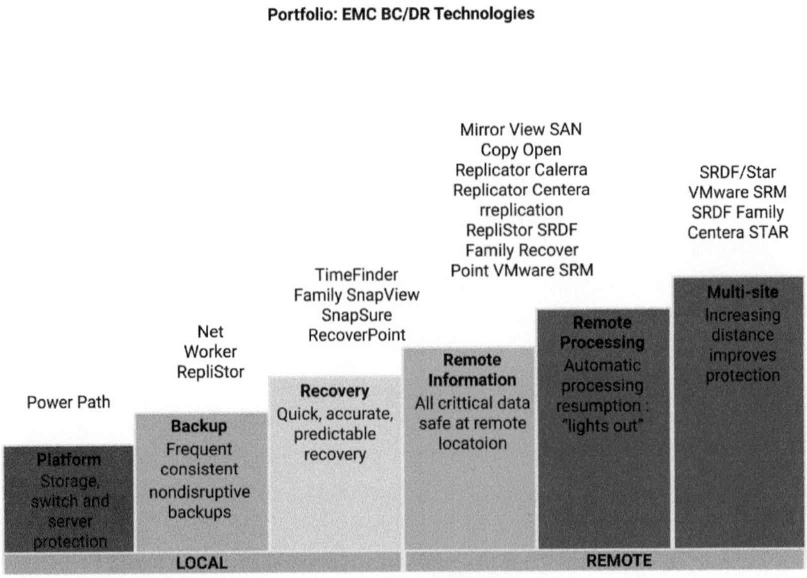

Portfolio: EMC BC/DR Technologies

Figure 3.61   Development of EMC BC/DR technologies.

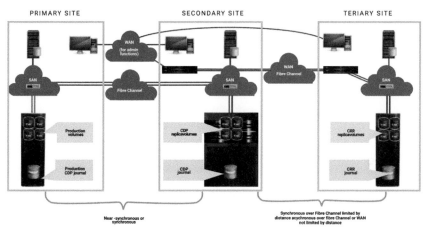

Figure 3.62   Features of the EMC approach.

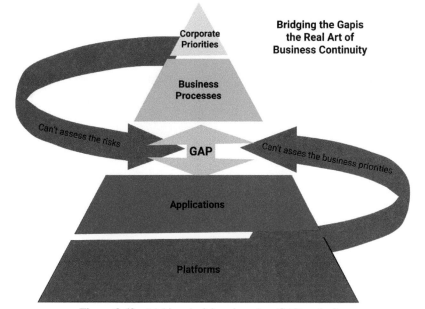

Figure 3.63   Making decisions based on GAP analysis.

(A. 17), ISO 20000-1, EMM/CMM, Compliance SOX, BCI and DRI, PMP, eTOM, COBIT 2019, RESILIA 2015, ITIL V4 and MOF 4.0, etc.

3.5.2   EMC Methodology

The main components of the BCM EMC methodology are shown in Figure 3.64.

Table 3.29   Methodological basis of EMC projects in the field of BCM.

No.	BSI and DRII standards
1)	Initiation and management
2)	Business impact analysis
3)	Risk evaluation and control
4)	Developing business continuity management strategies
5)	Emergency response and operations
6)	Developing and implementing business continuity and crisis management plans
7)	Awareness and training programs
8)	Maintaining and exercising business continuity and crisis management plans
9)	Crisis communications (company employees, contractors, partners, media, etc.)
10)	Coordination with external agencies

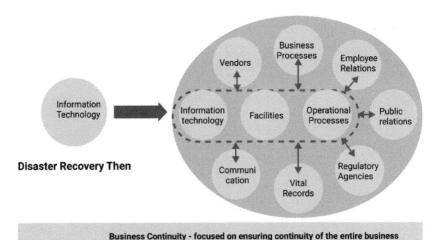

Figure 3.64   Main components of the BCM EMC methodology.

The following stages of ECP development are highlighted here:

- analysis of business requirements;
- business to process mapping (people, locations, procedures, applications, etc.);
- threat prioritization and risk assessment;
- business impact analysis;
- defining recovery time objective (RTO) and recovery point objects (RPO);

- current capability assessment;
- feasibility study of proposed solutions based on GAP analysis;
- BC strategy and recommendations.

Let us give a brief comment on the listed stages.

The first stage begins with the analysis of requirements and identification of critical business processes of the organization (Table 3.30), including providing IT services (Table 3.31). The main criteria of criticality of the business processes are:

- loss of income;
- transaction costs;
- marketing losses (loss of reputation, clients, partners, etc.).

Note that in the EMC methodology, quite a detailed attention is paid to mapping key business processes with a clear record of the IT component of each process.

Next, two main tasks are set and solved for each IT service:

Table 3.30 Example of identification of an organization's business processes.

Enterprise priority	Critical business process	Supporting critical application	Server platform
Profitability	Sales and activation	LEAPPlus	IBM AIX
		LEAPPlus	Win 2000
Customer retention (measured by Churn)		Provisioning	SunSolaris
		Mediation	HP-UX
		CASH	Win 2000
		MTelB	IBM AIX
		MtelInvoice	IBM AIX
		Mtel API	Win 2000
		Printing and Distribution	Win 2000
Improve overall operation to "industrial-grade"		IN (Pre-Paid)	Siemens
		IN Subscriber Mgmt.	Siemens
		SAP	BM AIX
		Document Archive	Win 2000
		PPP (SIM Card Mgmt)	
		Customer Care Call Center	IBM AIX

Table 3.31   Example of fixing the IT component of each process.

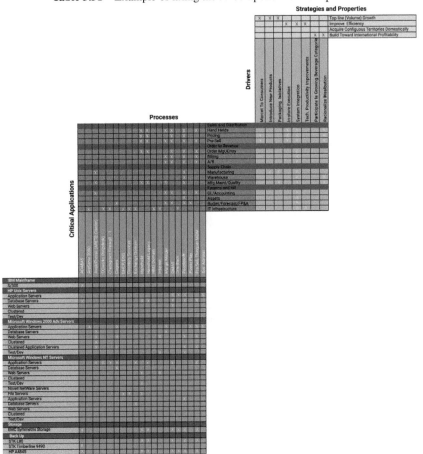

Ensuring the specified level of IT service availability parameters, considering:

- planned technological shutdowns;
- peak load;
- infrastructure upgrades, etc.

Providing a set level of parameters for restoring the functioning of key business processes in emergency situations, including:

- identification of possible risks (Table 3.32) and development of their management schemes (including risks of information security, natural disasters, terrorist attacks, terror, etc.);
- development of standard methods for restoring the functioning of IT services with minimal damage to the business (determining

Table 3.32   Identifying and ranking threats to business continuity.

Type of threat	Probability of incident	Property impact	Business impact	Sub-total	Ability to respond	Total score	Priority for mitigation
High = 5 Low = 1	High = 5 Low = 1	High = 5 Low = 1	High = 5		Poor=5 Good=1		
Environmental disasters:							
Tornado	4,0	3,0	4,0	11,0	4,0	15,0	#5
Hurricane	3,2	1,0	5,0	9,2	4,0	13,2	#7
Flood	5,0	4,0	3,5	12,5	4,0	16,5	#2
Snowstorm	5,0	4,0	5,0	14,0	2,0	16,0	#3
Earthquake	3,0	4,5	5,0	12,5	3,0	15,5	#4
Nuclear disaster	2,0	5,0	5,0	12,0	5,0	17,0	#1
Electrical storm	2,0	2,0	1,0	5,0	2,0	7,0	#8
Fire	4,0	4,0	4,0	12,0	2,5	14,5	#6
Landslide	1,0	4,0	3,5	8,5	4,0	12,5	#7
Loss of utilities and services							
Electrical power failure							
Loss of gas supply							
Coding system outage							
Communications services breakdown							
Equipment failure (excluding IT hardware)							
Serious information security incidents							
Cyber crime							
Computer virus							
IT system failure							
Human error							
Operator error							
Software bugs							
Errors in documenta-tion							
Deliberate disruption							
Act of terrorism							
Act of sabotage							
Act of war							

acceptable time characteristics of recovery, choosing a strategy, developing plans, selecting spare sites, etc.).

The next step is assessment of the business impact analysis. The main steps of this stage include the following:

- clarifying the characteristics of the company's main business processes;
- defining the composition and structure of supporting resources, structures, and operations;
- selection of criteria and indicators of criticality for business processes;
- ranking the organization's business processes by the degree of criticality;
- analysis of relationships between critical business processes;
- assessing the consequences of potential threats to business continuity;
- the definition of the goals and objectives of disaster recovery and recovery time;
- preparing the resulting BIA report;
- presentation of the main results of the BIA analysis to the organization's management.

Table 3.33 performs an example of BIA ratings that allow ranking applications by RTO (the next stage of the EMC methodology).

An example of a visual graphical representation of BIA analysis results is shown in Figure 3.65 and Table 3.34.

The next stages of the EMC methodology are devoted to developing an organization's business continuity management (BCM) strategy (Figures 3.66 and 3.67), as well as developing and implementing appropriate disaster recovery and business continuity plans (DRP and BCP) and appropriate DRP and BCP testing plans.

An approximate plan of the EMC project to create an enterprise BCM program is presented in Table 3.35. Examples of indicators of functional stability of the business processes are given in Table 3.36.

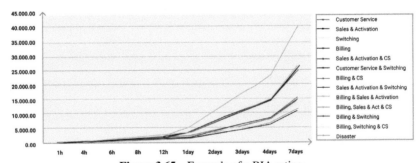

Figure 3.65   Example of a BIA rating.

Table 3.33   Example of a BIA rating.

Type	Event	1 Hour	6 Hours	12 Hours	24 Hours	3 Days	1 Week
Sales and activation	Any outage	64.451	386.708	773.415	1.546.830	4.640.491	10.827.813
Billing	Any outage	87.003	522.013	1.044.026	2.088.051	6.264.154	14.616.359
Customer care call center	Any outage	4.508	27.050	54.100	108.200	324.601	757.403
Combined sales and activation and CS	Any outage	68.960	413.758	827.515	1.655.031	4.965.093	11.585.216
Combined Billing and CS	Any outage	91.510	549.063	1.098.126	2.196.252	6.588.755	15.373.762
Combined sales and activation and switching	Any outage	147.685	886.111	1.772.222	3.544.443	10.633.330	24.811.103
Combined sales and activation and billing	Any outage	151.453	908.720	1.817.441	3.634.882	10.904.645	25.444.172
Combined sales and activation, billing, and CS	Any outage	155.962	935.771	1.871.541	3.743.082	11.229.246	26.201.575
Combined billing and switching	Any outage	170.236	1.021.416	2.042.832	4.085.664	12.256.992	28.599.649
Disaster	Any outage	239.196	956.782	2.870.347	5.740.695	17.222.085	40.184.865

Table 3.34   Application recovery time objective matrix.

Application	Number of business owners using application	RTO	LOB Owner:Trd HouEGT	LOB Owner: Trd Hou Pwt	LOB Owner: Trd PortFol	LOB Owner:Trd Hou EFT	LOB Owner:Trd SLC Gas	LOB Owner: Trd SLC Power	LOB Owner: Trd SLC Trade	LOB Owner:C:CommSvcs	LOB Owner: Trd SLC Admin	LOB Owner:Fin Controlership	LOB Owner:Fin TreasOps
AARC	1	120									120		
ACTI	1	48									48		
Adobe Acrobat	2	2				2		2					
Ancillary Sales Upload	1	1						1					
AOL Instant Messaging	2	0.25				2							
Bloomberg	7	0.25	0.25	0.25		2	8	1	1			48	
Business Objects	6	0.25		0.25	0.25	2	1	1	1	1			
Call A/S Bidding	1	1						1					
CAL ISO WENET	1	1						1					
CBS	1	3											
CCS	5	1					1	24	24	24		48	
Citrix	1	48								1			
Clarus	2	6										6	24
CPT	1	1						1					
CPT-CAL Plant Tracking	1	1				2							
COG	2	2	2										

Time to Recover and Loss by Solution Set

Business Continuity Solution Alternatives

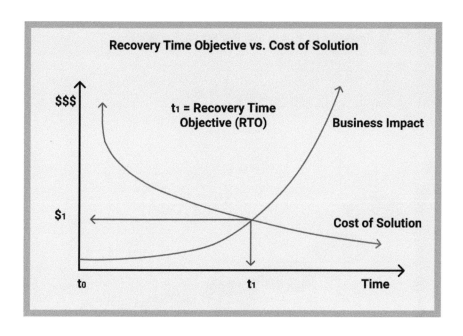

Figure 3.66   Defining BC strategies.

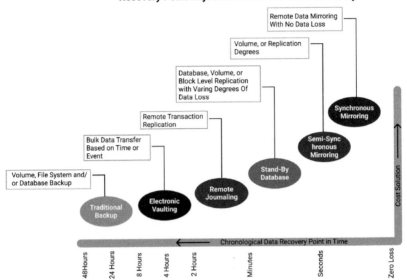

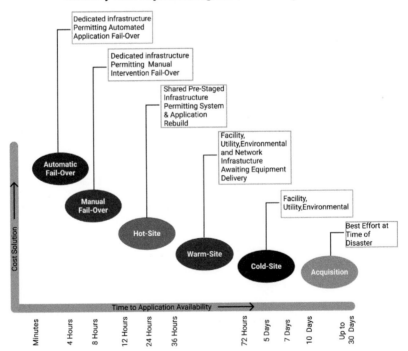

Figure 3.67   Example of defining a BC strategy.

Table 3.35 Sample plan of the EMC project for creating an enterprise BCM program.

Stage 1	Description	Results	Period
Project initiation	• Defining the project scope • Preparation of the project charter • Forming a project team • Conducting an introductory seminar on the goals and objectives of the project	• Approved list of the project team by the EMC and the customer • Project charter • Presentations and working materials of the introductory seminar	5 days

Stage 2	Description	• Results	Period
Assessment of the current state of the enterprise BCM program. **Clarification of the project framework (organizational and functional structure of the organization).**	• Analysis of the existing regulatory and methodological documentation of the organization in terms of business continuity (policies, regulations, standards, procedures, instructions, etc.) • Selective assessment of business continuity management, BCM • Interview transcripts, incident records, briefing logs, etc. • Total assessment of the maturity of the enterprise BCM program based on recommendations COSO, CobiT, ITIL, BS 25999, ISO/IEC 20000, ISO/IEC 27001, etc.	Documented assessment of the current state of the enterprise business continuity management program, BCM	15 days

Stage 3	Description	Results	Period
Business impact analysis, BIA	• Defining and describing key business process chains • Assessment of existing documented and undocumented business requirements for business continuity • Analysis of existing SLA and OLA agreements • Creating business loss scenarios in the event of business process interruptions • Determination of acceptable losses and development of requirements for emergency • restore it services (RTO and RPO)	A documented IA assessment containing the RPO/RTO for disaster recovery of IT services	20 days

Continued

Table 3.35   Continued

Stage 4	Description	Results	Period
Risk assessment, RA	• Identifying and ranking threats and vulnerabilities in business processes • Qualitative and quantitative assessment of business continuity risks • Reasonable choice of risk management methods • Development of draft risk management policies and regulations • Draft risk management policies and regulations • Recommendations for risk management	Documented business continuity risk assessment	20 days

Stage 5	Description	Results	Period
Development of a strategy for business continuity, the BCS	• Defining the specifics of business continuity and accident management recovery of IT services (office distribution, availability of staff involved in the business continuity process, language skills, differences, seismic stability, communication channels, availability of data providers center services, own platforms, etc.) • Documenting business continuity strategies	Documented business continuity management strategy, BCS	20 days

Stage 6	Description	Results	Period
Development and implementation of DRP and BCP plans	• Analysis of BCP and DRP requirements; • Development of draft BCP and DRP plans • Develop BCP and DRP test plans • The implementation of BCP and DRP; • Evaluating the effectiveness of BCP and DRP plans • Support and maintenance of BCP and DRP plans • Preparation of training programs (awareness)	• Draft BCP and DRP plans and related testing plans • Implementation plan • Evaluating the effectiveness of BCP and DRP plans • Regulations for supporting and supporting BCP and DRP plans • Awareness program	40 days

Table 3.36   Examples of indicators of functional stability of a certain business process.

Indicators of quality of functioning	
Stability indicator	**Computation method**
The number of incidents per month does not exceed the threshold value of the threshold number of incidents per month.	1) Maximum total downtime of the IT service, $T = (1 - A) \times N$. A: service availability,%; N: number of service hours* per month. 2) Threshold number of service incidents = T/ Average duration of a service incident *Service time: the time during which the service is available for use. Usually specified in the SLA
Consistent service availability	1) Acceptable downtime percentage, % = Acceptable financial losses during the specified period/Rated-financial losses during downtime during the specified period 2) Availability, $A = 1 - \%$
The current number of registered indicators* does not exceed the threshold value. *Indicators can be: the number of problems or known errors, the number of registered/ approved requests forchanges, the number of unsuccessful test implementations with a return plan, etc.; in short, all that does not directly reduce the level of service	Threshold number of indicators: 1) The 2-, 3-year trend in the number of registered indicators is being built during the period of stable business development. The average number of registered indicators is estimated $<Q> = (g1 + q2 + ...)$/Quantity of months, where $q1, q2, ...$ is the number of registered indicators in the first month, second month, and so on. 2) Estimated deviations $d1, d2, ...$ from the average in the first month, second month, and so on. 3) The average deviation is Estimated, $D = (d1 + d2 + ...)$/number of months. This deviation is declared a threshold. 4) Threshold number of indicators = $<Q>$ + threshold deviation. For frequent events, one can take a shorter time interval and consider trends on a daily basis instead of monthly. For more rare events, one can take quarterly reports.
Current number of active disaster recovery plans	Must be equal to zero for stable operation measure
Undisturbed SLA metrics	Are being measured

3.6   Microsoft Practice

In 1994, Microsoft Corporation adapted the ITIL library and developed a methodology for managing IT services to improve the quality of its own project activities. The mentioned methodology includes two libraries (areas) of knowledge:

- Microsoft Solutions Framework, MSF – to streamline the activities at the design and development stage of information systems;
- Microsoft Operations Framework, MOF – for improving operations at the stage of operation and maintenance of information systems.

As a result, -Microsoft has formed its own view of the IT service lifecycle model (Figure 3.68).

3.6.1   Characteristics of the Approach

At the same time, the best practices of information system design and development were focused on MSF knowledge:

- MSF Process Model;
- MSF Team Model;
- MSF Project Management Discipline;
- MSF Risk Management Discipline;
- MSF Readiness Management Discipline.

Accordingly, MOF's expertise includes best practices in operating and maintaining information systems:

- MOF Process Model for Operations;

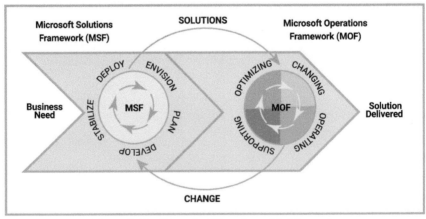

Figure 3.68   IT service lifecycle model, Microsoft.

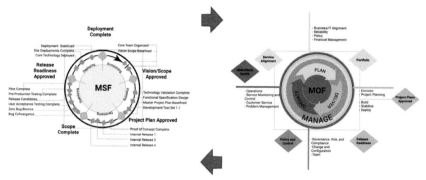

Figure 3.69   Interaction between MSF and MOF.

- MOF Team Model for Operations;
- Risk Management Discipline for Operations;
- Service Management Functions.

The interaction and distribution of authority between MSF and MOF is shown in Figure 3.69.

Note that the mentioned process model, MOF Process Model, is an extension of the standard processes of the ITIL V4 library (books: providing and supporting IT services) and consists of 20 IT management functions, Service Management Functions (SMF). According to Microsoft, if the processes of the ITIL library are abstract enough and need some detail in practice, then the MOF processes are more specific and adapted to practical application.

SMF functions were divided into related categories – quadrants (Figures 3.70 and 3.71), revealing the specific features of the lifecycle model of a certain IT service.

Currently, the fourth version of MOF has been released. The differences from previous versions of MOF are shown in Table 3.37.

3.6.2   ITCM Function

Among the marked processes and functions of MOF, an important place is occupied by the IT service continuity management function (Figure 3.72).

The main steps of the algorithm for implementing the "IT service continuity management" function include the following.

Analysis of requirements for the SMF function "Management of continuity of IT services":

- determining the content and structure of the plan;
- risk computation for each component of the IT service (Table 3.38).

Figure 3.70   MOF process model.

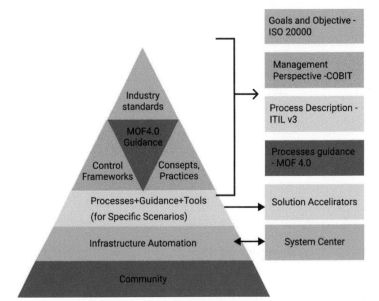

Figure 3.71   Ratio of MOF V4 methodology to COBIT 2019 and ITIL V4.

Table 3.37 Comparison of MOF v4 and MOF v3 versions.

Former name of SMF	MOF v3 quadrant	MOF v4 software area	New SMF name
Managing IT service levels	Optimization	Stage "Planning"	SMF function "business and IT Alignment"
Financial management	Optimization	Stage "Planning"	SMF function "Financial management"
Power control	Optimization	Stage "Planning"	SMF function "Reliability"
Availability management	Optimization	Stage "Planning"	SMF function "Reliability"
Managing the continuity of IT services	Optimization	Stage "Planning"	SMF function "Reliability"
Personnel management	Optimization	Level "Management"	SMF function "Operation team"
Security management	Optimization	Stage "Operation"	SMF function "Reliability"
Creation of infrastructure	Optimization	Stage "Implementation"	SMF function "Preliminary planning"
Change control	Change	Stage "Operation"	SMF function "Change and configuration"
Network administration	Change	Level "Management"	SMF function "Change and configuration"
Managing releases	Operation	Level "Management"	SMF function "Deployment"
System administration	Operation	Stage "Implementation"	SMF function "Operation"
Security administration	Operation	Stage "Operation"	SMF function "Operation"
Configuration management	Operation	Stage "Operation"	SMF function "Monitoring and service level control"
Monitoring and control of services	Operation	Stage "Operation"	SMF function "Operation"
Administration of directory services	Operation	Stage "Operation"	SMF function "Operation"
Storage management	Operation	Stage "Operation"	SMF function "Operation"
Scheduling tasks	Operation	Stage "Operation"	SMF function "Operation"

Continued

Table 3.37   Continued

Former name of SMF	MOF v3 quadrant	MOF v4 software area	New SMF name
Support service	Support	Stage "Operation"	SMF function "Customer service"
Incident management	Support	Stage "Operation"	SMF function "Customer service"
Problem management	Support	Stage «Operation»	SMF function "Problem management"
"Exploitation" by problems»		Level "Management"	SMF-function "Management, risk, and compliance with regulatory requirements"
Group model		Level "Management"	SMF function "Operation team"
		Stage "Planning"	SMF function "Policy"
		Stage "Implementation"	SMF function "Project planning"
		Stage "Implementation"	SMF function "Creation"
		Stage "Implementation"	SMF function "Stabilization"
Required changes assessment	Audit of operations management	Overview of the "Planning" stage	Management analysis "Portfolio"
Assessment of the required changes	Audit of operations management	Overview of the "Implementation" stage	Management analysis "Release ready"
Systematization of data obtained during operation	Audit of operations management	Overview of the "Operation" stage	Management analysis "Operational status"
Analysis of the service level agreement	Audit of operations management	Overview of the "Planning" stage	Management analysis "Service approval"
		Overview of the "Implementation" stage	Management analysis "Approving the project plan"
		Overview of the "Management" stage	Management analysis "Policy and control"

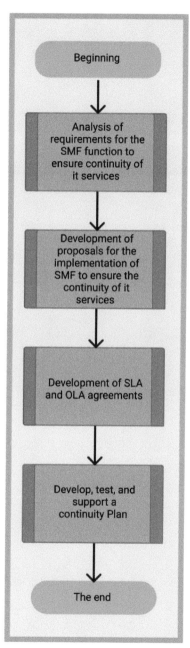

Figure 3.72 Algorithm for implementing the "IT service continuity management" function.

Table 3.38   Example of an IT service risk assessment.

Menaces	Exit	Rooms	Toils	Computer aids	Software system	DBMS, WWW technology, and so on	Approx. software
Fire		Average					
Flood		Low					
Virus attack			High		High		High
Termination of power supply	Average		Average				
Denial of service			Average	Average		Average	
Staff deficit			Average	Average	Average		Average
Terrorist act		Average	Average	Average			

Table 3.39 An example of emergency classification.

Level	Description	Sample
1)	All systems operate in normal mode.	All systems are functioning normally. - and- The client's search time for a phone number is less than 40 seconds.
2)	Slight deterioration in system performance. SLA execution is at risk.	Less than 5% of users cannot log in to the enterprise information system after two login attempts.
3)	The performance characteristics of the system does not allow to match the SLA.	More than 10% of users cannot log in to the enterprise information system after two login attempts. - or - The client's search time for a phone number is more than 60 seconds.
4)	The critical deterioration of the operational characteristics of the system. - or - Damage to more than one system.	More than 20% of users cannot log in to the enterprise information system after two login attempts. - or - The client's search time for a phone number is more than 90 seconds.
5)	Complete system failure.	100% of users cannot log in to the enterprise information system.

Develop proposals for the proper organization of the mentioned SMF function:

- development of an emergency response algorithm;
- the choice of disaster recovery scenarios.

Development of SLA and OLA agreements:
Develop, test, and support a continuity plan.

- defining criteria and indicators for the continuity of IT services;
- developing procedures for escalating problems and notifying employees;
- developing procedures for the initiation and decommissioning of IT services;
- choosing how to interact in emergency situations (Table 3.39);
- documenting countermeasures to ensure continuity.

4

ECP Development Samples

In this chapter, the key issues of enterprise continuity program (ECP) development are considered on the example of a large oil refining company *"Oil-2020"* (fictional name). The IT service *"Active Directory"* was chosen as the object of research and improvement; almost all the key business processes and IT services of the mentioned company depend on it, without exception.

The customer requested to conduct a risk assessment (RA) and business impact analysis (BIA), as well as develop BC strategies and *BCP/DRP* plans, based on the recommendations of the following BCM standards and practices:

- International standard *ISO 22301:2019 – Security and resilience – Business continuity management systems – Requirements)*[73];
- Practices of the *BCI Good Practice Guidelines (GPG2018)*[74];
- Disaster Recovery Institute International database *(DRI) (the Professional Practices for Business Continuity Management 2017 Edition)*[75];
- National standard *NIST SP 800-34 Rev 1 – Content Planning Guide for Federal Information Systems*[76];
- National standard *NIST SP 800-53 Rev 4 – Security and Privacy Controls for Federal Information Systems and Organizations*[77];
- Best practices *of COBIT 2019, RESILIA 2015, ITIL V4*, and *MOF 4.0* in the BCM *part.*

4.1   Characteristics of the Research Object

4.1.1   Current Active Directory Architecture

The current architecture of the Active Directory IT service was formed as a result of the merger of several independent information

---

[73] Available: https://www.iso.org/standard/75106.html

[74] https://www.thebci.org/training-qualifications/good-practice-guidelines.html

[75] www.drii.org

[76] https://nvlpubs.nist.gov/nistpubs/Legacy/SP/nistspecialpublication800-34r1.pdf

[77] https://nvlpubs.nist.gov/nistpubs/SpecialPublications/NIST.SP.800-53r4.pdf

Table 4.1   List of domains of the current Active Directory architecture.

Domain	Name	IP address	Roles
oil.ad	DC1-ROOT	10.1.10.10	GC, DNS, WINS, FSMO: Schema, Domain naming
oil.ad	DC2-ROOT	10.1.10.20	GC, DNS, WINS, FSMO: RID, PDC Emulator, Infra
co.oil.ad	DC0-CO	10.1.30.20	GC, DNS, WINS
co.oil.ad	DC2-CO	10.1.30.30	DNS, WINS, FSMO: Infra
co.oil.ad	DC3-CO	10.64.28.2	GC, DNS, WINS, DHCP,
co.oil.ad	DC4-CO	10.1.30.40	GC, DNS, WINS, FSMO: RID, PDC Emulator

platforms of subsidiaries and represents the so-called "one forest" of Windows 2016:

- Root domain "oil.ad" (does not contain resources and is the custodian of the administrator name and accounts).
- Eleven subdomains, including one subdomain in the Head Office "co. oil.ad."
- Active Directory service is activated on two root domain controllers and two subdomain controllers (3rd and 4th).

For a complete list of domains in the current Active Directory architecture, address Table 4.1.

The features of the current Active Directory architecture include:

- location of Active Directory database files on servers;
- refusing to use dedicated disks and/or storage;
- availability of an approved backup scheme;
- Computationcomputation of the temporal characteristics of the recovery process:

  ○ time to create a backup copy of any server is less than 15 minutes;
  ○ recovery time from a backup does not exceed 30 minutes;
  ○ minimum quantum of backup time is 8 hours.

General replication scheme in the *OIL* forest.*AD* looks as follows (Figure 4.1). The mentioned replication scheme was developed using System Center Operations Manager (SCOM) 2019.

Here, the full lines mark the replication paths that run inside the Moscow/Head Office site.

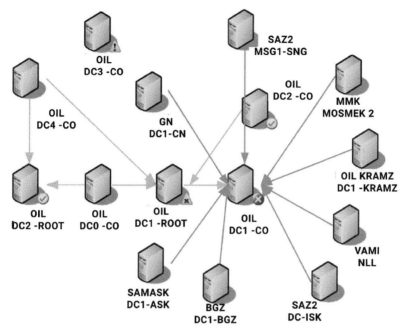

Figure 4.1   Current architecture of the *Active Directory* IT service.

4.1.2   Target Active Directory Architecture

The target architecture of Active Directory is a single-domain structure that is geographically distributed over several sites (according to the topology of narrow-band WAN channels). This highlights in particular the "one forest" of Windows 2016 and two domains:

- root domain "root.ad" for storing the names and accounts of company network administrators;
- subdomain – "hq.root.ad" to serve the company's employees and users.

The list of domains of the target Active Directory architecture is provided in Table 4.2.

It is assumed that each geographically remote site will use one or two domain controllers (depending on the number of users). All remote controllers will belong to the "hq.root" "domain.ad" and store information about the objects of its area of responsibility. In this case, replication will be implemented according to the "star" scheme.

Table 4.2   Role of domains of the current Active Directory architecture.

Domain	Name	Roles
root.ad	ROOTDC0	GC, DNS, WINS, FSMO: Schema, Domain naming
root.ad	ROOTDC3	GC, DNS, WINS, FSMO: RID, PDC Emulator, Infra
hq.root.ad	MSK-DC1	GC, DNS, WINS
hq.root.ad	MSK-DC2	GC, DNS, WINS, DHCP, FSMO: RID, PDC Emulator, Infra

Features of the current Active Directory architecture include:

- Location of Active Directory database files on servers; dedicated disks and/or storage are not used.
- SMOM 2019 server to monitor service continuity and notify of failures.
- Deploy a backup and restore system based on *Symantec Veritas Netbackup version 8.2* software. Before this event, the service data is backed up using the standard utility *wbadmin.exe* from the Windows 2016 tools.
- There are regulations for the normal mode of operation of the backup system. At the same time, the assessment of the temporary characteristics of the recovery process was made. The staff reservation scheme has been developed. The minimum quantum of time for backup is set at 8 hours.

4.2 BIA Example

4.2.1 Classification of Active Directory Processes and Services

The company distinguishes the following priorities for restoring IT services:

- *higher* (recovery period should not exceed 2 hours);
- *high* (recovery time – 4 hours);
- *standard* (recovery time – 8 hours);
- *low* (recovery time – 2 business days);
- *zero* (recovery period is 5 business days).

Therefore, the priority for restoring Active Directory IT services was defined as follows (Table 4.3). Following are the IT services:

- highest priority, highlighted in red;
- high priority, highlighted with a pink background;

Table 4.3 Classification of criticality of Active Directory IT services.

The IT service	Continuity requirements (RTO)	
	During business hours	**Outside of business hours**
Authenticate users	2 hours	4 hours
The Global Catalog Service	4 hours	4 hours
Storing policies and scripts	8 hours	16 hours
WINS and DNS service based on Active Directory for internal users	4 hours	16 hours

- standard priority, highlighted in yellow background;
- low or zero priority is not selected.

The survey identified a list of business services that depend on the functioning of the Active Directory service (Table 4.4).

4.2.2 Calculating RTO and RPO

The general model of the Active Directory IT service is presented in Figure 4.2.

Here, the Active Directory resources are allocated as providing resources:

- root domain controller(s);
- controller(s) of subdomains (including Head office);
- infrastructure for providing sites and environments;
- service personnel.

The following Active Directory services are highlighted:

- infrastructure of the enterprise local area network;
- enterprise WAN network infrastructure;
- power supply infrastructure.

The general matrix of resources and services is presented in Table 4.5.

The classification of criticality of *Active Directory* resources and services is shown in Table 4.6.

4.2.3 Active Directory Interrupt Scenarios

The analysis of the possible interruption scenarios was performed separately for uncorrelated and correlated scenarios (involving more than one resource in an incident). Internal and external threats [ISO/IEC

Table 4.4   List of business services that depend on Active Directory.

Business service	Support time	Users
Accounting information		
Admission order	9 × 5 9:00-18:00	All employees
Telephone directory	9 × 5 9:00-18:00	All employees
Workplace maintenance	9 × 5 9:00-18:00	All employees of the Central office
The Internet	9 × 5 9:00-18:00	Factories, Central office
Service desk	9 × 5 9:00-18:00	Factories, Central office
Documentum system	9 × 5 9:00-18:00	Facility Manager Divisions
Memos (Outlook)	9 × 5 9:00-18:00	All companies of the Company
Cognos BI system	9 × 5 9:00-18:00	All factories, Central office
Reporting to the Director-General	9 × 5 9:00-18:00	Administration company, enterprises of Central management
Emergency incident system	9 × 5 9:00-18:00	Safety department, DC, dispatchers, and security officers on duty
Management accounting base	9 × 5 9:00-18:00	HR
Electronic archive	9 × 5 9:00-18:00	Financial settlements center
Project committee	9 × 5 9:00-18:00	Top management
The system of personnel evaluation	9 × 5 9:00-18:00	All employees
Bank-client	9 × 5 9:00-18:00	Current Accounts Department
Almer	9 × 5 9:00-18:00	Management company, factories
Non-core assets of the company	9 × 5 9:00-18:00	The property administrative department
SAP S/4HANA	9 × 5 9:00-18:00	Factories, Central office
Financial and tax consolidation (OFA)	9 × 5 9:00-18:00	Factories, Central office
Bang. accounting and household management activity (1C)	9 × 5 9:00-18:00	Factories, Central office
Accounting for securities and bills of exchange (1C)	9 × 5 9:00-18:00	Management company, MC Divisions, etc. Companies
Accounting for payment tasks (1C-payments)	9 × 5 9:00-18:00	Management company, MC Divisions, etc. Companies
Payment tracking (1C)	9 × 5 9:00-18:00	Management company, MC Divisions, etc. Companies
Payroll and HR accounting (1C)	9 × 5 9:00-18:00	Management company, MC Divisions, etc. Companies

Continued

Table 4.4 Continued

Business service	Support time	Users
Accounting office (1C)	9 × 5 9:00-18:00	Management company, MC Divisions, etc. Companies
PlanDesigner budgeting system	9 × 5 9:00-18:00	All employees
Media overview	9 × 5 9:00-18:00	All employees
CEO's page	9 × 5 9:00-18:00	All employees
Online conference	9 × 5 9:00-18:00	All employees
The book order (internal)	9 × 5 9:00-18:00	All employees
Mail order (external)	9 × 5 9:00-18:00	All employees
Enterprise University	9 × 5 9:00-18:00	HR
Terms dictionary	9 × 5 9:00-18:00	All employees
Charity	9 × 5 9:00-18:00	All employees
Medical portal	9 × 5 9:00-18:00	All employees
Say thank you!	9 × 5 9:00-18:00	All employees
Order system for meeting rooms	9 × 5 9:00-18:00	All employees
Yandex search engine	9 × 5 9:00-18:00	All employees
Distance learning system	9 × 5 9:00-18:00	Factories, Central office
Electronic application system	9 × 5 9:00-18:00	All employees
WSS portal (internal)	9 × 5 9:00-18:00	All employees
Upstream	9 × 5 9:00-18:00	All employees
Intranet portal	24 × 7	All employees
The Bloomberg System	9 × 5 9:00-18:00	Employees of the Central office
System reuters	9 × 5 9:00-18:00	Employees of the Central office
Interfax system	9 × 5 9:00-18:00	Employees of the Central office
Consultant	9 × 5 9:00-18:00	Factories, Central office
Guarantor	9 × 5 9:00-18:00	Factories, Central office
Code	9 × 5 9:00-18:00	All employees
Citrix	9 × 5 9:00-18:00	Factories, Central office
Information record to external media	9 × 5 9:00-18:00	All employees of the Central office
Data recovery	9 × 5 9:00-18:00	All employees of the Central office
Network resources	9 × 5 9:00-18:00	All employees of the Central office
Print	9 × 5 9:00-18:00	All employees of the Central office
Scan	9 × 5 9:00-18:00	All employees of the Central office
Right fax	9 × 5 9:00-18:00	All employees of the Central office
Email	9 × 5 9:00-18:00	Factories, Central office

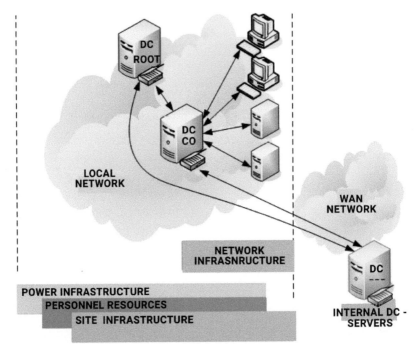

Figure 4.2   General architecture model for the *Active Directory* IT service.

Table 4.5   *Active Directory* matrix of resources and services.

	Authenticate users	The Global Catalog Service	Storing policies and scripts	WINS and DNS for internal users	WINS and DNS for external users
Resources					
Root domain controller(s);	+	+		+	+
Subdomain controller(s)	+	+	+	+	+
Infrastructure for providing sites and environments	+	+	+	+	+
Service personnel	+	+	+	+	+
Service					
The infrastructure of the enterprise local area network	+	+	+	+	+
Enterprise WAN network infrastructure	+	+		+	
Power supply infrastructure	+	+	+	+	+

Table 4.6 *Active Directory* criticality classificator of resources and services.

	RTO	Resource/*service*
0	**2 hours**	Root domain controller(s);
		Subdomain controller(s)
		Infrastructure for providing sites and environments;
		Service personnel.
		The infrastructure of the enterprise local area network;
		Enterprise WAN network infrastructure;
		Power supply infrastructure.
1	**4 hours**	
2	**8 hours**	
3	**>8 hours**	

27001:2013; ISO 22301:2019], natural, man-made, and anthropogenic (both unintended and intentional) impacts are considered.

Uncorrelated scenario of failure of the resource:
The uncorrelated scenarios mentioned (Table 4.7) assume that only one resource (most often one of its components) fails due to a technical failure or expiration of its service life. When using fault tolerance technologies such as *N*-fold redundancy and/or redundancy, it is assumed that only one *Active Directory* component that is being duplicated can fail.

Active Directory uncorrelated scenario of failure of the service:
Uncorrelated scenarios (Table 4.8) assume that only one service fails. Here, measures that significantly reduce the consequences of an interruption include measures to ensure fault tolerance of the service as well as measures to minimize the time to restore the service after an interruption.

Note: The Insert a tag "*Site and Environment Provision Infrastructure*" resource does not imply an uncorrected continuity violation scenario.

Active Directory correlated interrupt scenarios:
The correlated *Active Directory* scenarios mentioned are presented in Table 4.9.

Thus, in the course of the work performed, the following were identified and constructed:

- classification of criticality of services provided by the *Active Directory* service [ISO 22301: 2019];

428   ECP Development Samples

Table 4.7   *Active Directory* uncorrelated scenario of failure of the resource.

ID	1.1 (ROOT-HW)
Scenario	Failure of the root domain controller without damaging the drives
Affected resources	Root domain controller (0)
Factors that affect the likelihood of an incident	• Equipment life • Operating conditions of the equipment (stability of power supply and environmental parameters)
Factors affecting damage from an incident	• Availability of backup root domain controllers • Availability of similar backup equipment for the Company
ID	1.2 (ROOT-HDD)
Scenario	Failure of the root domain controller with damage to the drive that stores OS and program images
Affected resources	Root domain controller (0)
Factors that affect the likelihood of an incident	• Equipment life • Operating conditions of the equipment (stability of power supply and environmental parameters) • Level of protection against malicious attacks from inside the Company's network perimeter
Factors affecting damage from an incident	• Availability of backup root domain controllers • The application of the balancing technology queries • Availability of similar backup equipment for the Company • Availability of a backup copy of "OS+SOFTWARE" • The level of data backup
ID	1.3 (SUB-HW)
Scenario	Failure of the subdomain controller without damaging the drives
Affected resources	The controller subdomain (0)
Factors that affect the likelihood of an incident	• Equipment life • Operating conditions of the equipment (stability of power supply and environmental parameters)
Factors affecting damage from an incident	• Availability of backup domain controllers • Availability of similar backup equipment for the Company

ID	1.4 (SUB-HDD)
Scenario	Failure of the subdomain controller with damage to the drive that stores OS and program images
Affected resources	The controller subdomain (0)
Factors that affect the likelihood of an incident	• Equipment life • Operating conditions of the equipment (stability of power supply and environmental parameters) • Level of protection against malicious attacks from inside the Company's network perimeter
Factors affecting damage from an incident	• Availability of backup domain controllers • The application of the balancing technology queries • Availability of similar backup equipment for the Company • Availability of a backup copy of "OS+SOFTWARE" • The level of data backup

Note: The Insert a tag "*Site and Environment Provision Infrastructure*" resource does not imply an uncorrected continuity violation scenario.

ID	1.5 (PRSNL)
Scenario	Loss of personnel serving the IT service (dismissal, illness, accident, etc.)
Affected resources	Service personnel (0)
Note	*The scenario does not directly disrupt the continuity of the service, however, due to its significant impact on all aspects of ensuring continuity (including the development of other scenarios); it is considered an equal critical resource (GPG-2005, 1.1)*
Factors affecting damage from an incident	• Reserve of qualified personnel • Company personnel reserve • Completeness and relevance of documentation of enterprise technical information • Management policy of password information • Information security policy of the Company

Table 4.8   *Active Directory* uncorrelated scenario of failure of the service.

ID	2.1 (SRVC-LAN)
Scenario	The failure of the enterprise LAN
Affected resources	Enterprise LAN (0)
ID	2.1 (SRVC-WAN)
Scenario	The failure of the enterprise WAN network
Affected resources	Enterprise WAN (0)

- classification of criticality of *Active Directory* resources and related services (CobiT 2019, DSS04), which determines how to restore *Active Directory* in the event of an emergency;
- list of possible *Active Directory* interruption scenarios [ISO 22301:2019; GPG-2018].

4.3   Defining BC Strategies

The statement of the management (policy of business continuity):
The OIL Management company is aware of the need to ensure business continuity and cyber stability of its business processes and supporting IT services as the main obligation to its partners and customers. The OIL Management company is also aware of the importance of implementing BCM practices for the smooth functioning of the information infrastructure and providing services to its partners and customers in the event of emergencies. The policy of the OIL Management company is to create, implement, and support business continuity strategies, BCP/DRP plans, and corresponding BCP/DRP testing plans that are adequate to modern challenges and threats.

4.3.1   General Requirements

Typical threads information for the Active Directory IT service is provided in Table 4.10.

The typical business day for the *Active Directory* IT service is shown in Table 4.11.

Note that technical support for the company's branches that operate in other time zones is provided locally (also by administrators of the Active Directory IT service). The need to involve specialists of the Head Office technical service arises in the event of incidents that are located on the border of the responsibility of the technical services of the Head Office

Table 4.9 Active Directory correlated interrupt scenarios.

ID	3.1 (PWR)
Scenario	Failure of the power supply subsystem
Affected resources	Depending on the point of failure
Factors that affect the likelihood of an incident	• Stability of power supply from the current source • Agreements with the provider about the service level
Factors affecting damage from an incident	• Architecture of the power supply system • Duplicate sources available • Availability of backup sources • Availability of uninterruptible power supply systems • Availability of backup sites for placing equipment
ID	3.2 (ENVT)
Scenario	Failure of the subsystem for providing environmental parameters (air conditioning, etc.)
Affected resources	Depending on the point of failure
Factors that affect the likelihood of an incident	• Equipment life • The working conditions of the equipment (weather conditions) • Stability of power supply
Factors affecting damage from an incident	• Architecture of the environment parameter assurance system • Room characteristics that allow resources and services to function under certain conditions and when the system fails • Availability of backup sites for placing equipment
ID	3.3 (FIRE)
Scenario	Fire in the server room

Continued

Table 4.9   Continued

Affected resources	Depending on the nature of the damage
Factors that affect the likelihood of an incident	• Availability of conditions and/or equipment with increased fire hazard
Factors affecting damage from an incident	• Fire alarm system • Automatic fire extinguishing system • Availability of backup sites for placing equipment
ID	3.4 (FLOOD)
Scenario	Flooding of the server room
Affected resources	Depending on the nature of the damage
Factors that affect the likelihood of an incident	• Availability of conditions and/or equipment with increased fire hazard
Factors affecting damage from an incident	• Fire alarm system • Automatic fire extinguishing system • Availability of backup sites for placing equipment
ID	3.5 (MALWARE)
Scenario	Malicious software (or deliberate actions of an attacker) damages information on drives that store OS and program images
Affected resources	Depending on the nature of the damage
Factors that affect the likelihood of an incident	• Information security policy of the company • Anti-virus software
Factors affecting damage from an incident	• Intrusion detection system • Availability of a backup copy of "OS+SOFTWARE"
ID	3.6 (SRVROOM)
Scenario	Failure of the server room infrastructure with the inability to continue functioning in it for a long time
Affected resources	According to the territorial principle

Factors that affect the likelihood of an incident	
Factors affecting damage from an incident	• Presence of a geographically dispersed cluster solutions • Availability of remote OS backups+software and data storage • Availability of similar backup equipment at other company sites • Availability on other sites of the company, the infrastructure necessary for the operation of the service (connection to the Internet, connection to the WAN network, etc.)
ID	**3.7 (OFFBLOCK)**
Scenario	Failure of the entire building infrastructure (complete or partial destruction, terrorist attack, etc.)
Affected resources	According to the territorial principle
Factors that affect the likelihood of an incident	
Factors affecting damage from an incident	• Presence of a geographically dispersed cluster solutions • Availability of remote OS backups+software and data storage • Availability of similar backup equipment at other company sites • Availability on other sites of the company, the infrastructure necessary for the operation of the service (connection to the Internet, connection to the WAN network, etc.)
ID	**3.8 (SUBDOM)**
Scenario	Failure of the entire infrastructure of a subdomain
Affected resources	The controller subdomain (0)
Factors that affect the likelihood of an incident	• Information security policy of the company • Anti-virus software
Factors affecting damage from an incident	• Presence of a geographically dispersed cluster solutions • Availability of remote OS backups+software and data storage • Availability of similar backup equipment at other company sites • Availability on other sites of the company, the infrastructure necessary for the operation of the service (connection to the Internet, connection to the WAN network, etc.)

Table 4.10   Typical Active Directory information flows.

Flow of information	Character	Duration of transactions
User authentication requests	Around the clock (main – working hours, start of the working day)	Less than 1 minute
Requests for the Global Catalog Service	around the clock (main – working hours)	Less than 1 minute
Requests to get login and logoff scripts and update policies	Working hours (main – beginning of working day)	Less than 1 minute
Domain name service requests from the company network	Working hours	Less than 1 minute

Table 4.11   Typical Active Directory business day.

Time	Action
1:00–1:15	Creating backups of domain controllers
9:00	Getting started with the technical support group
9:00	The beginning of the work of the service administrators
9:00	Start of the working day in Moscow time zone offices (start of the main load)
18:00	Start of the working day in Moscow time zone offices (start of the main load)
18:00	The beginning of the work of the service administrators
20:00	Getting started with the technical support group

and branches of the company. At the same time, only single incidents that required the departure of specialists from the technical support service of the Head Office to the branch during non-working hours were recorded.

4.3.2   Detailed Reading of RTO and RPO

We suggest using the following typical incident management scheme performed in Figure 4.3 in order to determine the necessary and sufficient measures for emergency recovery of Active Directory.

Phase 0 – the incident detection:
It includes the time interval between the incident and the time when the responsible person was informed. An incident can be detected (depending on the nature of the damage) by the service users, administrators, or

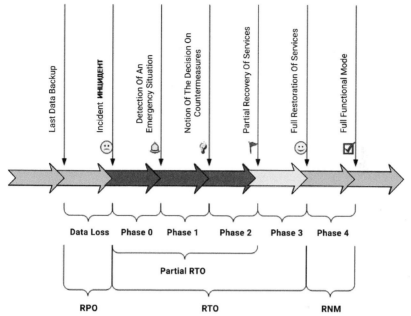

Figure 4.3   Typical event schema for Active Directory.

technical support in manual mode (during planned or unplanned events), administrators, or technical support using automatic monitoring systems. For some resources and services, detection occurs almost immediately after the service fails due to its demand by users. For other resources and services that use them, minimizing the duration of the discovery phase is an important task because operations and transactions rejected during phase 0 increase the amount of data loss; long phase 0 increases the RTO and partial RTO parameters even if there is no data loss (for example, during night hours or during weekends).

The duration of the phase (T_0) is proposed to be equal to the time period from the actual occurrence of the incident to its detection by the stuff of the company's employees ($T_0 - A$):

- for a scenario where an emergency situation is detected by a service user – the average time between consumer requests to the AD service;
- for the automatic monitoring system detection scenario – the AD service survey period;
- for the detection scenario when performing scheduled events (for example, daily or weekly log file viewing) – the period of implementation of the above measures.

The period of time from the incident detection (in most cases, from the receipt of information on the company's ServiceDesk) to informing the responsible person or a substitute specialist ($T_0 - S$):

- if an incident occurs during the shift of the responsible person – 5 minutes;
- if an incident occurs outside the work shift:

  o if the responsible person needs to be present – the time required for their appearance at the workplace or at a remote workplace with the required access to the company's network;
  o if it could be managed by phone – 10 minutes.

$$T_0 = T_{0-A} + T_{0-S}.$$

In this regard, attention should be focused on the need for a qualitative and adequate analysis of the incident detection phase, the introduction of automatic monitoring of resource/service availability in required cases, and the development of interaction processes with ServiceDesk for detecting and partial identification of an incident.

Phase 1 – identification and decision on countermeasures:
This includes the time interval between receiving information about the incident to the responsible person and deciding about the scenario of measures taken to restore IT services and resume business. The duration of the phase largely depends on:

- duration of the destructive effect:
  o for pinhole incidents (hardware failure, power loss, etc.), the response scenario is usually developed fairly quickly;
  o on the contrary, for long-term exposure incidents (fire, viral epidemic, etc.), decision-making may be more difficult;
- incident is included in the list of pre-considered (typical) scenarios, which carries out most of the identification work and, possibly, decision – made by the technical support service and ServiceDesk;
- qualification (training) of ServiceDesk staff for the preliminary identification of an incident and involvement of the specialists in activities according to the service continuity plan.

The duration of the phase (T_1) for typical uncorrelated incidents is proposed to be estimated based on the following time values:

- *10 minutes* – the time required to identify the resource and the specific component of the resource that led to the emergency as well as the damage rates;

- *10 minutes* – the time required to check the workability (unaffected by an accident) of the related service resources.

Accordingly, due to the ability to simultaneously perform the operations, the time is equal to *10 minutes* if there are *two or more* qualified employees and *20 minutes* if there is only one qualified.

Phase 2 – partial service restoration:
The partial service recovery phase is separated from the full one in the following cases:

- making a decision about a scenario with partial service recovery (for example, for consumers that business functions are largely dependent on the IT service functioning);
- ability to continue service functioning without data, which was lost during the time period between the last backup and the point of incident – in this case, the service is considered partially functioning after restoring resources and data backup and fully restored-after data re-entering.

In general case (in the absence of a duplicate resource), the duration of the phase (T_2) is summarized (if necessary, including the possibility of parallel execution of some activities) from the time:

- hardware recovery/replacement ($T_{2-\text{HW}}$);
- operating system and application software installations ($T_{2-\text{SW}}$);
- restore a backup of the operating system and application software ($T_{2-\text{REC}}$);
- restore data storage (possibly partial-only for the most critical users) ($T_{2-\text{DATA}}$).

If there is a duplicate resource, a decision must be made on whether the service with a failed but duplicated resource is fully functional or whether the service is in *Phase 3* until the failed resource is restored. This decision is usually made based on the performance margin of duplicate devices, whether it is sufficient for a business-acceptable quality of service operation. For example, in a redundancy scheme with passive (pending) hardware ("*AA-P,*" "*AAA-P,*" etc.), the service is, in most cases, recognized as fully functioning, and in redundancy schemes without passive hardware ("*AA,*" "*AAA,*" etc.), the solution depends on the load on the hardware before and after the reallocation of functions.

When duplicating with manual switching (or, alternatively, *i*th manual disabling a faulty resource), the duration of the phase (T_2) is assumed to be equal to the duration of the system reconfiguration (T_{2-C}) (usually no more than 20 minutes) and the duration of transients in the system, associated with moving functions from the failed resource to the duplicate (T_{2-A}) (if there is no need to restore backups on the duplicate device, no more than *10 minutes*).

When duplicating with automatic switching, the duration of the phase (T_2) is assumed to be equal to (T_{2-A}).

Phase 3 – partial restoration of services:
When separated from phase 2, the full recovery phase differs in the following points:

- Parameters T_{3-HW}, T_{3-SW}, T_{3-REC}, and T_{3-DATA} for typical scenarios that have a value similar to the parameters T_{2-HW}, T_{2-SW}, T_{2-REC}, and T_{2-DATA} correspond to activities for the remaining (not the most critical) resources.
- Total duration of the phase (T_3) is added to the duration of re-entering data lost from the last recovery point as well as accumulated during the HW-(I/SW)-S (T_{3-R}) phases; in the first approximation, T_{3-R} can be estimated as $(T_{3-R} < RPO)$; however, in practice, the data recovery procedure is usually less than the *RPO* duration. In addition, if appropriate resources are available, the data recovery process can occur in parallel with some other operations.

In the absence of phase 2, the duration (T_3) includes the total costs for the parameters T_{3-HW}, T_{3-SW}, T_{3-REC}, T_{3-DATA}, and the costs for the parameter T_{3-R}.

Phase 4 – switching to normal operation mode:
The recovery phase of normal operation is separated from *Phase 3* in cases where the service infrastructure contains redundant elements and can provide full functionality of business services based on them. However, until the damaged service component is restored, it cannot be recognized as functioning in the normal mode (design-mode) because the redundancy options are exhausted, and in the event of a repeat incident, the calculated fault tolerance characteristics cannot be provided. The duration of the phase (T_4) corresponds to the time of bringing the system architecture to the state corresponding to the design, i.e., complete restoration of the disabled components.

The computation is complex and correlated scenarios.
Correlated scenarios are calculated separately for each damaged resource, considering the priorities defined at the risk analysis stage [OIL-BCP-AD-1]. The features of this computation include the following:

- possibility of parallel implementation of activities, in cases, when they have different resources as the impact object;
- resource constraints that may limit both the parallelization of recovery processes and the overall recoverability of the service (for example, if there is a lack of backup hardware);
- main types of resources that restrictions can affect the recovery process are [ISO 22301:2019, RESILIA 2015]:
  ○ personnel with sufficient qualifications;
  ○ hardware;
  ○ accessories (such as mounting hardware) and infrastructure elements (such as power supplies or fasteners).

The computation of the time parameters of scenarios with a long-term impact (fire, virus epidemic, etc.) is influenced by the uncertainty generated by the source of the destructive impact. For example, an adequate evaluation of damage and selection of solution on the complex of rehabilitation measures, in case of fire, is possible only after the fire suppression; in a viral epidemic with unknown vulnerability – only after blocking the ways to spread a malicious code; in the deliberate actions of the attacker – only after identifying him or blocking his means of influence on the service.

RPO computation:
The RPO value describes the duration of the data loss period in the event of an incident and is usually determined by the time that has elapsed since the last state backup. For resources that do not have a state (stateless), the RPO value is not calculated, and backups are formed after changes in the OS or application software. In some cases, by the decision of business unit representatives, the RPO may include the value partial RTO based on the fact that during phase 3, employees of business units may be forced to re-enter data that could not be processed by the IT service during phases *0-1-2*.

4.3.3 Selection of Technical Solutions

The well-known technical methods for ensuring business continuity and disaster recovery of IT services include the following:

- creation of data backups;
- preparation of a hardware reserve (some percentage of the currently functioning hardware);
- proactive monitoring of the quality of hardware operation (SMART technology, daily memory test, and similar);

- duplicate network cards with connections to different active network hardware;
- application of fault-tolerant data storage and processing technologies in RAM (error correction, memory mirroring, etc.);
- application of fault-tolerant data storage technologies on storage devices (RAID and similar);
- use of network protocols with delivery confirmation (receipt) and session control;
- accumulation of reserve (redundancy) in the cooling system of system units;
- feedback from power supply infrastructure elements and environmental monitoring sensors in order to correctly end sessions and switch to the disabled state and/or reallocate the load to undamaged backup/ backup components;
- cross-training of personnel.

Description of private technical solutions:
Clusterization of servers:
Server clustering provides redundancy and/or hot redundancy options at the application software level. The advantages of a cluster solution include:

- automatic switching to backup resource components in case of hardware failure;
- automatic switching to backup resource components in case of failure of application and system software.

The limitation of cluster solutions is the fact that they provide application layer resilience and are abstracted from the stored data, which requires a separate consideration of the continuity of the data warehouse. In addition to the above advantages, clustering technology will restore the system to its normal state after an incident when a backup component is installed, i.e., it does not require data migration or load regrouping.

Volume Shadow Copy technology:
The Volume Shadow Copy technology, implemented in Microsoft Windows 2016, significantly reduces the recovery time of large-volume backups and, as a result, the RTO value by restoring the last copy, which allows us to recommend its use for backup of large data stores in the service.

Synchronous and asynchronous replication:
In case of implementing a remote backup storage or geographically distributed cluster systems, there is a need to choose the type of replication: synchronous (with remote party confirmation of recording) or asynchronous

(with deferred recording). The advantages and disadvantages of these types of replication include the following types:

- for synchronous replication: replication:
  - 100% reliable data transfer – identity of both data copies;
  - reduced performance due to waiting for transaction confirmation;
  - use of more simultaneous read/write procedures due to waiting for replication confirmation;
  - requires an average reduction of the permissible load from the design-4 times.
- for asynchronous replications:
  - increase of the probability of irreversible loss of unreplicated data if the original is damaged;
  - slight decrease in the speed of the service.

Note that synchronous data replication is recommended for critical data ($RPO = 0$), e.g., for services and resources of the 0th critical category.

Duplication of channels for interaction between resources and services:
Continuity of the service depends on the correct formation of relationships between resources that maximize the use of backup and duplicate resource components in automatic mode. In particular, there is a recommendation to automatically switch to a component that was not damaged by an incident:

- reserving interaction with domain controllers (it is recommended to have two domain controllers for each region connected to other WAN networks);
- reserving interaction with global directory servers (it is recommended to have two global directory servers for each region connected to other WAN networks);
- for subscribers with static local network settings – static specifying two domain name servers;
- for subscribers with DHCP configuration – specify two domain name servers in the DHCP configuration package.

4.3.4 Possible Recovery Strategies

We suggest using the following recovery strategies (Table 4.12) to ensure the continuity of resources used by the Active Directory service.

Here, the parameters (RPO, $pRTO$, RTO, and RNM) for correlated incident scenarios with damage to multiple resources and/or service components with local damage (number of damaged resources: 2–5; expected

Table 4.12   Possible strategies for restoring the Active Directory.

ID	1.1 (ROOT-HW), 1.3 (SUB-HW)			
Scenario	Failure of the root domain controller without damaging the drives			
Affected resources	Root domain controller (0)			
Current parameters	Hardware – standard; an unallocated cold reserve exists.			
	pRTO	**RTO**	**RNM**	**Additional resources**
Strategy 1 *(minimum)* • cold reserve AO	T_0 = 10 min. T_1 = 10 min. T_{2-H} = 15 min. **35 min.**	**35 min.**		• *Cold reserve AO (Unallocated.)*
Strategy 2 • *(current, target)* • duplication ("AA") • cold reserve AO	T_{2-A} = 10 min. **10 min.**	**10 min.**	T_0 = 10 min. T_1 = 10 min. T_{2-H} = 15 min. **35 min.**	• *Duplication* • *Cold reserve AO (Unallocated.)*
ID	1.2 (ROOT-HDD), 1.4 (SUB-HDD)			
Scenario	Failure of the root domain controller with damage to the drive that stores OS and program images			
Affected resources	Root domain controller (0)			
Current parameters	Data backup The time to restore a copy is 25 minutes; after the server is restored to non-authoritarian mode, replication from the undamaged controller will occur automatically.			
	pRTO	**RTO**	**RNM**	**Additional resources**
Strategy 1 *(minimum)* • cold reserve AO	T_0 = 10 min. T_1 = 10 min. T_{2-H} = 15 min. T_{2-B} = 25 min. **1 hour**	**1 hour**		• *Cold reserve AO (Unallocated.)*
Strategy 2 • *(current, target)* • duplication ("AA") • cold reserve AO • backup copying	T_{2-A} = 10 min. **10 min.**	**10 min.**	T_0 = 10 min. T_1 = 10 min. T_{2-H} = 15 min. T_{2-B} = 25 min. **1 hour**	• *Duplication* • *Cold reserve AO (Unallocated.)* • *Backup copying*

Continued

Table 4.12 Continued

ID	1.5 (PRSNL)	
Scenario	Loss of personnel serving the IT service (dismissal, illness, accident, etc.)	
Affected resources	Service personnel (0)	
	Specifications	**Additional resources**
Strategy 1 *(current)*	Duplication of functions, processes of training, and retraining of service personnel by two (lead/backup) or more specialists.	• *Expenses for training and training of reserve personnel*
Strategy 2 *(recommended)*	Creating a knowledge base for existing IT infrastructure	• *Allocation of working time for preparing materials*
Strategy 3 *(current)*	Partial transfer of IT resources management rights to the technical support group, with the possibility of the temporary transfer of full rights in emergency cases.	• *Policy for storing and providing authorization information in emergency cases*
Strategy 4 *(recommended)*	Territorial office of reserve specialists or part of the technical support group in another building or in another city.	• *Duplicate room* • *Policy for storing and providing authorization information in emergency cases*
Strategy 5 *(option)*	Outsourcing of service personnel	• *The cost of outsourcing*
Note	Currently, the global computer networks department of the Company is launching a project to create a 24-hour on-call network technical support service based on the company's branch in Krasnoyarsk.	

characteristics of service recovery: do not exceed business requirements ($pRTO = 2$ hours; $RPO = 1$ day)) calculated, according to the above method with additional restrictions required:

- on the resources of service personnel when performing several recovery tasks in parallel;
- on the backup hardware fleet;

- on the performance of the system restore from backup during the parallel execution of several tasks on the restoration.

4.3.5   Restoring the IT Service

The IT service continuity strategy describes options for restoring the backup infrastructure of the service, while the *business continuity strategy* considers options for the functioning of business processes when it is completely impossible to restore the service.

Current continuity characteristics of related services:
The mentioned continuity characteristics of related services are presented in Table 4.13.

Factors that have a significant impact on the choice of strategy:
The main factors that influence the choice of a backup infrastructure recovery strategy are the following:

- Nature of damage to the server room/building (**SITE**):
  - ("**0**") platform infrastructure is not damaged (for example, the "3.5 MALWARE scenario");

Table 4.13   Current continuity characteristics of related services.

ID	2.1 (SRVC-LAN)
Scenario	The failure of the enterprise LAN
Service	Enterprise LAN (0)
Current state	LAN Head Office is built on the full redundancy principle; the basic units are three server rooms (two in one building of the Head Office; one in the other); the server room is fully connected (3 GB/s); there is an access to the switchboards on the floors of buildings, which allowitching cabinets with active equipment ("AP"), each of the devices connected to the server two optical channels; each server has two network interfaces.
ID	2.1 (SRVC-WAN)
Scenario	The failure of the enterprise WAN network
Service	Enterprise WAN (0)
Current state	Channels enterprise WAN obtained from two independent providers (RTCOMM and TransTK); the "last mile" channels routed independently; boundary active equipment independently; AS a raised area; SLAs with providers have not been concluded; the great divisions of the company have the backup WAN link; the average recovery time of the communication channel in the Central regions – up to 6 hours; in some regions – up to 24 hours.

○ ("**3**") one server room was damaged, but its infrastructure will be restored in the near future (less than 5 days);

○ ("**9**") one server room is damaged, and its infrastructure will be unavailable for a long time (more than 5 days);

○ ("**99**") significantly damaged the building and all available server rooms ("3.7 OFFBLOCK scenario").

■ Extent of damage to the domain/forest infrastructure (**DOMAINS**):

○ ("**2-0**") both root controllers are damaged; the controllers are damaged;

○ ("**0–2**") both subdomain controllers are damaged;

○ ("**2–2**") entire service infrastructure is damaged.

Options ("1–0") and ("0–1") refer to scenarios of damage to the resource level.

Description of private technical solutions:
Minimal service architecture:
Current architecture of the Active Directory IT service is as follows:

• root domain controller ("root.ad");
• controller subdomain ("hq.root.ad").

If necessary, the server data can be installed within a single hardware complex – the subdomain controller "hq.root.ad" + virtual server of the root domain controller on the same hardware in order to minimize the cost of providing auxiliary infrastructure (power supply, computer network, etc.).

Tables 4.14 and 4.15 provide an approximate computation of the cost of hardware for the service in a minimal architecture.

Organization of "last mile" wireless channels:
In case of failure of the "last mile" wire channel due to mechanical damage, it is possible to use pairs of wireless modems or connect to the Internet provider via wireless equipment in order to minimize the time to restore the service (including on a new or backup site).

Organization of backup WAN channels via virtual private network (VPN):
The presence of separate points of interface with the global Internet in large divisions of the company uses protocols for creating cryptographically

Table 4.14 Calculating the cost of hardware.

Components	Equipment	Number of units	Costs, $
Root domain controller (0)	Dell PowerEdge	1	3.400
The controller subdomain	Dell PowerEdge	1	3.400
	Subtotal		**6.800**

Table 4.15   Summary table of the decision characteristics.

Specifications	Value
Full cost of equipment	6.800 $
Share of IT services provided	100% – full range; there is no reduction in performance
The share of the user population	100%
Labor costs at the preliminary stage	The definition of a server room, setting up infrastructure components – 30 min; hardware installation – 15 min; testing the backup platform – 15 minutes. (creating backups – as part of an existing backup process)
The time of restoration of service (partial) – pRTO	From the moment the full set of hardware is available in the server room: • hardware installation – 20 minutes; • restoring OS from an image (Acronis) – 30 minutes in parallel; • recovery from backups (one LTO3 drive) – 2 × 30 minutes = 1 hour. Total: 1 hour 50 min. if there are two specialists.
Service recovery time (full) – RTO	// – //

protected tunnels in emergency cases to form a temporary infrastructure of the WAN network or its individual elements.

The disadvantages of VPN channels over the Internet are dynamically changing total bandwidth, unpredictable delays on the route of packets, and, as a result, the lack of technical ability to provide the specified quality of service.

Analysis of options for continuity strategies for the service:
We suggest using the following strategies (Table 4.15) (MOF-SCM) ("+": the strategy is applicable for this factor value; "–": the strategy is not applicable for this factor value; "": such actions are not required) to ensure the continuity of the Active Directory IT service in case of serious and global incidents.

4.3.6   The Business Recovery

The business continuity strategy regulates measures to partially replace the functionality of a failed service if the duration of service failure exceeds business requirements.

Table 4.16   Analysis of strategy options.

	Room ("SITE")				Domains ("DOMAINS")		
	"0" – not damaged	"3" – one, up to 3 hours	"9" – one, long-term	"99" – both, long-term	"2-0" – root	"0-2" – subdomains	"2-2" is the entire infrastructure
Strategy 1 Moving a site inside a building *Note: the existing architecture allows the service to function on the basis of a single server platform*			+	–	+	+	+
Strategy 2 Creating an ad platform in another company building in Moscow *Note: possibly based on the principles of minimal architecture (p. 5. 3. 2)*			?	+	+	+	+
Strategy 3 Authoritative subdomain recovery, inclusion in the root domain	+	+	+	+	–	+	+
Strategy 4 Authoritative restore of the root domain, connection subdomain	+	+	+	+	+	–	+

Main services provided to businesses:
The main IT services of the Active Directory service are:

- user authentication;
- the Global Catalog Service;
- storing policies and scripts;
- WINS and DNS services based on Active Directory.

Description of private technical solutions:
Third-party LDAP server:
Providing access services to information stored in the Active Directory database is possible using a third-party LDAP server, provided that the information stored in the controllers is periodically replicated (RPO = 1 day or RPO = 4 hours).

Note that some applications may not be fully compatible with a third-party LDAP server. This, together with the fact that there is no already deployed infrastructure for third-party LDAP servers, does not recommend a third-party LDAP server as having advantages over a Microsoft Windows-based server.

Third-party DNS server:
Domain name resolution services can be provided using a third-party DNS server, provided that the information stored in the controllers is periodically replicated (RPO = 1 day or RPO = 4 hours).

The servers of the global networks department running FreeBSD can be used as the DNS server.

Switching to other cryptographic authentication schemes:
If there is a need to authenticate users during a service failure, one can temporarily switch to other authentication schemes if the application software supports them.

In particular, if there is no critical identification of each individual user, and it is only necessary to restrict access to a particular application, it is possible to temporarily grant all users who have access rights to the application a single group account.

Here, when deciding about applying this strategy, one has to:

1. make sure that authentication is technically possible in this application using a different authentication protocol;
2. choose a cryptographic protocol from the available options;
3. decide about the possibility and feasibility of introducing a group account.

Transferring authentication to the network layer:
User authentication can be transferred from the application layer to the network layer of the OSI model through:

- secure network connection settings (for example, over the IPSEC Protocol);
- introduce a policy for connecting to a resource only via a secure channel;
- disable authentication on the resource.

It should be noted that there are significant costs to service personnel for this strategy, which recommends it only for VIP users.

Analysis of options for continuity strategies for the service:
We suggest using the following strategies (Table 4.17) ("+": the strategy is applicable to replace this functionality; "?": the strategy partially replaces this functionality) to partially replace the functionality of the Active Directory IT service in business processes.

Table 4.17 Analysis of options for Active Directory recovery strategies.

	Authorization	Global Catalog	Policies and scripts	DNS	WINS
Strategy 1 Third-party LDAP server		?			
Strategy 2 Third-party DNS server				+	
Strategy 3 Other authentication schemes	?				
Strategy 4 Authentication at the network level	?				

The analysis of strategy options, for ensuring the continuity of the service, revealed a high specificity of the service, which does not allow replacing its functionality to a sufficient extent with any other service. In this regard, we decided to restore the Active Directory service in the event of incidents that significantly damage its architecture using the same architecture scheme as the current one.

The current load level on domain controllers will temporarily restore the service based on two servers (one is the root domain controller "root. ad" and the other, subdomain controller "hq.root.ad"). Service restoration in such a limited hardware solution will, if necessary, launch the service, including outside the server room, providing a sufficient level of measures to protect information and ensure infrastructure (stability of power supply, local network, environmental parameters, etc.).

The service has the highest recovery priority (no more than 2 hours). Alternative options for ensuring the continuity of business services (if it is impossible to restore full or partial IT service) are considered low-acceptable and equal priority:

• third-party LDAP server;
• third-party DNS server;
• other authentication schemes;
• authentication at the network level.

We recommend switching to them only when the recovery process of the main IT service is critical.

Thus, during the development stage of continuity strategies for the Active Directory IT service, the following data were generated and analyzed:

• minimum, current, and target option strategies to ensure the continuity of the individual resources IT service;

- options for strategies to ensure continuity (backup recovery) of the IT service as a whole in the event of serious or global incidents;
- options for strategies for partial replacement of IT service functionality in business processes if it is impossible to restore it in the time required by the business.

A list of factors that influence the choice of a particular continuity strategy is formed, recommendations for decision-making are formulated, and a preliminary assessment of additional resources required for the implementation of strategies (including at the stage of preliminary activities) is made.

Developing a strategy to ensure the continuity of the Active Directory IT service will allow the company to develop and implement a set of measures to reduce potential financial losses caused by violations in the operation of this service.

4.4 BCP Example

The main goals and objectives of BCP are:

- informational and organizational support for the process of managing the restoration of the regular mode of operation of the Active Directory IT service in the implementation of a scenario that falls within the scope of the above ones;
- informational and organizational support for partial replacement of the service functionality in cases of serious and/or global incidents that prevent the service from being restored within the time required by business processes;
- ensuring the availability of the service and its components;
- definition of the preventive measures required to implement strategies to ensure continuity;
- protection of data and service resources in emergency situations;
- reduction of the impact of service downtime on business.

4.4.1 Requirements Analysis

In accordance with the contract for the provision of consulting services dated April 10, 2019, the "Ernst & Young" company provided consulting services in relation to the assessment of the internal control and risk management system of OIL Management company. The purpose of the services was to assist the company's management in evaluating and submitting proposals for improving approaches to the internal control system and enterprise risk management methods. The services were provided in accordance with the *COSO Internal Control Integrated Framework* and *the Combined Code on Enterprise Governance* standards, the *COSO Enterprise Risk Management Integrated Framework* methodology, *the CobIT 2019, and ITSM* methodologies, as well as the requirements of *article 404 and article 302 of the Sarbanes–Oxley Act (SOA) and the Public Company Accounting Oversight Board (PCAOB)*.

In terms of IT risk assessment, Ernst & Young recommends:

".... Develop and implement a comprehensive plan to ensure the continuity of information systems and related business processes of the Company...." It is recommended that the plan reflect the following:

- prioritization of information systems, IT resources, and services, according to the degree of importance and the amount of risk (the amount of financial losses in case of unavailability);
- composition, authority, and responsibility of the management team, as well as functional disaster recovery teams;
- issues of mobilization of the required personnel;
- main contact persons (including name, position, office, mobile, and home phone numbers) and ways to notify them;
- issues of information interaction both within the company and with equipment, suppliers, clients, state control structures, mass media, etc.;
- implementation of critical business processes and placement of required staff to alternative premises;
- list of resources that are minimal for disaster recovery, including archived information, primary documents, forms, templates, etc.;
- disaster recovery requirements for core business processes and IT infrastructure, including priorities and recovery time;
- procedures for restoring functional infrastructure.

Please note that all employees responsible for its implementation must be familiar with this plan, and the plan itself must be approved by the company's top management. "...Copies of the business continuity plan for the company's information systems and related business processes (both printed and electronic) should be available to all employees involved in the process. The effectiveness of the activities included in the plan should be periodically tested (at least on an annual basis). The plan should be reviewed and modified with all changes in operations, organizational structure, business processes, and IT systems that affect recovery capabilities and strategies...."

Results of the audit of the "ensuring continuity of services" sub-process:

Risk P29. Interruption of critical business processes (production process, shipment of finished products, payment processing, financial reporting, etc.) in the event of an emergency.

KR29. 1. A "matrix of services" has been developed, which describes the areas of responsibility of departments for ensuring business continuity, in the context of IT services:

- Department of applied IS.
- Department of hardware and office systems.
- SAP s/4HANA Department.
- Service Desk Project.
- Electronic document management project.
- Internet/intranet portal project.

KR29. 2. The use of a remote data center (city of Beblingen, Germany), connected via two independent providers.

KR26. 2. The use of fault-tolerant systems (geographically distributed clusters, backup systems): for example, for SAP R/3 applications.

KR26. 3. Plans developed to eliminate the consequences of failures.

KR48. 3. Centralized backup system (*Veritas*).

The Risk of R30. There are no alternative ways to perform critical business processes in the event of failure of supporting IT services.

KR30. 1. Radio modems (SkyLink provider) provide communication in the event of cable failure/disconnection of the main provider.

KR30. 2. The use of technology "thin client" to preserve the possibility of remote connection and perform sensitive operations.

KR30. 3. An independent backup pool of email addresses (based on Orange Business Services) is reserved for Management.

KR30. 4. OpenText RightFax System can be used as an alternative to the existing MS Exchange email system.

KR29. 2. The use of a remote data center (city of Beblingen, Germany), connected via two independent providers.

4.4.2 BCP Content and Structure

The continuity plan for the Active Directory IT service was developed, based on the recommendations [ISO 22301:2019; GPG-2018; NIST-34; RESILIA 2015; MOF-SCM] (Figures 4.4 and 4.5) and contains:

- roles of employees involved in recovery (and prevention) operations and their responsibilities before, during, and after an incident;

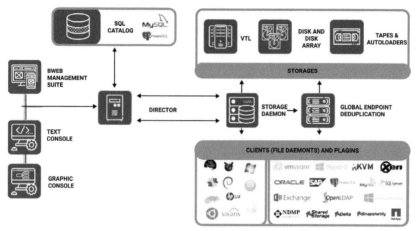

Figure 4.4 Active Directory backup automation scheme.

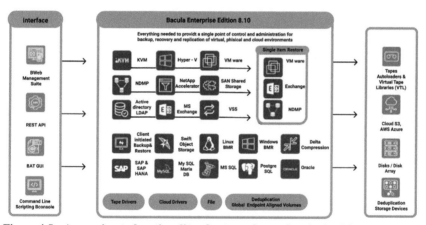

Figure 4.5 **Approximate functionality of automation tools** – Active Directory backup.

- recovery procedures specifying the roles of responsible persons and a sequence of actions graph;
- contact information and procedures for mobilizing responsible employees;
- procedure of information cooperation in the remediation of the procedure of information interaction in the elimination of incidents;
- list of resources minimally required to restore the service;
- test plan of measures for ensuring continuity;
- examples of the tests.

BCP support:

The company's IT Department (Table 4.18) is responsible for maintaining the relevance and availability of the BCP Plan.

Table 4.18   Frequency of updating continuity documentation.

Documentation component	Refresh rate
Risk assessment (BIA and RA)	6 months
Strategy to ensure continuity of service	12 months
Roles and responsibilities of the recovery team	12 months
Emergency procedure	6 months
Recovery team members and contact information	3 months

Maintaining awareness:
The list of people to send the BCP plan includes:

- management of the company's IT Department;
- leadership of the company security service;
- recovery team personnel.

Copies of the set of documents are stored in electronic form at the following address:
http://prometey.int.oil.ru/FD/IT/
and in printed form at the following points:

Moscow, Kamskaya str., 13, p. 1 (Head Office), Forum no. 2, of. 602
Moscow, kasimovskaya str., 1, of. 205.

Roles and responsibilities of the organization's employees:
Emergency recovery team:
It is necessary to create a permanent emergency recovery team to maintain readiness to respond in the event of threats affecting the continuity of the service. The disaster recovery team roles listed below should be assigned to two or more employees (under the "lead/backup" scheme):

- head of the emergency recovery team;
- coordinator of the continuity planning process;
- responsible for backup;
- responsible for remote backup storage (company security);
- responsible for allocating/reserving work sites (including providing them with power supply infrastructure and environmental parameters) (administrative facility department);
- responsible for the operation of the local network;
- responsible for the operation of data transmission channels (WAN, Internet, etc.) (global networks department);
- responsible for the provision of hardware;

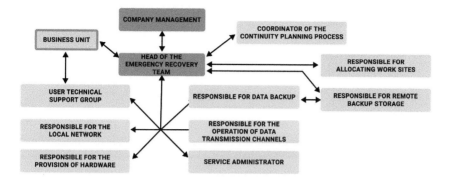

Figure 4.6 Interaction diagram of the organization's emergency recovery team.

- service administrator;
- technical support specialists for users in the event of a violation of the normal operation of the service (technical support group).

The interaction of specialists of the emergency recovery team is shown in Figure 4.6.

Responsibilities of emergency recovery team staff:
Emergency recovery team manager:
Responsibilities before the accident:

- know and understand the entire emergency management process;
- transfer experience to new group members;
- participate in the activities on the revision of policies, plans, and procedures;
- participate in exercises to test the plan.

Responsibilities during the emergency period:

- assess the situation and decide to declare an emergency and authorize plan activation using the emergency management process;
- organize evacuation of people from the building, if required;
- inform the involved *emergency recovery team staff* about the decisions made;
- based on information about damage, make a decision about using the building, placing equipment, restoring, or moving it to another location;
- be the only official source of information about the accident;
- maintain contact with the parent organization and cooperate with local authorities if necessary;

- provide timely information to *the company's management*, public relations department, and *company business divisions*;
- hold regular meetings with *emergency recovery staff* to exchange information on the status of individual tasks and the overall status of recovery efforts;
- authorize the necessary costs to ensure recovery (equipment, materials, temporary third-party assistance in installing, configuring hardware and software, transportation costs, etc.); authorize bypassing regular procurement procedures to speed up the delivery of necessary materials.

Responsibilities after an accident:

- provide details related to the accident and recovery process:
  - ○ *company's management* and authorized persons to conduct an audit of the completed recovery actions;
  - ○ *company's security service* to conduct an investigation;
  - ○ *continuity planning coordinator* to review and improve existing strategies, plans, and procedures.

Continuity planning coordinator:
Responsibilities before the accident:

- initiate the process of initial development of strategies and a continuity plan for the most critical IT services;
- maintain acuteness of strategies and continuity plans when changes occur in the company's business goals, business processes, environment, and the nature and structure of supporting IT processes;
- maintain awareness of all company employees involved in ensuring the continuity of the service;
- initiate periodic testing of continuity procedures.

Responsibilities during the emergency period:

- provide information support to the *emergency recovery manager* on issues related to making decisions about the choice of a particular recovery strategy.

Responsibilities after an accident:

- initiate the process of reviewing strategies, plans, and continuity procedures if their actual implementation during the incident differed from the expected one in certain parameters (duration of operations, quality of partial service restoration, and so on).

Responsible for backup:
Responsibilities before the accident:

- support the backup process in accordance with the company's strategy;
- notify the *emergency recovery manager* of significant changes in backup parameters (volume, duration of copying/restoring, etc.) that require modification of procedures, plans, or continuity strategies.

Responsibilities during the emergency period:

- provide the *emergency recovery manager* with information (type, expiration date, location, estimated recovery time, etc.) about available backups of damaged data;
- participate in the process of restoring backups;
- coordinate the backup recovery process in cases where multiple *emergency recovery specialists* or a *technical support team* perform the recovery process;
- monitor the correctness of the backup recovery process (identity of the original data).

Responsibilities after an accident:

- provide details related to the accident and recovery process to audit actions taken and improve existing strategies, plans, and procedures.

Responsible for remote backup storage:
Responsibilities before the accident:

- provide secure remote storage of backup sets received from the backup manager or the emergency recovery manager.

Responsibilities during the emergency period:

- provide the required set of backup copies of data at the request of the emergency recovery managers within the time limits specified in the regulations.

Responsible for allocating/reserving work sites:
Responsibilities before the accident:

- maintain a list of potential premises for organizing worksites in all company buildings with the following information:
  - free space for equipment placement;
  - free space for staff accommodation;
  - quality of power supply and control of environmental parameters (air conditioning);

○ equipment of the premises with a local network.
- coordinate the selection of the primary backup site with the *emergency recovery manager* and the continuity planning coordinator;
- monitor the compliance of the primary backup site status (in accordance with the continuity strategy) with the required standards.

Responsibilities during the emergency period:

- provide support for the process of moving the service infrastructure to the primary backup site if the *emergency recovery manager* makes such a decision;
- provide informational support for the process of selecting a different backup work site from the prepared list of potential premises, if, for some reason, the primary backup site cannot be used or does not allow providing the required quality of service to the service being moved.

Responsibilities after an accident:

- in case of long-term relocation of the service to the primary backup site in the main status select a new primary backup site;
- provide details related to the process of moving the service to a backup work site to audit the actions taken and improve existing strategies, plans, and procedures.

Responsible for the operation of the local network:
Responsibilities before the accident:

- maintain the state of the local network, in accordance with the company's strategy, for ensuring continuity of services;
- notify the *emergency recovery manager* of significant changes in local network parameters (bandwidth, topology, and ability to automatically redirect traffic) that affect the quality of service continuity.

Responsibilities during the emergency period:

- provide the *emergency recovery manager* with information about the nature of damage to the local network of the Head Office, if it occurred, and an assessment of the parameters of the partial/full network recovery process;
- participate in the process of restoring backups;
- coordinate the backup recovery process in cases where multiple *emergency recovery specialists* or a *technical support team* perform the recovery process;

Responsibilities after an accident:

- provide details related to the accident and recovery process to audit actions taken and improve existing strategies, plans, and procedures.

Responsible for the operation of data transmission channels:
Responsibilities before the accident:

- maintain the state of the local network in accordance with the company's strategy for ensuring continuity of services;
- interact with data transfer service providers to improve the quality and fault tolerance of the service, as well as formalize these parameters;
- notify the *emergency recovery manager* of significant changes in local network parameters (bandwidth, topology, and ability to automatically redirect traffic) that affect the quality of service continuity.

Responsibilities during the emergency period:

- provide the *emergency recovery manager* with information about the nature of damage to the local network of the Head Office, if it occurred, and an assessment of the parameters of the partial/full network recovery process;
- interact with specialists of data transmission service providers in order to restore the normal operation of channels;
- participate in the process of activating backup data channels (including using wireless technologies and/or the Internet), if such a decision is made.

Responsibilities after an accident:

- provide details related to the accident and recovery process to audit actions taken and improve existing strategies, plans, and procedures.

Responsible for the hardware:
Responsibilities before the accident:

- maintain a hardware reserve in accordance with the service continuity strategy;
- maintain a list of equipment that can be exceptionally repurposed to perform the service support function (equipment for secondary services, test equipment, equipment used during the development of the new IT systems, etc.), including in other company buildings.

Responsibilities during the emergency period:

- support the process of replacing failed hardware with backup hardware;

- provide support for the second-level reserve conversion process;
- provide support for emergency purchase of new hardware in required cases.

Responsibilities after an accident:

- fill in the hardware reserve.

The service administrator:
Responsibilities before the accident:

- be aware of all options for service continuity strategies and plans;
- keep the service knowledge base up-to-date (architecture, topology, manual settings, parameters, etc.);
- participate in testing continuity strategies, plans, and procedures.

Responsibilities during the emergency period:

- identify and provide the emergency recovery team manager with information about the nature of damage to the service architecture and assessment of parameters of partial/full restoration process of the service on the basis of its own analysis of the situation and information received from the emergency recovery personnel;
- provide information support for the decision-making process on choosing a service recovery strategy;
- participate in the process of restoring the service, including informing the *emergency recovery manager* of updated estimates of parameters for partial/full service restoration;
- coordinate the backup recovery process in cases where multiple *emergency recovery specialists* or a *technical support team* perform the recovery process;
- monitor the integrity and quality of the restored service.

Responsibilities after an accident:

- provide details related to the accident and recovery process to audit actions taken and improve existing strategies, plans, and procedures.

Technical support specialists for users:
Responsibilities before the accident:

- be aware of all options for service continuity strategies and plans;
- participate in testing continuity strategies, plans, and procedures.

Responsibilities during the emergency period:

- provide information support to users at the service recovery stage according to the scheme chosen by the *emergency recovery manager.*

Responsibilities after an accident:

- provide details related to the accident and recovery process to audit actions taken and improve existing strategies, plans, and procedures.

4.4.3 Management Procedure

Backup control centers:
If it is not possible to manage the restoration of service in the Head Office premises that are normally used for the accommodation of service person-nel, a reservation is provided (possibly unallocated-with re-profiling for the time of restoration) of the premises:

a. in the Head Office building;
b. in a building outside the company's Head Office.

Requirements for a backup control center:

- availability of 24 hours for several days in a row;
  working places (including with LAN support) for seven people;
  city phone;
- fax machine in the immediate vicinity;
- internet connection (possible via modem or wireless connection);
- stock of ordinary office stationery.

Backup control center at the Head Office:
The backup control center in the Head Office is a room _____,
re-profiling is required – yes/no.

Backup control center outside the Head Office building:
The backup control center in the Head Office is a room _____,
re-profiling is required – yes/no.

Phase 0 – incident detection:
During phase 0 (Table 4.19 and Figure 4.7), there is an initial detection of the incident, possibly by persons who are not experts in this matter, an initial assessment of the severity of the damage, and informing the head of the emergency recovery.

Phase 1 – incident identification and strategy selection:
During phase 1 (Figure 4.8 and Table 4.20), the emergency recovery team identifies the nature of the damage, assesses the severity of the damage in detail, and selects a recovery strategy.
    Following is the Phase 1 diagram.

Table 4.19   List of typical actions when detecting a security incident.

Subject	Action	Date–time
Any person	If the incident affects the **safety of people, call special services** – 01 or 112	
Unqualified person	**Call** **Service Desk** – _____ or _____ **Security service** – _____ or _____	
Technical support group	Estimate the duration of service failure, **if it exceeds 1 hour,** to inform the head of the emergency recovery, **otherwise** inform the service administrator or take actions yourself (contact information is in Appendix 1)	
Monitoring system specialist	Estimate the duration of service failure, **if it exceeds 1 hour,** to inform the head of the emergency recovery, **otherwise** inform the administrator of the service or technical support group (contact information is in Appendix 1)	
ServiceDesk service	Inform the service administrator or technical support team (contact information is in Appendix 1)	
The service administrator	Estimate the duration of service failure, **if it exceeds 1 hour,** to inform the head of the emergency recovery, **otherwise** take actions yourself or assign them to the technical support team (contact information is in Appendix 1)	

Phases 2 and 3 – partial and full restoration of the service:

During phases 2 and 3 (Figure 4.9 and Table 4.21), the partial and full service recovery processes are performed, respectively. At the same time, it is possible to identify unforeseen circumstances that may lead to a change in the level of the incident and return to the stage of strategy selection, possibly with the transition to a higher-level plan.

A depiction of the algorithm of Phases 2 and 3 is as follows:

Phase 4 – switching to normal operation mode:

Phase 4 is characterized by providing a full range of services for business processes on the part of the IT service (Table 4.22); however, its internal architecture differs from the regular one (for example, there is no duplication of components).

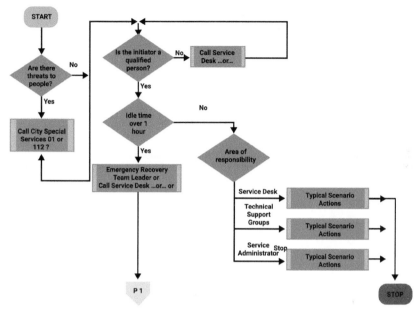

Figure 4.7 Phase 0 algorithm.

Incident recording and analyzing:
At the end of the recovery process (possibly before the end of phase 4), a detailed documentation of the incident and the process of restoring the service from its consequences is required. In addition, analysis of the recovery process may lead to changes in the strategy and plans for ensuring continuity of services (Table 4.23).

4.4.4 BCP Testing

The purpose of the test:
The effectiveness of a business continuity plan can only be verified by a regular testing. Recovery procedures, emergency procedures that are used when the main infrastructure and resources are unavailable, contact information, etc., cannot be considered usable if they have not been tested in the appropriate test conditions.

Testing tasks:
Testing the plan:

- updates your business continuity plans to meet changing business requirements and business conditions;
- allows staff to practice using emergency procedures;

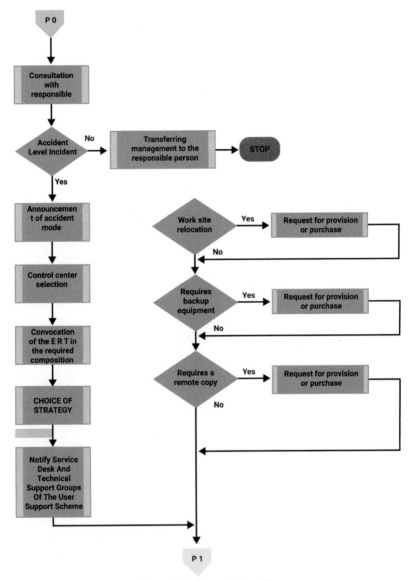

Figure 4.8   Phase 1 algorithm.

- verifies the correctness of BCP documentation and the accuracy of specific procedures;
- demonstrates readiness for a possible accident and gives internal confidence to all interested parties;
- meets the requirements of industry standards and state regulation;
- unites people assigned to the emergency recovery team.

Table 4.20 Order of actions in case of a security incident.

Subject	Action	Date–time
Emergency recovery manager specialist responsible for the damaged component (+Backup solutions specialist)	Analyze the nature of damage Estimate the time of service recovery **If the damage is of the nature of an accident,** then make a decision to declare an accident, **otherwise** transfer incident management to the specialist responsible for the damaged component	
Emergency recovery team manager	Declare about the accident Select the control center to use Announce the meeting of emergency recovery team in the required number of people (contact information is in Appendix 1)	
Emergency recovery team manager	Select a recovery strategy based on [OIL-BCP-NP-2] and based on available resources (Appendix 4). Notify emergency recovery team staff about it	
Emergency recovery team manager	**If the strategy requires moving the work site,** then request the responsible person to allocate a backup work site, as well as identify/get personnel and/or transport to move the equipment	
Emergency recovery team manager	**If the strategy requires using a reserve of equipment,** then request the responsible person to allocate a hardware reserve and/or initiate an emergency hardware purchase process	
Emergency recovery team manager	**If the strategy requires recovery from a deleted backup,** then request the responsible person to deliver the corresponding backup copy of data from the remote storage	
Emergency recovery team manager	Notify ServiceDesk and the technical support team of the adopted recovery strategy and their action plan in case of user requests	

Types of plan tests:

There are four main types of tests used:

- test for checking call lists and inventory;
- desktop test;
- system, application, and data recovery test;
- full test.

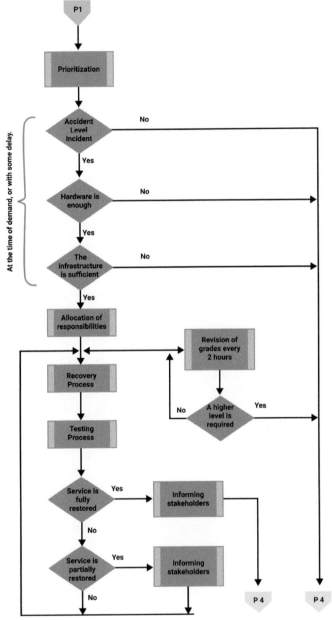

Figure 4.9   The algorithm of phases 2 and 3.

Table 4.21 List of typical actions when restoring the service.

Subject	Action	Date–time
Emergency recovery team manager	Determine recovery priorities	
The service administrator	Check whether the hardware is sufficient at the time of demand (Appendix 4)	
Responsible for backup	Check whether the backup material is sufficient at the time of demand (Appendix 4)	
Emergency recovery team manager	Check whether the work site is provided with infrastructure (power supply, air conditioning, local area network, global network, etc.) (see Appendix 4)	
Emergency recovery team manager	Allocate responsibilities between the staff of emergency recovery team. Start the recovery process	
Emergency recovery team manager	Provide a process for documenting steps and problem situations that occur during the recovery process	
Service administrator; emergency recovery team responsible persons	Recovery process The process of testing the integrity and performance of the restored system (contact information in Appendixes 2 and 3)	
Emergency recovery team manager	Ensure that the recovery time parameters are reviewed at 2-hour intervals. *If the recovery process exceeds the limits of this recovery level,* then make the escalation of measures to ensure continuity *(moving to phase 1 at the "strategy selection" point). If the recovery process has reached the point where the service is partially restored,* then go to phase 3. *If the recovery process has reached the point where the service is partially restored,* then go to phase 3. Ensure that management, ServiceDesk, and technical support are informed of changes.	

Table 4.22   List of typical actions when entering normal operation mode.

Subject	Action	Date–time
Emergency recovery team manager	Determine differences from the standard architecture. Determine component recovery priorities. Get the missing components from the reserve or order them.	
Responsible person; the service administrator	Bring the service architecture to the standard one.	
The service administrator	Check the integrity and fault tolerance of the service.	
Emergency recovery team manager	Notify interested parties of the full transition of the service to normal operation.	

Table 4.23   Documenting the security incident and recovery process.

Subject	Action	Date–time
Emergency recovery team manager; responsible person; the service administrator	Documenting the security incident and recovery process	
Emergency recovery team manager; the plan coordinator	Analysis of the recovery process. Search for discrepancies between expected and observed characteristics	
Emergency recovery team manager	Familiarizing emergency recovery team staff with the results of the incident analysis	
Plan coordinator	Making adjustments to the strategy and continuity plans. Familiarizing emergency recovery team staff with updated versions of the strategy and continuity plans	

Test for checking call lists and inventory:
Definition:

- the BCP sections containing inventory lists, address, and contact information are checked separately.

Goal:

- check and confirm the accuracy and completeness of the information on the date of verification.

Features and limitations:

- It is recommended to run together with the desktop test and not as a separate test.

- It requires minimal effort and resources for planning, preparation, and implementation.

Desktop test:
Definition:

- The entire emergency team gathers together and steps through the plan processes in accordance with the specific scenario of the simulated accident.

Goal:

- demonstration of the effectiveness of the BCP by applying IT to the selected scenario and possibly identify inconsistencies in the plan's procedures;
- check the processes and connections within the emergency team;
- training staff in the application of the plan.

Features and limitations:

- When called, the emergency processes are activated, but the procedures themselves are not executed.
- The scope of testing can include the entire BCP, or any part of IT, process, or division within the organization.
- Testing is safe for the operation of the main production processes and economical use of resources.
- The result has a limited value.

Partial test:
Definition:

- The implementation of the plan's procedures for the rehabilitation of infrastructure ensures the life of the individual subsystems, applications, and their data.

Goal:

- Check the procedures for restoring critical subsystems and backup data within the time allotted by the plan and in accordance with known dependencies between applications and services.

Features and limitations:

- focuses on restoring the IT infrastructure and IT and not business processes;
- requires the participation of IT staff for execution and representatives of business users for verification;
- it can be costly and interfere with normal business activity, with inadequate planning and resource allocation.

Full test:
Definition:

- The emergency team performs BCP processes in accordance with the accepted realistic scenario in full.

Goal:

- determining how well the plan works out the selected accident scenario internally and externally;
- training of personnel in real time and in close to real conditions.

Features and limitations:

- focuses on the plan as a whole, rather than on individual departments, functions, or processes;
- all BCP processes are activated and procedures are performed in full from the beginning to the end (i.e., production systems are actually switched to emergency mode, buildings are evacuated, etc.);
- the scope of testing may include participation/interaction with external parties (suppliers, state, and city emergency services);
- an expensive, time-consuming, and resource-intensive event;
- it could be dangerous for the functioning of the organization if resources are not properly planned and allocated.

Testing program:
The plan testing program is determined by the importance of the business process covered by the plan, its complexity, the complexity of the IT infrastructure supporting the business process, the history of emergencies, and test events.

Depending on these conditions, a sequence of tests is compiled that make up the testing program for a certain period, usually 1–2 years.

Testing process:
Test planning:
Careful test planning is a prerequisite for successful (effective) testing:

- during this phase, the scope, tasks, scenarios, expected results, and initial and final test conditions must be defined and documented.
- the test plan must be agreed with the affected parties and approved by the appropriate level of management.

Preparing the test:
This phase includes the following:

- notification and involvement of all personnel involved in the testing plan;

- conducting a workshop with participants and observers to discuss the testing plan and its rules;
- preparing the necessary resources, permissions, transport, etc.;
- notification of the test date/time to all affected internal departments and external contacts within a reasonable and sufficient time.

Testing:
This phase includes the following:

- performing BCP procedures in accordance with the plan and approved scenario;
- monitoring the procedure execution time and documenting the results obtained;
- documentation of problems encountered and ways to avoid them, and any deviations from the pre-defined plan procedures.

Documenting results:
Fixing the results is very important in the testing process. All participants must submit reports on their work or actions.

The main methods for collecting results are the following:

- identification of problems when performing BCP procedures (including automatic/automated logging of actions);
- documenting the effectiveness of efforts after testing (what was good/bad? what prevented? what deviations from the procedure, and so on).

Report and offers:

- After testing, the test results should be analyzed for problems, omissions, intersections, and interactions with other processes.
- The solution to the identified problems and omissions should be entrusted to the appropriate personnel.
- The solutions found and possible measures to improve the plan's processes and procedures should be collected in the test report and submitted to the emergency team and management for review.
- The BCP plan must be adjusted in accordance with the change procedure.
- The components of the plan that caused the recovery failure (problems) should be considered for re-testing.

Recommendations on the choice of test scenarios:

- Scenarios must correspond to the actual operating conditions.
- Scenarios should be selected according to the BIA results.
- At the beginning, light and partial tests should be planned. If they are successfully completed, one can proceed to gradually complicate and expand the tests up to the full plan test.

- The minimum interval between tests should not exceed 3 months.
- Specialized (focus) testing of BCP can also be used, especially for significant changes in production processes, technologies, and personnel-members of emergency teams.
- Tests and scenarios should be diverse.
- The tests should test documented emergency procedures, not the professionalism, resourcefulness, and dedication of the staff.
- The participation of all participants involved is key factor. All group members should participate, not just the leaders. Introducing key personnel unavailability to the scenario and performing procedures by backup team members helps identifying weaknesses in documenting procedures and sharing experience.
- Tests should not interfere with normal work and, if possible, should be planned for a period of minimal business activity or during off-hours.
- Some tests may be performed without prior notice to staff.
- We recommend including the plan's external interfaces (suppliers, partners, and city emergency services) in testing.

Thus, during the development stage of continuity strategies for the Active Directory IT service, the following data were generated and analyzed:

- content of emergency repair teams;
- roles and responsibilities of company employees involved in recovery operations;
- contact information and procedures for the mobilization of members of the rehabilitation team;
- service and infrastructure recovery procedures that implement IT process and business continuity strategies developed in [OIL-BCP-AD-2];
- issues of information interaction during the recovery process;
- list of the minimum resources required to restore the functionality of the IT service according to the developed strategies;
- principles and procedures for managing and testing a continuity plan.

The development and implementation of the process for ensuring the continuity of the Active Directory IT process and business processes that use its services will allow the company to reduce possible financial losses caused by service failures due to various reasons.

Example 1. List of minimum required resources:
The list of minimum necessary resources describes the following:

- volume of preliminary operations at the planning stage to ensure continuity;

- possible costs in the process of restoring the service.

The overall responsibility for providing the recovery process with the minimum necessary resources is borne by the head of the emergency recovery team. The responsibility for maintaining/reserving/purchasing specific components is borne by specific emergency recovery team specialists.

The resource – level accident:

- hardware (server or its interchangeable part, active or passive network hardware);
- a backup copy of the data if the data storage device fails;
- resource of the personnel (one or two specialists) responsible for the failed component.

Moving a work site (inside a building, to another building, to another city):
The option using a minimal hardware architecture is focused on restoring the service without moving the source code:

- server for the subdomain controller and virtual root domain controller;
- backup copy (disk image) of "subdomain OS" + "virtual server software" + "root domain virtual OS" *(note: the image cannot be obtained from the current architecture – a single additional installation is required)*;
- backup of Active Directory databases (using OS tools) (from the current architecture);
- working platform for hosting one server;
- power supply infrastructure;
- infrastructure for ensuring environmental conditions *(not required under acceptable conditions in the room-ventilation, absence of high humidity, and dust)*;
- infrastructure of the enterprise local area network;
- access to the company's WAN network *(not required for partial service restoration without branch network support)*;
- personnel capable of performing roles:
  - service administrator;
  - responsible for backup;
  - network administrator;
- one staff workplace (no further than 100 m from the work site), equipped with a personal computer connected to a local network, possibly mobile, or a console for temporary work directly on the server being restored.

Moving a work site (inside a building, to another building, to another city):
An option of using an architecture similar to the current one is focused on moving the current hardware:

- server for the root domain controller;
- backup copy of the OS domain controller + Active Directory database (using OS tools);
- server for the subdomain controller;
- backup copy of the OS subdomain controller + Active Directory database (using OS tools);
- work area for hosting two servers and active network equipment;
- power supply infrastructure;
- infrastructure for ensuring environmental conditions *(not required under acceptable conditions in the room-ventilation, absence of high humidity, and dust)*;
- infrastructure of the enterprise local area network;
- access to the company's WAN network *(not required for partial service recovery without support for branch network mailboxes)*;
- personnel capable of performing roles:
  ○ service administrator;
  ○ responsible for backup;
  ○ network administrator;
- two working places for staff (no further than 100 m from the work site), equipped with personal computers connected to a local network, possibly mobile.

Third-party LDAP server:

- ready-to-operate hardware and software complex (one server) with up-to-date replication of the Active Directory database;
- working platform for hosting one server;
- power supply infrastructure;
- infrastructure for ensuring environmental conditions *(not required under acceptable conditions in the room-ventilation, absence of high humidity, and dust)*;
- infrastructure of the enterprise local area network;
- access to the company's WAN network *(not required for partial service restoration without branch network support)*.

Third-party DNS server:

- current replication of the Active Directory domain name database;
- working platform for hosting one server;
- personnel capable of performing roles: network administrator.

Example 2. Desktop test:

«Approved»
Management of "OIL-UK" LLC
Position, Full Name, date.
«Agreed»
Division 1, Full Name, signature, date
Division 2, Full Name, signature, date
Division 1, Full Name, signature, date

The test plan:
Test volume:
Desktop modeling of personnel actions to overcome an accident in the Active Directory IT service that occurred due to the impact of malicious software code on domain controllers "hq.root.ad."

Test tasks:

- estimation of time to complete recovery procedures;
- assessment of compliance with the time characteristics and quality of recovery for the service;
- checking the availability of documentation.

The conditions for the start of the test – scenario:
The accident occurred on July 04 on Tuesday at 12: 30. As a result of exposure to malicious software code launched by someone from the company's employees on the local network, and using an unpublished vulnerability in the implementation of the Kerberos Protocol, on both domain controllers "hq.root.ad," the operating system was damaged, but the servers themselves rebooted, and the Active Directory database files were not damaged. At the time of the accident, most of the employees were in the cafeteria. The last backup copy of the data was made in accordance with the regulations.

Additional details of the situation:
The lead service administrator is unavailable until the end of the week due to this issue. All other employees are at their jobs.

Test completion criteria:
The test ends with a discussion of the procedure for commissioning the system to users.

Resources – participants and roles:
All available members of the emergency recovery team.

Test management tools:

- strategy to ensure continuity of service;
- recovery procedure;
- call list.

Expected test results for the scenario:
It is assumed that all necessary documents and procedures will be available, recovery actions will be simulated, and user service will be restored within 60 minutes.

Place, date, and time of the test:
Meeting room No. _____; "___" _____ 2020 g.; 15:00–16:30

Preparation
Ordering a meeting room: _____.
Copies of the plan and documented procedures: all participants
Notification of test participants by email (1 week in advance):_____.

Conducting the test:
Communicating the scenario to participants: _____.
Discussion of recovery progress: all participants in accordance with the roles defined by the plan.
Responsible for documenting issues that arose during the discussion: _____.

Analysis and conclusions:
Analysis and report on test results: _____
Assign those responsible for solving identified problems:_____.
Preparing the resulting test report: _____.
Updating the plan in accordance with the solutions found: _____.

Example 3 – partial test:

«Approved»
Management of "OIL-UK" LLC
Position, Full Name, date.
«Agreed»
Division 1, Full Name, signature, date
Division 2, Full Name, signature, date
Division 1, Full Name, signature, date

The test plan:
Test volume:
Modeling of personnel actions to overcome a service accident that occurred due to the failure of the air conditioning system of one of the server rooms.

Test tasks:

- checking availability and sufficiency of backup hardware configurations;
- assessment of the degree of documentation of recovery procedures;
- estimation of time to complete recovery procedures;
- checking contacts with external service and support providers;
- assessment of compliance with the quality parameters of the recovery process in emergency conditions;
- checking the availability of documentation.

The conditions for the start of the test – scenario:
The accident occurred on July 04 on Tuesday at 12: 30. The temperature in Moscow is +28°C; by 14:30, it is expected to be +34°C. As a result of overload, the air conditioning system of one of the server rooms failed. The backup air conditioners could not cope with the load due to lack of power. The server room temperature has exceeded the set threshold. The work of several servers in the server terminated abnormally. Restarting the system after identifying the crash and starting the air conditioners (30 minutes from the start of the scenario), an error occurred in the system data area on the hard disk of one of the root domain controllers. The last backup copy of the data was made in accordance with the regulations.

Additional details of the situation:
The lead administrator of the service is on another vacation and is only available by mobile phone. The lead responsible backup specialist is out of the office and unavailable by phone for the first 60 minutes of the test. All other employees are at their jobs.

Test completion criteria – the test ends:

- testing the restored system by a support user or
- if the system cannot be restored without violating the established scenario or
- hours after the start, if the restoration of normal mode was unsuccessful.

Resources – participants and roles:
All available members of the emergency recovery team.

Test management tools:

- continuity strategy and plan;
- recovery procedure;
- call list;
- monitoring and timing of procedures.

Expected test results for the scenario:
Automatic switching of all affected resources and related services to backup components – 10 minutes.

Receiving a signal from the automatic monitoring system about the failure of several servers – 5 minutes.

Notification of responsible persons – 5 minutes

Identifying the problem and deciding on a counter-action strategy – 10 minutes

Preparing backup hardware for the server – 15 minutes

Restoring a backup copy for the server – 25 minutes

Testing the integrity of the restored system – 10 minutes

Performing a functional test (DNS query and test authentication) – 5 minutes

It is assumed that the automatic switching to backup components will occur within 10 minutes, and the return to normal operation (with duplication of all system components) – after 1 hour and 30 minutes.

Place, date, and time of the test:
"___" _____ 2020, beginning at 10:00
Recovery management office _____
Objects-server side _____

Preparation:
Notification of test participants by email (1 week in advance): _____
Providing access to the room for test participants: _____
Copies of the plan, documented procedures: all participants
Ensuring availability of backup equipment: _____

Seminar:
Before the test "___" _____ 2020, from 11 a.m. to 12 a.m. in the room no. _____ a seminar is held for test participants.
Responsible for conducting the seminar (invitations, premises): _____
Discussion of the scope and tasks of the test: _____
Discussion of the test time schedule: _____
Discussion of the script: _____
The discussion of the rules of the test: _____
Questions and answers
Clarification of the test conditions based on the results of the workshop

Conducting the test:
Communicating the scenario to participants: _____

Discussion of recovery progress: all participants in accordance with the roles defined by the plan

Discussion of recovery progress: all participants in accordance with the roles defined by the plan

Monitoring and timing of procedures: _____

Analysis and conclusions:

Report on test results: all participants in accordance with the roles defined by the plan.

The collection of reports of participants and storage of results:_____

Analysis and report on test results: _____

Assign those responsible for solving identified problems:_____

Preparing the resulting test report: _____

Updating the plan in accordance with the solutions found:_____

Set a date for retesting if the recovery was unsuccessful:_____

Conclusion

Dear Reader,

We hope that our book was interesting and beneficial to you!

The relevance of creating and implementing an *enterprise sustainability management program* is explained by the need to maintain and develop your business in the face of emergencies that have a negative impact on the company's operations (and even business suspension). Especially in the transition to the sixth technological order (digital economy) and *Industry 4.0* technologies: *Artificial intelligence (AI), cloud and foggy computing, 5G+, Internet of Things (IoT)/industrial Internet of Things (IIoT), Big Data and ETL, Q-computing, blockchain, virtual and augmented reality (VR/AR)*, etc.

Indeed, in most technologically advanced companies today, business development priorities include:

- minimizing the cyber risks of doing business by protecting the interests in the information sphere and cyberspace;
- provision of safe, trusted, and adequate management of their company;
- planning and supporting business sustainability;
- improving the quality of cybersecurity (cyber stability) activities;
- reducing costs and improving the efficiency of investment in cybersecurity (cyber resilience);
- increasing the level of trust in the company from shareholders, potential investors, business partners, professional participants in the securities market, authorized state bodies, other interested parties, etc.

On the other hand, the requirements for business optimization impose additional requirements and imply the construction of a *cyber-resilient information infrastructure* in the following conditions:

- lack of credit funds for business development;
- review of long-term investments in the company's shares;
- stopping high-risk IT and cybersecurity projects;
- focus on quick impact projects;
- staff reductions of IT services and information security;
- saving and reducing operating and capital costs.

481

Therefore, against this background, the following tasks are relevant in the first order:

- analysis of the effectiveness of the company's business lines, divisions, and employees;
- adequate reallocation of budget funds;
- strengthening control and management tools;
- strengthening financial control.

What would I recommend to responsible persons? First of all, we must maintain prudence and calmness and develop and implement an enterprise business sustainability management program in a timely manner. It is appropriate to recall the famous saying of Winston Churchill: *"Any crisis is a new opportunity."* Indeed, modern challenges and threats of business interruption and the difficult economic situation, among other things, open up new opportunities for cybersecurity and IT services and allow us to implement truly effective and effective measures to ensure cyber stability.

In conclusion, here are some practical recommendations for the development and implementation of effective enterprise business sustainability management programs.

1. In the developed business sustainability management programs, the business needs to consider equally: *regulatory, economic, technological, technical, organizational, and managerial aspects of planning and managing business sustainability*. Only in this case, *a reasonable balance* can be achieved between the *cost and effectiveness* of the planned organizational and technical measures *to ensure cyber stability*.
2. Business sustainability management *programs must not contradict the regulatory framework,* including regulatory documents *(Federal laws, Presidential decrees, Government resolutions, etc.) and regulatory and technical documents (state standards and regulatory guidelines* in the field *of* cybersecurity and cyber stability).
3. In *business sustainability management programs,* it is desirable to consider the following guidelines and recommendations:
   - requirements and recommendations *of NIST Special Publication 800-160 Volume 2. Systems Security Engineering. Cyber Resiliency Considerations for the Engineering of Trustworthy Secure Systems – (Draft), March 2018, etc.*[78];
   - MITRE guidelines *"Cyber Resilience Engineering Aid – the Updated Cyber Resilience Engineering Framework and Guidance*

---

[78] https://www.nist.gov/publications/

on Applying Cyber Resilience Techniques," MTR140499R1, PR 15-1334 (May 2015), etc.[79];

- best practice *in business continuity management in the ISO* series of standards *22301, 22313, 22317, 22318, 22330, and 22331*[80].
- requirements and recommendations of *national standards and practices* for *ensuring business resilience* and building a cyber-resilient *information infrastructure.*

4. When reflecting the economic approach, based on the concept of cyber risk management in the developed business sustainability programs, it is recommended to pay attention to the following methods:
   - application of information analysis, *Applied Information Economics* (AIE);
   - computation of the consumer index, *Customer Index* (CI);
   - computation of the added economic value, *Economic Value Added* (EVA);
   - determining the initial economic value, *Economic Value Sourced* (EVS);
   - asset portfolio management, *Portfolio Management* (PM);
   - assessment of the real opportunities, *Real Option Valuation* (ROV);
   - support for the lifecycle of artificial systems, *System Lifecycle Analysis* (SLCA);
   - computation of the *Balanced Scorecard* (BSC);
   - computation of the *Total Cost of Ownership* (TCO);
   - functional cost analysis, *Activity Based Costing* (ABC), etc.

5. When developing a detailed plan to ensure the business continuity and sustainability of BCP and DRP, it is advisable to use the recommendations and guidelines of two well-known American institutions, SANS Institute[81],[82]. The recommendations of the NIST US 800[83] series standards will also be useful – CIS[84], NSA[85], and others. This will allow, in particular, the following:
   - correctly define the goals of creating a technical architecture for a cyber-resilient business information infrastructure;
   - develop an effective system for sustainability management of businesses through the management of these risks;

---

[79] http://www.mitre.org/sites/default/files/publications
[80] https://www.iso.org
[81] www.sans.org
[82] www.drii.org
[83] www.nist.gov
[84] www.cisecurity.org
[85] www.nsa.gov

- calculate a set of detailed not only qualitative but also quantitative indicators of cyber stability to assess the compliance of the enterprise business sustainability management program with the stated goals;
- select and use the required tools to ensure business sustainability and assess its current state;
- implement the required methods for monitoring and managing cyber resiliency with a sound system of metrics and measures to ensure business sustainability, which will allow objective assessment of the stability of critical business processes, business processes and IT services, and manage business sustainability in the face of growing threats to cybersecurity.

I wish you success in your difficult but interesting work on creating and implementing *enterprise programs for managing business resilience* and building a *cyber-resilient* information infrastructure.

Professor Sergei A. Petrenko
S.Petrenko@innopolis.ru
Russia-Germany
January 2021

References

[1] Z. A. Collier, M. Panwar, A. A. Ganin, A. Kott and I. Linkov, "Security Metrics in Industrial Control Systems," in *Cyber Security of Industrial Control Systems, Including SCADA Systems*, New York, NY, Springer, 2016.

[2] Vorobiev, E. G., Petrenko, S. A., Kovaleva, I. V., Abrosimov, I. K. Analysis of computer security incidents using fuzzy logic, In Proceedings of the 20th IEEE International Conference on Soft Computing and Measurements (24–26 May 2017,), SCM 2017, pp 349–352, St. Petersburg, Russia, 2017.

[3] Vorobiev, E. G., Petrenko, S. A., Kovaleva, I. V., Abrosimov, I. K. Organization of the entrusted calculations in crucial objects of informatization under uncertainty, In Proceedings of the 20th IEEE International Conference on Soft Computing and Measurements (24–26 May 2017). SCM, pp 299–300. DOI: 10.1109/SCM.2017.7970566, St. Petersburg, Russia, 2017.

[4] Abdelzaher T, Kott A. Resiliency and robustness of complex systems and networks. Adaptive, Dynamic and Resilient Systems 2013. P. 67–86.

[5] Appliance of information and communication technologies for development. Resolution of the General Assembly of the UN. Document A / RES / 65/141 dated December 20, 2010 [Electronic resource]. - Access mode: http://www.un.org/en/ga/search/view_doc.asp?symbol=A/RES/65/141.

[6] Bakkensen, L. A., Fox-Lent, C., Read, L. K. and Linkov, I. (2017), "Validating Resilience and Vulnerability Indices in the Context of Natural Disasters". Risk Analysis, 37: 982–1004. doi:10.1111/risa.12677.

[7] Barabanov A.V., Markov A.S., Tsirlov V.L. Methodological Framework for Analysis and Synthesis of a Set of Secure Software Development Controls, *Journal of Theoretical and Applied Information Technology*, 2016, vol. 88, No 1, pp. 77–88.

[8] Barabanov A., Markov A., Tsirlov V. Procedure for Substantiated Development of Measures to Design Secure Software for Automated Process Control Systems. In Proceedings of the *12th International Siberian Conference on Control and Communications* (Moscow, Russia, May 12–14, 2016). SIBCON 2016. IEEE, 7491660, 1–4. DOI: 10.1109/SIBCON.2016.7491660.

[9] Barabanov A., Markov A. Modern Trends in the Regulatory Framework of the Information Security Compliance Assessment in Russia Based on Common Criteria. In *Proceedings of the 8th International Conference on Security of Information and Networks* (Sochi, Russian Federation, September 08–10, 2015). SIN '15. ACM New York, NY, USA, 2015, pp. 30–33. DOI: 10.1145/2799979.2799980.

[10] Barabanov A.V., Markov A.S., Tsirlov V.L. Statistics of Software Vulnerability Detection in Certification Testing. *Journal of Physics: Conference Series*. 2018. V. 1015. P. 042033.

[11] Barabanov A.V., Markov A.S., Tsirlov V.L. Information Security Controls Against Cross-Site Request Forgery Attacks On Software Application of Automated Systems. *Journal of Physics: Conference Series*. 2018. V. 1015. P. 042034.

[12] Boccia, F. (Ed), Leonardi, R. (Ed) (2016). The Challenge of the Digital Economy. Springer Nature Switzerland AG. Part of Springer Nature, 148 p.

[13] Bostick, T. P., Holzer, T. H., & Sarkani, S. (2017). Enabling stakeholder involvement in coastal disaster resilience planning. *Risk Analysis, 37*(6), 1181–1200.

[14] Bostick, T. P., Connelly, E. B., Lambert, J. H., & Linkov, I. (2018). Resilience Science, Policy and Investment for Civil Infrastructure. Reliability Engineering & System Safety *175*:19–23.

[15] Committee on Payments and Market Infrastructures, Bank for International Settlements, "Cyber resilience in financial market infrastructures," November 2014. [Online]. Available: http://www.bis.org/cpmi/publ/d122.pdf.

[16] Committee on Payments and Market Infrastructures, Board of the International Organization of Securities Commissions, "Guidance on cyber resilience for financial market infrastructures - consultative report,"November 2015. [Online]. Available: http://www.bis.org/cpmi/publ/d138.pdf.

[17] Connelly, E. B., Allen, C. R., Hatfield, K., Palma-Oliveira, J. M., Woods, D. D., & Linkov, I. (2017). Features of resilience. *Environment Systems and Decisions, 37*(1), 46–50.

[18] Cyber Resilience and Response, Department of Homeland Security (DHS) 2018, https://www.dhs.gov/sites/default/files/publications/2018_AEP_Cyber_Resilience_and_Response.pdf

[19] Cyber-resilience: Range of practices, Basel Committee on Banking Supervision (December 2018), https://www.bis.org/bcbs/publ/d454.pdf

[20] *Cyber-resilience: the key to business security,* https://www.panda-security.com/mediacenter/src/uploads/2018/05/Cyber-Resilience-Report-EN.pdf

[21] CYBER RESILIENCE ALLIANCE, A Science and Innovation Audit Report sponsored by the Department for Business, Energy and Industrial Strategy, https://swlep.co.uk/docs/default-source/strategy/industrial-strategy/a-science-and-innovation-audit-for-the-cyber-resilience-alliance.pdf?sfvrsn=d1ee7f92_4

[22] Cyber resilience, Specia Report (2017), https://www.acs.org.au/content/dam/acs/acs-documents/ACS%20-%20Cyber%20Resilience%20Special%20Report%20-%2021.06.pdf

[23] Cyber resilience in the digital age. Implications for the GCC region, EY (2017), https://www.ey.com/Publication/vwLUAssets/ey-cyber-resilience-inthe-digital-age-implications-for-the-gcc-region/$File/ey-cyber-resilience-inthe-digital-age-implications-for-the-gcc-region.pdf

[24] D. J. Bodeau, "Analysis Through a Resilience Lens: Experiences and Lessons-Learned (PR 15–1309) (presentation)," in *5th Annual Secure and Resilient Cyber Architectures Invitational*, McLean, VA, 2015.

[25] D. Bodeau and R. Graubart, "Cyber Resiliency Assessment: Enabling Architectural Improvement (MTR 120407, PR 12-3795)," May 2013. [Online]. Available: http://www.mitre.org/sites/default/files/pdf/12_3795.pdf.

[26] Dan Goodin. (27 January 2016). Ars Technica. "Israel's Electric Authority Hit by 'Severe' Hack Attack." Last accessed on 19 April 2017, http://arstechnica.com/security/2016/01/israels-electric-grid-hit-by-severe-hack-attack/.

[27] Department of Homeland Security. (6 December 2016). ICS-CERT. "Advisory (ICSA-16-231-01) Locus Energy LGate Command Injection Vulnerability." Last accessed on 19 April 2017, https://ics-cert.us-cert.gov/advisories/ICSA-16-231-01-0.

[28] DHS, "Assessments: Cyber Resilience Review (CRR)," US-CERT, [Online]. Available: https://www.uscert.gov/ccubedvp/assessments.

[29] DHS, "Cyber Resilience Review (CRR): NIST Cybersecurity Framework Crosswalk," February 2014. [Online]. Available: https://

www.us-cert.gov/sites/default/files/c3vp/csc-crr-nist-framework-crosswalk.pdf.

[30] DHS, "Cyber Resilience Review Fact Sheet," 26 September 2014. [Online]. Available: https://www.dhs.gov/sites/default/files/publications/Cyber-Resilience-Review-Fact-Sheet-508.pdf.

[31] Dorofeev A.V., Markov A.S., Tsirlov V.L. Social Media in Identifying Threats to Ensure Safe Life in a Modern City, *Communications in Computer and Information Science*, 2016, vol. 674, pp. 441–449. DOI: 10.1007/978-3-319-49700-6_44.

[32] E. L. F. Schipper and L. Langston, "A comparative overview of resilience measurement frameworks: analyzing indicators and approaches (ODI Working Paper 422)," July 2015. [Online]. Available: http://www.odi.org/sites/odi.org.uk/files/odi-assets/publications-opinion-files/9754.pdf.

[33] Jerry Rozeman, Roberta Witty, David Gregory, "16 Tips to Enhance Your IT Disaster Recovery Program", 28 February 2020, [Online]. Available: https://www.gartner.com/en/documents/3981595

[34] Roberta Witty, David Gregory, "Leading Through COVID-19: Pandemic Preparedness Requires Strong Business Continuity Management", February 2020, [Online]. Available: https://www.gartner.com/en/webinars/3982036/pandemic-preparedness-requires-strong-business-continuity-manage

[35] Roberta Witty, L Akshay, "Gartner's Business Continuity Management Program Methodology", 12 January 2016, [Online]. Available: https://www.gartner.com/en/documents/3185019

[36] Roberta Witty, "Toolkit: Job Description for the Business Continuity Management Program Leader", 25 October 2016, [Online]. Available: https://www.gartner.com/en/documents/3491618

[37] David Gregory, Roberta Witty, "2020 Strategic Road Map for Business Continuity Management", 21 February 2020, [Online]. Available: https://www.gartner.com/en/documents/3981203

[38] Enterprise Risk Management Research Team, "Introduction to Business Continuity Management", 08 May 2019, [Online]. Available: https://www.gartner.com/en/documents/3913381

[39] John Morency, Roberta Witty, Robert Rhame, Donna Scott, Dave Russell, "Predicts 2016: Business Continuity Management and IT Service Continuity Management", 20 November 2015, [Online]. Available: https://www.gartner.com/en/documents/3170527

[40] Chloe Demrovsky President & CEO, DRI International, "Stronger Together: Deploying cross-sectoral collaboration to overcome the

global preparedness gap", Global Platform for Disaster Risk Reduction Geneva, [Online]. Available: https://www.preventionweb.net/files/globalplatform/5cd5a91dd5fd3Stronger_Together-_deploying_cross-sectoral_collaboration_to_overcome_the_global_preparedness_gap.pdf, https://www.youtube.com/watch?v=ZbThj1HVbJY

[41] Chloe Demrovsky President & CEO, DRI International, "Watch Out For These Global Business Risks In 2020", Jan 15, 2020, [Online]. Available: https://www.forbes.com/sites/chloedemrovsky/2020/01/15/watch-out-for-these-global-business-risks-in-2020/#53eba3883b6f

[42] Petrenko Sergei. Big Data Technologies for Monitoring of Computer Security: A Case Study of the Russian Federation, ISBN 978-3-319-79035-0 and ISBN 978-3-319-79036-7 (eBook), https://doi.org/10.1007/978-3-319-79036-7 © 2018 Springer Nature Switzerland AG, part of Springer Nature, 1st ed. 2018, XXVII, 249 p. 93 illus.

[43] Petrenko Sergei. Cyber Security Innovation for the Digital Economy: A Case Study of the Russian Federation, ISBN: 978-87-7022-022-4 (Hardback) and 978-87-7022-021-7 (Ebook) © 2018 River Publishers, River Publishers Series in Security and Digital Forensics, 1st ed. 2018, 490 p. 198 illus.

[44] Petrenko, A.S., Petrenko S.A., Makoveichuk, K.A., Chetyrbok, P.V. The IIoT/IoT device control model based on narrow-band IoT (NB-IoT), 2018 IEEE Conference of Russian Young Researchers in Electrical and Electronic Engineering (EIConRus), 2018, pp. 950–953.

[45] Petrenko, A.S., Petrenko, S.A., Makoveichuk, K.A., Chetyrbok, P.V. Protection model of PCS of subway from attacks type «wanna cry», «petya» and «bad rabbit» IoT, 2018 IEEE Conference of Russian Young Researchers in Electrical and Electronic Engineering (EIConRus), 2018, pp. 945–949.

[46] Petrenko, S.A., Stupin, D.D. Assignment of semantics calculations in invariants of similarity. 2017 IVth International Conference on Engineering and Telecommunication (EnT), 2017, pp. 127–129.

[47] Petrenko, A.S., Petrenko, S.A., Makoveichuk, K.A., Chetyrbok, P.V. About readiness for digital economy. 2017 IEEE II International Conference on Control in Technical Systems (CTS), 2017, pp. 96–99.

[48] Petrenko, S.A., Makoveichuk, K.A. Ontology of cyber security of self-recovering smart Grid. CEUR Workshop. 2017, pp. 98–106.

[49] Petrenko, S.A., Makoveichuk, K.A. Big data technologies for cybersecurity. CEUR Workshop. 2017, pp. 107–111.

[50] Petrenko, S.A., Petrenko, A.S., Makoveichuk, K.A. Problem of developing an early-warning cybersecurity system for critically important governmental information assets. CEUR Workshop. 2017, pp. 112–117.

[51] Petrenko, S.A. The concept of maintaining the efficiency of cybersystem in the context of information and technical impacts, Proceedings of the ISA RAS, Risk management and safety, Vol. 41, pp.175–193, Russia, 2009.

[52] Petrenko, S.A., Methods of detecting intrusions and anomalies of the functioning of cybersystem, Risk management and safety, Vol. 41, pp. 194–202. Russia- 2009.

[53] Petrenko, S.A. Stability problem of the cybersystem functioning under the conditions of destructive effects, Proceedings of the ISA RAS, Risk management and security, Vol. 52. pp. 68–105, Russia, 2010.

[54] Petrenko, S.A., Methods of ensuring the stability of the functioning of cybersystems under conditions of destructive effects, Proceedings of the ISA RAS, Risk management and security Vol. 52, pp. 106–151, Russia, 2010.

[55] Petrenko, S.A. The Cyber Threat model on innovation analytics DARPA, Trudy SPII RAN, Issue. 39, pp. 26–41, Russia, 2015.

[56] Petrenko, S. A., Petrenko, A. S. New Doctrine of Information Security of the Russian Federation, Information Protection, Inside. No. 1 (73). - pp. 33–39, Russia, 2017.

[57] Petrenko, S. A., Petrenko, A. S. New Doctrine as an Impulse for the Development of Domestic Information Security Technologies // Intellect & Technology, No. 2 (13), pp. 70–75, Russia, 2017.

[58] Petrenko, S.A., Kurbatov, V.A., Bugaev, I.A., Petrenko, A.S. Cognitive system of early cyber - attack warning, Protection of information, Inside, No. 3 (69), pp. 74–82, Russia, 2016.

[59] Petrenko, S. A., Petrenko, A. S. Big data technologies in the field of information security, Protection of information, Inside, No. 4 (70), pp. 82–88, Russia, 2016.

[60] Petrenko, A.S., Bugaev, I.A., Petrenko, S.A. Master data management system SOPKA, Information protection, Inside. No. 5 (71), pp. 37–43, Russia, 2016.

[61] Petrenko, S. A., Petrenko, A. S. Designing the corporate segment SOPKA, Protection of information, Inside. No. 6 (72), pp. 47–52, Russia, 2016.

[62] Petrenko, S. A., Petrenko, A.S. Practice of application the GOST R IEC 61508, Information protection, Insider, No 2 (68), pp. 42–49, Russia, 2016.

[63] Petrenko, S. A., Petrenko, A.S. From Detection to Prevention: Trends and Prospects of Development of Situational Centers in the Russian Federation, Intellect & Technology, No. 1 (12), pp. 68–71, Russia, 2017.

[64] Petrenko, S.A., Stupin, D.D. (2017). National Early Warning System on Cyber - attack: a scientific monograph [under the general editorship of SF Boev] "Publishing House" Athena ", University of Innopolis; Innopolis, Russia, p. 440.

[65] Petrenko, S. A., Petrenko, A. S. Creation of a cognitive supercomputer for the cyber - attack prevention, Protection of information. Inside, No. 3 (75), pp. 14–22, Russia, 2017.

[66] Petrenko, S.A., Asadullin, A. Ya., Petrenko, A.S. Evolution of the von Neumann architecture, Protection of information. Inside. No. 2 (74), pp. 8–28, Russia, 2017.

[67] Petrenko, S. A., Petrenko, A. S. Super-productive monitoring centers for security threats, Part 1, Protection of information. Inside, No. 2 (74), pp. 29–36, Russia, 2017.

[68] Petrenko, S. A., Petrenko, A.S. Super-productive monitoring centers for security threats, Part 2, Protection of information, Inside, No. 3 (75), pp. 48–57, Russia, 2017.

[69] Petrenko, S. A., Petrenko, A. S. Profile of the security of the mobile operating system, Tizen, Information security. Inside, No. 4 (76), pp. 33–42, Russia, 2017.

[70] Petrenko, S.A., Shamsutdinov, T.I., Petrenko, A.S. Scientific and technical problems of development of situational centers in the Russian Federation, Information protection, Inside, No. 6 (72). pp. 37–43, Russia, 2016.

[71] Petrenko, S. A., Petrenko, A. S. The first interstate cyber-training of the CIS countries: "Cyber-Antiterror-2016", Information protection, Inside, No. 5 (71), pp. 57–63, Russia, 2016.

[72] Petrenko, S. A., Petrenko, A. A. Ontology of the cyber-security of self-healing SmartGrid, Protection of information, Inside, No. 2 (68), pp. 12–24, Russia, 2016.

[73] Petrenko, S. A., Petrenko, A. A. The way to increase the stability of LTE-network in the conditions of destructive cyber – attacks, Questions of cybersecurity, No. 2 (10), pp. 36–42, Russia, 2015.

[74] Petrenko, S. A., Petrenko, A.A., Cyberunits: methodical recommendations of ENISA, Questions of cybersecurity, No. 3 (11), pp. 2–14, Russia, 2015.

[75] Petrenko, S. A., Petrenko, A. A. Research and Development Agency DARPA in the field of cybersecurity, Questions of cybersecurity, No. 4 (12), Russia, pp. 2–22, 2015.

[76] Petrenko, S.A. Methods of Information and Technical Impact on Cyber Systems and Possible Countermeasures, Proceedings of ISA RAS, Risk Management and Security, pp. 104–146, Russia, 2009.

[77] Petrenko, S. A., Petrenko, A.A. (2002). Intranet Security audit (Information technologies for engineers), DMK Press, Moscow, Russia, p 416.

[78] Petrenko, S.A., Simonov, S.V. (2004). Management of Information Risks, Economically justified safety (Information technology for engineers), DMK-Press, Moscow, Russia, p.384.

[79] Petrenko, S.A., Kurbatov, V.A. (2005). Information Security Policies (Information Technologies for Engineers), DMK Press, p. 400, Russia, Moscow.

[80] Petrenko, S. A., Petrenko, A.S. (2016). Lecture 12, Perspective tasks of information security, Intelligent Information Radiophysical Systems, MSTU, N. E Bauman; [ed. S.F. Boev, D.D. Stupin, A.A. Kochkarov], Moscow, Russia, pp. 155–166.

[81] Petrenko, S. A., Petrenko, A. S. The task of semantics of partially correct calculations in similarity invariants, Remote educational technologies, Materials of the II All-Russian Scientific and Practical Internet Conference, pp. 365–371, Russia, 2017.

[82] Petrenko, S. A., Petrenko A. A. Information Security Audit Internet/ Intranet (Information Technologies for Engineers), 2 nd ed, DMK-Press, p. 314, Moscow, Russia, 2012.

[83] Petrenko Sergei. Cyber Resilience, ISBN: 978-87-7022-11-60 (Hardback) and 877-022-11-62 (Ebook) © 2019 River Publishers, River Publishers Series in Security and Digital Forensics, 1st ed. 2019, 492 p. 207 illus.

[84] Petrenko Sergei. La Administraciyn De La Ciberseguridad. Industria 4.0: ISBN: 978-84-17445-28-7, EAN: 9788417445287 © 2019, Lugar de ediciyyn: ESPACA, Servicio de Publicaciones de la Universidad de Oviedo; Ediciyn: 1 (13 de noviembre de 2019), 276 p.

[85] Sergei Petrenko [0000-0003-0644-1731] and Khismatullina Elvira [0000-0002-8765-1097]. Cyber-resilience concept for Industry

4.0 digital platforms in the face of growing cybersecurity threats. Software Technology: Methods and Tools, 51st International Conference, TOOLS 2019, Innopolis, Russia, October 15–17, 2019, Proceedings. Editors: Mazzara, M., Bruel, J.-M., Meyer, B., Petrenko, A. (Eds.), eBook ISBN 978-3-030-29852-4, DOI 10.1007/978-3-030-29852-4, Softcover ISBN 978-3-030-29851-7, 420 p. URL: https://www.springer.com/gp/book/9783030298517.

[86] Sergei Petrenko [0000-0003-0644-1731] and Khismatullina Elvira [0000-0002-8765-1097]. Method of improving the Cyber Resilience for Industry 4.0. Digital platforms. Software Technology: Methods and Tools, 51st International Conference, TOOLS 2019, Innopolis, Russia, October 15–17, 2019, Proceedings. Editors: Mazzara, M., Bruel, J.-M., Meyer, B., Petrenko, A. (Eds.), eBook ISBN 978-3-030-29852-4, DOI 10.1007/978-3-030-29852-4, Softcover ISBN 978-3-030-29851-7, 420 p. URL: https://www.springer.com/gp/book/9783030298517.

[87] Petrenko Sergei. Developing a Cybersecurity Immune System for Industry 4.0, ISBN: 978-87-7022-188-7 (Hardback) 978-87-7022-187-0 (Ebook) © 2020 River Publishers, River Publishers Series in Security and Digital Forensics, 1st ed. 2020, 456 p. 297 illus.

[88] Business Resiliency, Ensuring employee safety, maintaining operational continuity, and meeting customer expectations, [Online]. Available: https://www.cisco.com/c/en/us/about/business-continuity.html

[89] Susie Wee, Developer, Business Continuity Planning Today, March 10, 2020, [Online]. Available: https://blogs.cisco.com/developer/business-continuity

[90] Global Business Resiliency (GBR) Program Policy, [Online]. Available: https://www.cisco.com/c/dam/en_us/about/business-continuity/global-business-resiliency-program-policy.pdf

[91] Mario Ruiz –Technical Solutions Architect, Design of Robust Business Continuity Solutions to satisfy Regulatory Guidelines, July 10–14 2016, CiscoLive, Las Vegas, NV, [Online]. Available: https://www.ciscolive.com/c/dam/r/ciscolive/us/docs/2016/pdf/BRKDCN-2300.pdf

[92] Tony Savoy, Enhance IT resiliency and business continuity, IBM Services, ©2020 IBM Corporation, [Online]. Available: https://www.ibm.com/account/reg/at-en/signup?formid=urx-43756

[93] Enhance IT resiliency and business continuity, **©2020 IBM Corporation,** IBM Resiliency Services, [Online]. Available: https://www.ibm.com/uk-en/services/business-continuity?p1=

Search&p4=43700052448792406&p5=b&cm_mmc=Search_
Google-_-1S_1S-_-EP_RU-_-%2Bibm%20%2Bresiliency_b&cm_
mmca7=71700000064861972&cm_mmca8=aud-311016886972:k-
wd-47088086215&cm_mmca9=EAIaIQobChMIldic-
qKOS6QIVAc53Ch08pAPNEAAYASAAEgIs9_D_BwE&cm_
mmca10=426169590952&cm_mmca11=b&gclid=EAIaIQob-
ChMIldicqKOS6QIVAc53Ch08pAPNEAAYASAAEgIs9_D_
BwE&gclsrc=aw.ds

[94] Building the business case for resiliency Linking resiliency to business objectives can help IT professionals make a case for corporate
investment, IBM Global Technology Services, White Paper, ©2015
IBM Resiliency Services, [Online]. Available: https://www.ibm.
com/downloads/cas/WRZGRKKM

[95] Charlotte Brooks Clem Leung Aslam Mirza Curtis Neal Yin Lei Qiu
John Sing Francis TH Wong Ian R Wright, IBM System Storage
Business Continuity: Part 1 Planning Guide, RedBooks, ©2007
IBM Corporation, [Online]. Available: https://www.redbooks.ibm.
com/redbooks/pdfs/sg246547.pdf

[96] Smartha Guha Thakurta, Business Continuity Planning In an
Organisation, EMC Proven Profesional Knowledge Sharing 2009,
[Online]. Available: https://education.dellemc.com/content/dam/
dell-emc/documents/en-us/KS2009_Thakurta-Business_Continuity_
Planning_in_an_Organization.pdf

[97] HPE Helion and Veritas Continuity. Managed Disaster Recovery-
as-a-Service to the HPE Helion OpenStack Managed Cloud, ©2015
Veritas, [Online]. Available: https://www.karma-group.ru/upload/
iblock/518/HP-Helion-and-Veritas-Continuity-EN.pdf

[98] Learn about protecting your cloud topology against disasters, ©2020 Oracle, Architecture Center, Solution PlayBooks,
[Online]. Available: https://docs.oracle.com/en/solutions/design-dr/
#GUID-0F9BB5FE-E49D-4132-BB29-FD5903B5C2D4

[99] Oracle Optimized Solution for Secure Disaster Recovery Highest
Application Availability with Oracle SuperCluster, ©2015 Oracle,
[Online]. Available: http://www.oracle.com/us/solutions/oos/oos4dr-
solbrief-10-2015-final-2737434.pdf

[100] Business Continuity Preparedness Handbook, Managing risk through
proactive planning, ©2016 AT&T, https://www.att.com/Common/
about_us/pdf/business_continuity_handbook.pdf

[101] Petrenko Sergei. Cyber Resilience, ISBN: 978-87-7022-11-60
(Hardback) and 877-022-11-62 (Ebook) © 2019 River Publishers,

River Publishers Series in Security and Digital Forensics, 1st ed. 2019, 492 p. 207 illus.

[102] Beraud P., Cruz A., Hassell S. and Meadows S., "Using Cyber Maneuver to Improve Network Resiliency," in *MILCOM*, Baltimore, MD, 2011.

[103] Colbert, E. J., Kott, A., Knachel III, L., & Sullivan, D. T. (2017). *Modeling Cyber Physical War Gaming* (Technical Report No. ARL-TR-8079). US Army Research Laboratory, Aberdeen Proving Ground, United States.

[104] Collier, Z. A., Linkov, I., DiMase, D., Walters, S., Tehranipoor, M., & Lambert, J. (2014a). Risk-Based Cybersecurity Standards: Policy Challenges and Opportunities. Computer 47:70–76.

[105] Collier, Z. A., Panwar, M., Ganin, A. A., Kott, A., & Linkov, I. (2016). Security metrics in industrial control systems. In *Cybersecurity of SCADA and other industrial control systems* (pp. 167–185). Cham: Springer International Publishing.

[106] Collier, Z. A., Walters, S., DiMase, D., Keisler, J. M., & Linkov, I. (2014b). A semi-quantitative risk assessment standard for counterfeit electronics detection. *SAE International Journal of Aerospace, 7*(1), 171–181.

[107] Connelly, E. B., Allen, C. R., Hatfield, K., Palma-Oliveira, J. M., Woods, D. D., & Linkov, I. (2017). Features of resilience. *Environment Systems and Decisions, 37*(1), 46–50.

[108] Petrenko Sergei. Cyber Resilience, ISBN: 978-87-7022-11-60 (Hardback) and 877-022-11-62 (Ebook) © 2019 River Publishers, River Publishers Series in Security and Digital Forensics, 1st ed. 2019, 492 p. 207 illus.

[109] Sergei Petrenko [0000-0003-0644-1731] and Khismatullina Elvira [0000-0002-8765-1097]. Cyber-resilience concept for Industry 4.0 digital platforms in the face of growing cybersecurity threats. Software Technology: Methods and Tools, 51st International Conference, TOOLS 2019, Innopolis, Russia, October 15–17, 2019, Proceedings. Editors: Mazzara, M., Bruel, J.-M., Meyer, B., Petrenko, A. (Eds.), eBook ISBN 978-3-030-29852-4, DOI 10.1007/978-3-030-29852-4, Softcover ISBN 978-3-030-29851-7, 420 p. URL: https://www.springer.com/gp/book/9783030298517.

[110] Sergei Petrenko [0000-0003-0644-1731] and Khismatullina Elvira [0000-0002-8765-1097]. Method of improving the Cyber Resilience for Industry 4.0. Digital platforms. Software Technology: Methods and Tools, 51st International Conference, TOOLS 2019, Innopolis, Russia,

October 15–17, 2019, Proceedings. Editors: Mazzara, M., Bruel, J.-M., Meyer, B., Petrenko, A. (Eds.), eBook ISBN 978-3-030-29852-4, DOI 10.1007/978-3-030-29852-4, Softcover ISBN 978-3-030-29851-7, 420 p. URL: https://www.springer.com/gp/book/9783030298517.

[111] Petrenko, S.A. Methods of Information and Technical Impact on Cyber Systems and Possible Countermeasures, Proceedings of ISA RAS, Risk Management and Security, pp. 104–146, Russia, 2009.

[112] Petrenko Sergei. Cyber Resilience, ISBN: 978-87-7022-11-60 (Hardback) and 877-022-11-62 (Ebook) © 2019 River Publishers, River Publishers Series in Security and Digital Forensics, 1st ed. 2019, 492 p. 207 illus.

[113] The Cyber Resilience Blueprint: A New Perspective on Security, https://www.symantec.com/content/en/us/enterprise/white_papers/b-cyber-resilience-blueprint-wp-0814.pdf

[114] The MITRE Corporation (ed.), "Fourth Annual Secure and Resilient Cyber Architectures Invitational," 2015. [Online]. Available: http://www.mitre.org/sites/default/files/pdf/2014-Secure-Resilient-Cyber-Architectures-Report-15-0704.pdf.

[115] The MITRE Corporation (ed.), "Third Annual Secure and Resilient Cyber Architectures Workshop," December 2013. [Online]. Available: http://www.mitre.org/sites/default/files/publications/13-4210.pdf.

[116] The MITRE Corporation (ed.), "2nd Secure and Resilient Cyber Architectures Workshop: Final Report," 2012. [Online]. Available: https://registerdev1.mitre.org/sr/2012_resiliency_workshop_report.pdf.

[117] NIST Special Publication 800-160 VOLUME 2. Systems Security Engineering. Cyber Resiliency Considerations for the Engineering of Trustworthy Secure Systems – (Draft), March 2018, https://insidecybersecurity.com/sites/insidecybersecurity.com/files/documents/2018/mar/cs03202018_NIST_Systems_Security.pdf

[118] NIST Special Publication 800-160 VOLUME 3. Systems Security Engineering. Software Assurance Considerations for the Engineering of Trustworthy Secure Systems – (Draft), December 20, 2019

[119] NIST Special Publication 800-160 VOLUME 4. Systems Security Engineering. Hardware Assurance Considerations for the Engineering of Trustworthy Secure Systems – (Draft), December 20, 2020.

[120] NIST SP 800-34. Rev. 1: Contingency Planning Guide for Federal Information Systems Paperback – February 18, 2014 https://www.amazon.com/NIST-Special-Publication-800-34-Rev/dp/1495983706

[121] NIST, Framework for improving critical infrastructure cybersecurity, version 1.1, draft 2, 16 April 2018, https://www.nist.gov/publications/framework-improving-critical-infrastructure-cybersecurity-version-11, or https://doi.org/10.6028/NIST.CSWP.04162018

[122] NIST, "Framework for Improving Critical Infrastructure Security, Version 1.0," 12 February 2014. [Online]. Available: http://www.nist.gov/cyberframework/upload/cybersecurity-framework-021214.pdf.

[123] Logan O. Mailloux, Engineering Secure and Resilient Cyber-Physical Systems (2018), Systems Engineering Cyber Center for Research, US Air Force, https://www.caecommunity.org/sites/default/files/symposium_presentations/Engineering_Secure_and_Resilient_Cyber-Physical_Systems.pdf

[124] Lomako, A. G., Petrenko, S. A., Petrenko, A. S. Model of the Immune System of Stable Computations, In: Information Systems and Technologies in Modeling and Control. Materials of the all-Russian scientific-practical conference, pp. 250–254, Russia, 2017.

[125] Lomako, A. G., Petrenko, S. A., Petrenko, A. S., Representation of perturbation dynamics for the organization of computations with memory, In: Remote educational technologies, Materials of the II All-Russian Scientific and Practical Internet Conference, pp. 355–359, 2017.

[126] Lomako, A. G., Petrenko, S. A., Petrenko, A. S. Realization of the immune system of the stable computations organization, In: Information systems and technologies in modelling and management, Materials of the All-Russian scientific and practical conference, pp. 255–259, Russia, 2017.

[127] IBM Corporation (2018), James Boyles, «Cybersecurity and YOU!! CYBER RESILIENCE - PREPARE FOR WHEN, NOT IF», https://files.nc.gov/ncdit/documents/files/2018%20NCSAM%20Symposium%20-%20Cyber%20Resilience%20-%20IBM.pdf

[128] IBM Corporation (2018), Felicity March, «Cyber Resilience», https://www-05.ibm.com/dk/think-copenhagen/assets/pdf/Studie3_Session2_Speaker4_Felicity_March_IBM.pdf

[129] IBM Corporation (2018), Jean-Michel Lamby Associate Partner - IBM Security, «Cyber Resiliency. Minimizing the impact of breaches on business continuity», https://www-05.ibm.com/be/think-brussels/assets/pdf/Minimizing_the_impact_of_breaches_on_business_continuity_by_Jean_Michel_Lamby.pdf

[130] IBM Corporation (2018), ARNE JACOBSEN, «IBM RESILIENT: INTELLIGENT ORCHESTRATION THE NEXT GENERATION OF

INCIDENT RESPONSE», https://www-05.ibm.com/se/securitysummit/ assets/pdf/IBM_Resilient-Arne_Jacobsen.pdf

[131] IBM "IBM's Smarter Cities Challenge: Boston—Report." Last accessed on 12 April 2017, https://www.smartercitieschallenge.org/ assets/cities/boston-united-states/documents/boston-united-states-full-report-2012.pdf.

[132] IDC (June 2018), Phil Goodwin, Sean Pike, «Five Key Technologies for Enabling a Cyber-Resilience Framework», https://cdn2.hub-spot.net/hubfs/4366404/QRadar/QRadar%20Content/Five%20 Key%20Technologies%20for%20Enabling%20a%20Cyber%20 Resilience%20Framework.pdf?t=1535932423907

[133] Ian Johnson, (15 June 2013). The New York Times. "China's Great Uprooting: Moving 250 Million into Cities." Last accessed on 12 April 2017, http://www.nytimes.com/2013/06/16/world/ asia/chinas-great-uprooting-moving-250-million-into-cities. html?pagewanted=all&_r=0.

[134] INCOSE, "Resilience Engineering," in *INCOSE Systems Engineering Handbook: A Guide for System Life Cycle Processes and Activities, Fourth Edition*, Hoboken, NJ, John Wiley & Sons, 2015, pp. 229–231.

[135] ISO/IEC. (2015). "Smart Cities: Preliminary Report 2014." Last accessed on 12 April 2017, https://www.iso.org/files/live/sites/ isoorg/files/developing_standards/docs/en/smart_cities_report-jtc1.pdf.

[136] Guzik, V. F., Kalyaev, I. A., Levin, I. I. (2016). Reconfigurable computing systems; [under the Society. ed. I.A. Kalyayeva], Publishing house SFU, Rostov-on-Don, p. 472.

[137] Abdelzaher T, Kott A. Resiliency and robustness of complex systems and networks. Adaptive, Dynamic and Resilient Systems 2013. P. 67–86.

[138] Abramov, S. M. History of development and implementation of a series of Russian supercomputers with cluster architecture, History of domestic electronic computers, 2 nd ed., Rev. and additional; color. Ill, Publishing house "Capital Encyclopedia", Moscow, Russia, 2016.

[139] Beraud P., Cruz A., Hassell S. and Meadows S., "Using Cyber Maneuver to Improve Network Resiliency," in *MILCOM*, Baltimore, MD, 2011.

[140] Abramov, S. M. Research in the field of supercomputer technologies of the IPS RAS: a retrospective and perspective. Proc, Proceedings

of the International Conference "Software Systems: Theory and Applications", Publishing house "University of Pereslavl", vol. 1. pp. 153–192. Russia, Pereslavl, 2009.

[141] Biryukov, D. N., Lomako, A. G. Approach to Building a Cyber Threat Prevention System. Problems of Information Security. Computer systems, Publishing house of Polytechnic University, vol. 2, pp. 13–19, St. Petersburg, Russia, 2013.

[142] Biryukov, D. N., Lomako, A. G., Sabirov, T. R. Multilevel Modeling of Pre-Emptive Behavior Scenarios. Problems of Information Security. Computer systems, Publishing house of Polytechnic University, vol. 4, pp. 41–50. St. Petersburg, Russia, 2014.

[143] Biryukov, D. N., Glukhov, A. P., Pilkevich, S. V., Sabirov, T. R. Approach to the processing of knowledge in the memory of an intellectual system, Natural and technical sciences, No. 11, pp. 455–466, Russia, 2015

[144] Biryukov, D. N., Lomako, A. G., Rostovtsev, Yu. G. The appearance of anticipatory systems to prevent the risks of cyber threat realization, Proceedings of SPIIRAS, Issue. 2 (39), pp. 5–25, Russia, 2015.

[145] Biryukov, D. N., Lomako, A. G., Petrenko, S. A. Generating scenarios for preventing cyber – attacks, Protecting information, Inside, No. 4 (76).- 2017.

[146] Biryukov, D. N., Rostovtsev, Y. G. Approach to constructing a consistent theory of synthesis of scenarios of anticipatory behavior in a conflict. Proc. SPIIRAS. 1(38), pp 94–111, Russia, 2015.

[147] DiMase, D., Collier, Z. A., Heffner, K., & Linkov, I. (2015). Systems engineering framework for cyber physical security and resilience. *Environment Systems and Decisions, 35*(2), 291–300.

[148] Eisenberg, D. A., Linkov, I., Park, J., Bates, M., Fox-Lent, C., & Seager, T. (2014). Resilience metrics: Lessons from military doctrines. *Solutions, 5*(5), 76–87.

[149] Ganin, A. A., Massaro, E., Gutfraind, A., Steen, N., Keisler, J. M., Kott, A., Mangoubi, R., & Linkov, I. (2016). Operational resilience: Concepts, design and analysis. *Scientific Reports, 6*, 19540.

[150] Logan O. Mailloux, Engineering Secure and Resilient Cyber-Physical Systems (2018), Systems Engineering Cyber Center for Research, US Air Force, https://www.caecommunity.org/sites/default/files/symposium_presentations/Engineering_Secure_and_Resilient_Cyber-Physical_Systems.pdf

[151] Petrenko, S.A. Stability problem of the cybersystem functioning under the conditions of destructive effects, Proceedings of the ISA

RAS, Risk management and security, Vol. 52. pp. 68–105, Russia, 2010.

[152] Petrenko, S.A. The concept of maintaining the efficiency of cybersystem in the context of information and technical impacts, Proceedings of the ISA RAS, Risk management and safety, Vol. 41, pp.175–193, Russia, 2009

[153] Petrenko, S.A., Methods of detecting intrusions and anomalies of the functioning of cybersystem, Risk management and safety, Vol. 41, pp. 194–202. Russia- 2009.

[154] Petrenko, S.A., Methods of ensuring the stability of the functioning of cybersystems under conditions of destructive effects, Proceedings of the ISA RAS, Risk management and security Vol. 52, pp. 106–151, Russia, 2010

[155] Petrenko, S.A., Stupin, D.D. Assignment of semantics calculations in invariants of similarity. 2017 IVth International Conference on Engineering and Telecommunication (EnT), 2017, pp. 127–129.

[156] Petrenko Sergei. Big Data Technologies for Monitoring of Computer Security: A Case Study of the Russian Federation, ISBN 978-3-319-79035-0 and ISBN 978-3-319-79036-7 (eBook), https://doi.org/10.1007/978-3-319-79036-7 © 2018 Springer Nature Switzerland AG, part of Springer Nature, 1st ed. 2018, XXVII, 249 p. 93 illus.

[157] Petrenko, S.A., Stupin, D.D. (2017). National Early Warning System on Cyber - attack: a scientific monograph [under the general editorship of SF Boev] "Publishing House" Athena ", University of Innopolis; Innopolis, Russia, p. 440.

[158] Petrenko Sergei. Cyber Security Innovation for the Digital Economy: A Case Study of the Russian Federation, ISBN: 978-87-7022-022-4 (Hardback) and 978-87-7022-021-7 (Ebook) © 2018 River Publishers, River Publishers Series in Security and Digital Forensics, 1st ed. 2018, 490 p. 198 illus.

[159] Petrenko Sergei. Cyber Resilience, ISBN: 978-87-7022-11-60 (Hardback) and 877-022-11-62 (Ebook) © 2019 River Publishers, River Publishers Series in Security and Digital Forensics, 1st ed. 2019, 492 p. 207 illus.

[160] Petrenko Sergei. Developing a Cybersecurity Immune System for Industry 4.0, ISBN: 978-87-7022-188-7 (Hardback) 978-87-7022-187-0 (Ebook) © 2020 River Publishers, River Publishers Series in Security and Digital Forensics, 1st ed. 2020, 456 p. 297 illus.

[161] The MITRE Corporation (ed.), "Fourth Annual Secure and Resilient Cyber Architectures Invitational," 2015. [Online]. Available: http://www.mitre.org/sites/default/files/pdf/2014-Secure-Resilient-Cyber-Architectures-Report-15-0704.pdf.

[162] The MITRE Corporation (ed.), "Third Annual Secure and Resilient Cyber Architectures Workshop," December 2013. [Online]. Available: http://www.mitre.org/sites/default/files/publications/13-4210.pdf.

[163] The MITRE Corporation (ed.), "2nd Secure and Resilient Cyber Architectures Workshop: Final Report," 2012. [Online]. Available: https://registerdev1.mitre.org/sr/2012_resiliency_workshop_report.pdf

[164] NIST Special Publication 800-160 VOLUME 2. Systems Security Engineering. Cyber Resiliency Considerations for the Engineering of Trustworthy Secure Systems – (Draft), March 2018, https://insidecybersecurity.com/sites/insidecybersecurity.com/files/documents/2018/mar/cs03202018_NIST_Systems_Security.pdf

[165] NIST Special Publication 800-160 VOLUME 3. Systems Security Engineering. Software Assurance Considerations for the Engineering of Trustworthy Secure Systems – (Draft), December 20, 2019

[166] NIST Special Publication 800-160 VOLUME 4. Systems Security Engineering. Hardware Assurance Considerations for the Engineering of Trustworthy Secure Systems – (Draft), December 20, 2020.

[167] NIST SP 800-34. Rev. 1: Contingency Planning Guide for Federal Information Systems Paperback – February 18, 2014 https://www.amazon.com/NIST-Special-Publication-800-34-Rev/dp/1495983706

[168] Johnson, P. 2017. "With The Public Clouds Of Amazon, Microsoft And Google, Big Data Is The Proverbial Big Deal." Forbes, Jun 15. Web access: https://www.forbes.com/sites/johnsonpierr/2017/06/15/with-the-public-clouds-of-amazonmicrosoft-and-google-big-data-is-the-proverbial-big-deal/#2a37a76b2ac3

[169] Kenney, M., & Zysman, J. (2016). "The Rise of the Platform Economy." Issues in Science and Technology, 32(3), 61–69.

[170] Mamaev, M. A., Petrenko, S.A. Technologies of information protection on the Internet. - St. Petersburg.: publishing house "Peter", p. 848, Russia, St.Petersburg, 2002.

[171] Marz, N., Warren, J. Big data. Principles and practice of building scalable data processing systems in real time, Williams, p. 292, Moscow, Russia, 2016.

[172] Ovidiu Vermesan, Peter Friess (ed.) (*2016*). Digitising the Industry - Internet of Things Connecting the Physical, Digital and Virtual Worlds. River Publishers.

[173] Petrenko, S. A., Petrenko, A. S. New Doctrine of Information Security of the Russian Federation, Information Protection, Inside. No. 1 (73). - pp. 33–39, Russia, 2017.

[174] Ramjee Prasad and Leo P. Ligthart (ed.) (2018). Towards Future Technologies for Business Ecosystem Innovation. River Publishers.

[175] Richard Graubart, The MITRE Corporation, Cyber Resiliency Engineering Framework, The Secure and Resilient Cyber Ecosystem (SRCE) Industry Workshop Tuesday, November 17, 2015, [Online]. Available:https://secwww.jhuapl.edu/SRCE-Workshop/past-events/2015/docs/abstracts/Abstract_Graubart_MITRE.pdf

[176] S. Noel, J. Ludwig, P. Jain, D. Johnson, R. K. Thomas, J. McFarland, B. King, S. Webster and B. Tello, "Analyzing Mission Impacts of Cyber Actions (AMICA), STO-MP-AVT-211," 1 June 2015. [Online]. Available: http://csis.gmu.edu/noel/pubs/2015_AMICA.pdf.

[177] The BCI Cyber Resilience Report, [Online]. Available: https://www.b-c-training.com/img/uploads/resources/BCI-Cyber-Resilience-Report-2018.pdf

[178] Petrenko Sergei. Big Data Technologies for Monitoring of Computer Security: A Case Study of the Russian Federation, ISBN 978-3-319-79035-0 and ISBN 978-3-319-79036-7 (eBook), https://doi.org/10.1007/978-3-319-79036-7 © 2018 Springer Nature Switzerland AG, part of Springer Nature, 1st ed. 2018, XXVII, 249 p. 93 illus.

[179] Petrenko Sergei. Cyber Security Innovation for the Digital Economy: A Case Study of the Russian Federation, ISBN: 978-87-7022-022-4 (Hardback) and 978-87-7022-021-7 (Ebook) © 2018 River Publishers, River Publishers Series in Security and Digital Forensics, 1st ed. 2018, 490 p. 198 illus

[180] Petrenko Sergei. Big Data Technologies for Monitoring of Computer Security: A Case Study of the Russian Federation, ISBN 978-3-319-79035-0 and ISBN 978-3-319-79036-7 (eBook), https://doi.org/10.1007/978-3-319-79036-7 © 2018 Springer Nature Switzerland AG, part of Springer Nature, 1st ed. 2018, XXVII, 249 p. 93 illus.

[181] Petrenko Sergei. Cyber Security Innovation for the Digital Economy: A Case Study of the Russian Federation, ISBN: 978-87-7022-022-4 (Hardback) and 978-87-7022-021-7 (Ebook) © 2018

River Publishers, River Publishers Series in Security and Digital Forensics, 1st ed. 2018, 490 p. 198 illus.

[182] Petrenko, S. A., Petrenko, A.S. (2016). Lecture 12, Perspective tasks of information security, Intelligent Information Radiophysical Systems, MSTU, N. E Bauman; [ed. S.F. Boev, D.D. Stupin, A.A. Kochkarov], Moscow, Russia, pp. 155–166.

[183] Petrenko Sergei. Cyber Security Innovation for the Digital Economy: A Case Study of the Russian Federation, ISBN: 978-87-7022-022-4 (Hardback) and 978-87-7022-021-7 (Ebook) © 2018 River Publishers, River Publishers Series in Security and Digital Forensics, 1st ed. 2018, 490 p. 198 illus.

[184] Petrenko Sergei. Developing a Cybersecurity Immune System for Industry 4.0, ISBN: 978-87-7022-188-7 (Hardback) 978-87-7022-187-0 (Ebook) © 2020 River Publishers, River Publishers Series in Security and Digital Forensics, 1st ed. 2020, 456 p. 297 illus.

[185] Petrenko Sergei. Cyber Security Innovation for the Digital Economy: A Case Study of the Russian Federation, ISBN: 978-87-7022-022-4 (Hardback) and 978-87-7022-021-7 (Ebook) © 2018 River Publishers, River Publishers Series in Security and Digital Forensics, 1st ed. 2018, 490 p. 198 illus.

[186] Cyber Security Innovation for the Digital Economy: A Case Study of the Russian Federation, ISBN: 978-87-7022-022-4 (Hardback) and 978-87-7022-021-7 (Ebook) © 2018 River Publishers, River Publishers Series in Security and Digital Forensics, 1st ed. 2018, 490 p. 198 illus.

[187] Rus, D. 2015. "The Robots Are Coming: How Technological Breakthroughs Will Transform Everyday Life." In: Rose, G (ed.) The Fourth Industrial Revolution: A Davos Reader. Council on Foreign Relations.

[188] Sergei Petrenko [0000-0003-0644-1731] and Khismatullina Elvira [0000-0002-8765-1097]. Cyber-resilience concept for Industry 4.0 digital platforms in the face of growing cybersecurity threats. Software Technology: Methods and Tools, 51st International Conference, TOOLS 2019, Innopolis, Russia, October 15–17, 2019, Proceedings. Editors: Mazzara, M., Bruel, J.-M., Meyer, B., Petrenko, A. (Eds.), eBook ISBN 978-3-030-29852-4, DOI 10.1007/978-3-030-29852-4, Softcover ISBN 978-3-030-29851-7, 420 p. URL: https://www.springer.com/gp/book/9783030298517.

[189] Petrenko Sergei. Cyber Resilience, ISBN: 978-87-7022-11-60 (Hardback) and 877-022-11-62 (Ebook) © 2019 River Publishers, River Publishers Series in Security and Digital Forensics, 1st ed. 2019, 492 p. 207 illus.

[190] *Arnold V.I., "Theory of Catastrophes," Edition Three, supplemented by M., Science, 1990 - 128 p. 84,000 copies. ISBN 5-02-014271-9*

[191] René Tom, Structural stability and morphogenesis, Logos Publishing House, Year of publication: 2002, ISBN: 5-8163-0032-6, https://www.livelib.ru/book/1000554483-strukturnaya-ustojchivost-i-morfogenez-rene-tom.

[192] Petrenko Sergei. Cyber Resilience, ISBN: 978-87-7022-11-60 (Hardback) and 877-022-11-62 (Ebook) © 2019 River Publishers, River Publishers Series in Security and Digital Forensics, 1st ed. 2019, 492 p. 207 illus.

[193] Mesarovich M., Mako D., Takahara I., Theory of Hierarchical Multi-Level Systems, Publishing House "Mir" 1973, 344 pp.

[194] ISO 22301:2019. Security and resilience — Business continuity management systems — Requirements, [Online]. Available: https://www.iso.org/standard/75106.html

[195] ISO 22313:2020. Security and resilience — Business continuity management systems — Guidance on the use of ISO 22301, [Online]. Available: https://www.iso.org/standard/75107.html

[196] ISO/TS 22317:2015. Societal security -- Business continuity management systems – Guidelines for business impact analysis (BIA), [Online]. Available: https://www.iso.org/standard/50054.html

[197] ISO/TS 22318:2018. Security and resilience — Business continuity management systems — Guidelines for people aspects of business continuity, [Online]. Available: https://www.iso.org/standard/65336.html

[198] ISO/TS 22330:2018, Security and resilience – Business continuity management systems – Guidelines for people aspects of business continuity, [Online]. Available: https://www.iso.org/standard/50067.html

[199] ISO/TS 22331:2018, Security and resilience – Business continuity management systems – Guidelines for business continuity strategy, [Online]. Available: https://www.iso.org/standard/50068.html

[200] ISO 22301:2012. Societal security – Business continuity management systems – Requirements, [Online]. Available: https://www.iso.org/standard/50038.html

[201] Petrenko Sergei. Cyber Resilience, ISBN: 978-87-7022-11-60 (Hardback) and 877-022-11-62 (Ebook) © 2019 River Publishers, River Publishers Series in Security and Digital Forensics, 1st ed. 2019, 492 p. 207 illus

[202] Cyber Security Innovation for the Digital Economy: A Case Study of the Russian Federation, ISBN: 978-87-7022-022-4 (Hardback) and 978-87-7022-021-7 (Ebook) © 2018 River Publishers, River Publishers Series in Security and Digital Forensics, 1st ed. 2018, 490 p. 198 illus.

[203] Petrenko Sergei. Cyber Resilience, ISBN: 978-87-7022-11-60 (Hardback) and 877-022-11-62 (Ebook) © 2019 River Publishers, River Publishers Series in Security and Digital Forensics, 1st ed. 2019, 492 p. 207 illus

[204] How to Build Resiliency through Business Continuity Management, © 2017 Accenture, [Online]. Available: https://www.accenture.com/t20170113T003242Z__w__/us-en/_acnmedia/PDF-40/Accenture-InsideOps-Business-Continuity-Management

[205] Continuity in Crisis April 2020 How to run effective business services during the COVID-19 pandemic, © 2020 Accenture, [Online]. Available: https://www.accenture.com/_acnmedia/PDF-120/Accenture-COVID-19-Continuity-in-Crisis-Effective-Business-Services.pdf

[206] Attilio Di Lorenzo, Accenture Methodologies and Framework, Disaster Recovery Workshop – *Third Edition*, Rome, 8th April 2013 © 2013 Accenture, [Online]. Available: https://www.dis.uniroma1.it/~ciciani/files/Universit_%20La%20Sapienza%20-%20IT%20Disaster%20Re%20covery%20-%20Workshop%20Accenture_08042013.pdf

[207] OPERATIONAL RESILIENCE IS FINANCIAL RESILIENCE, Fordham./Accenture Compliance Series, Operational Resiliency Industry Perspectives and Compliance Considerations, Fordham Law School Skadden Conference Center 150 West 62nd Street, New York City, Wednesday, 13 November 2019 © 2019 Accenture, [Online]. Available: https://www.fordham.edu/download/downloads/id/13995/fordham_accenture_operational_resiliency_2019_course_materials.pdf

[208] Business Continuity Management, © 2020 Ernst & Young, China. [Online]. Available: https://www.ey.com/Publication/vwLUAssets/ey-business-continuity-management/$File/ey-business-continuity-management.pdf

[209] Business Continuity Management. Current trends © 2011 Ernst & Young, China. [Online]. Available: https://www.eyjapan.jp/services/advisory/global-contents/pdf/Insights-on-it-risk-2011-08-en.pdf

[210] Ensuring Business Continuity, © 2020 KPMG, [Online]. Available: https://home.kpmg/at/de/home/services/advisory/risk-consulting/corporate-risk-services/ensuring-business-continuity.html

[211] Business Continuity Management, © 2020 KPMG, [Online]. Available: https://home.kpmg/de/de/home/dienstleistungen/advisory/consulting/security-consulting/business-continuity-management.html

[212] Building a Continuity Culture A survey of Canadian decision makers on Business Continuity Planning, © 2006 KPMG, [Online]. Available: http://www.dcag.com/images/BusinessContinuity.pdf

[213] Business Continuity Planning Solutions © 2020 PWC. [Online]. Available: https://www.pwc.com/us/en/services/risk-assurance/risk-compliance-and-governance/business-continuity-planning.html

[214] Business Continuity Management © 2020 PWC. [Online]. Available: https://www.pwc.com/gh/en/services/advisory/bcm.html

[215] Ed Matley, Marie Lavoie Dufort, Cyber Security and Business Continuity Management, EPICC, October 2016, © 2016 PWC. [Online]. Available: http://www.epicc.org/uploadfiles/documents/PwC%20-%20Cyber%20Security%20and%20Business%20Continuity%20Management.pdf

[216] Business Continuity Management services, © 2020 deloitte. [Online]. Available: https://www2.deloitte.com/jp/en/pages/risk/solutions/bcm-bcp/bcm.html

[217] Business continuity and recovery management. Confidence and reliability, © 2020 deloitte. [Online]. Available: https://www2.deloitte.com/ru/en/pages/risk/articles/business-continuity-and-recovery-management.html

[218] IBM Corporation (2018), James Boyles, «Cybersecurity and YOU!! CYBER RESILIENCE - PREPARE FOR WHEN, NOT IF», https://files.nc.gov/ncdit/documents/files/2018%20NCSAM%20Symposium%20-%20Cyber%20Resilience%20-%20IBM.pdf

[219] IBM "IBM's Smarter Cities Challenge: Boston—Report." Last accessed on 12 April 2017, https://www.smartercitieschallenge.org/assets/cities/boston-united-states/documents/boston-united-states-full-report-2012.pdf

[220] IBM Corporation (2018), ARNE JACOBSEN, «IBM RESILIENT: INTELLIGENT ORCHESTRATION THE NEXT GENERATION

OF INCIDENT RESPONSE», https://www-05.ibm.com/se/securitysummit/assets/pdf/IBM_Resilient-Arne_Jacobsen.pdf

[221] IBM Corporation (2018), Jean-Michel Lamby Associate Partner - IBM Security, «Cyber Resiliency. Minimizing the impact of breaches on business continuity», https://www-05.ibm.com/be/think-brussels/assets/pdf/Minimizing_the_impact_of_breaches_on_business_continuity_by_Jean_Michel_Lamby.pdf

[222] IBM Corporation (2018), Felicity March, «Cyber Resilience», https://www-05.ibm.com/dk/think-copenhagen/assets/pdf/Studie3_Session2_Speaker4_Felicity_March_IBM.pdf

[223] Business Resiliency, Ensuring employee safety, maintaining operational continuity, and meeting customer expectations, [Online]. Available: https://www.cisco.com/c/en/us/about/business-continuity.html

[224] HPE Helion and Veritas Continuity. Managed Disaster Recovery-as-a-Service to the HPE Helion OpenStack Managed Cloud, ©2015 Veritas, [Online]. Available: https://www.karma-group.ru/upload/iblock/518/HP-Helion-and-Veritas-Continuity-EN.pdf

[225] Smartha Guha Thakurta, Business Continuity Planning In an Organisation, EMC Proven Profesional Knowledge Sharing 2009, [Online]. Available: https://education.dellemc.com/content/dam/dell-emc/documents/en-us/KS2009_Thakurta-Business_Continuity_Planning_in_an_Organization.pdf

[226] Charlotte Brooks Clem Leung Aslam Mirza Curtis Neal Yin Lei Qiu John Sing Francis TH Wong Ian R Wright, IBM System Storage Business Continuity: Part 1 Planning Guide, RedBooks, ©2007 IBM Corporation, [Online]. Available: https://www.redbooks.ibm.com/redbooks/pdfs/sg246547.pdf

[227] Building the business case for resiliency Linking resiliency to business objectives can help IT professionals make a case for corporate investment, IBM Global Technology Services, White Paper, ©2015 IBM Resiliency Services, [Online]. Available: https://www.ibm.com/downloads/cas/WRZGRKKM

[228] Enhance IT resiliency and business continuity, ©2020 IBM Corporation, IBM Resiliency Services, [Online]. Available: https://www.ibm.com/uk-en/services/business-continuity?p1=Search&p4=43700052448792406&p5=b&cm_mmc=Search_Google-_-1S_1S-_-EP_RU-_-%2Bibm%20%2Bresiliency_b&cm_mmca7=71700000064861972&cm_mmca8=aud-311016886972:kwd-47088086215&cm_

mmca9=EAIaIQobChMIldicqKOS6QIVAc53Ch08pAP-
NEAAYASAAEgIs9_D_BwE&cm_mmca10=426169590952&cm_
mmca11=b&gclid=EAIaIQobChMIldicqKOS6QIVAc53Ch08pAP-
NEAAYASAAEgIs9_D_BwE&gclsrc=aw.ds

[229] Tony Savoy, Enhance IT resiliency and business continuity, IBM Services, ©2020 IBM Corporation, [Online]. Available: https://www.ibm.com/account/reg/at-en/signup?formid=urx-43756

[230] Mario Ruiz –Technical Solutions Architect, Design of Robust Business Continuity Solutions to satisfy Regulatory Guidelines, July 10–14 2016, CiscoLive, Las Vegas, NV, [Online]. Available: https://www.ciscolive.com/c/dam/r/ciscolive/us/docs/2016/pdf/BRKDCN-2300.pdf

[231] Global Business Resiliency (GBR) Program Policy, [Online]. Available: https://www.cisco.com/c/dam/en_us/about/business-continuity/global-business-resiliency-program-policy.pdf

[232] Susie Wee, Developer, Business Continuity Planning Today, March 10, 2020,[Online]. Available: https://blogs.cisco.com/developer/business-continuity

Index

B

Business continuity (BC) 1,
    97–98, 121–22, 147, 251–252,
    323–326, 368, 466
Business continuity management
    systems (BCMS) 3, 9, 10,
    97–98, 151, 251–252, 338
Business continuity management
    (BCM) 1, 55, 97–98, 121–
    122, 147, 251–252, 323–326,
    338, 466
BCM Lifecycle (PDCA Model
    ISO) 56, 97–98, 147, 148–
    149, 165–166, 251–252, 397,
    452, 473–474
    BCP Development 147,
        165–166, 424, 454
    BCP Testing 172, 428–430,
        439, 452
    BCM Aspects 147, 168, 381,
        423–424, 428
    BCM Controls (ISO/IEC
        (A. 14 or A. 17) 191, 250,
        257, 338
    Business impact analysis
        (BIA) 16, 21, 47, 169,
        97–98, 306, 338, 426, 431
        RTO – target recovery
            time 16, 21, 426, 426,
            429–430, 439

        RPO – target recovery
            point 16, 21, 426,
            426–427, 429–430, 439
    BCP Strategy 17, 47, 50–51,
        53, 170, 313, 322, 343,
        437, 444, 447
    Business continuity plan
        (BCP) 147, 323–326, 346,
        449, 452
    BCP Testing 323–326, 449,
        452, 473
    Support for BCP and DRP
        plans 147, 323–326, 452,
        473
    Approving and implementing
        BCP and DRP plans
        323–326, 346, 452, 473
BCM Best Practices 121–122,
    137, 141, 151, 205, 323–326,
    329, 348, 368, 375, 381,
    380–381, 398, 412, 466
    Principles for Business
        Continuity (ISO
        22301) 137, 141, 145, 153
    Good Practice Guidelines
        Edition (BCI) 137, 153,
        154, 161
    The Professional Practices
        for Business Continuity
        Management Edition
        (DRII) 137, 156, 161,

DSS process (COBIT®
2019) 137, 219, 222, 415
DSS Maturity Levels
(COBIT® 2019) 137, 219,
220
ITIL practices 137, 218, 237,
240, 243, 415
ITCM Function (MOF 4.0.)
137, 243, 412, 415, 417
RESILIA 2015 137
NIST SP 800-34 137
NFPA 1600:2019 137
ASIS ORM.1-2017 137
BCM Technologies 26, 28, 34,
36–37, 64, 69, 77, 84, 87, 91,
368, 375, 381, 398, 444
Elastic clustered NAS head
systems 26, 28, 69, 84
Quantum StorNext for CIFS
26, 28, 69
NFS–based access via a block
storage system running
VPLEX™ 26, 28
Converged solutions (Cisco,
NetApp, IBM, Pure,
etc.) 26, 69, 77, 82
Application of hypercon-
verged systems for
HyperFlex virtualization
platforms, etc. 26, 67, 77,
82, 87
Integrated systems for Big
Data and predictive
analytics 30–31, 67, 73,
77, 91
Ceph cluster vSphere
Metro Storage Cluster
(vMSC) 30–31, 33
Clustering databases (Oracle
RAC) 77, 91

Integrated backup and copy
storage solutions (Veeam,
Commvault, Coherence,
etc.) 77, 80, 82, 87
Integrated platforms for storing
archives and secondary
data (Coherence, Scality,
SwiftStack, etc.) 77, 87, 91
SDS platform (Ceph, Gluster,
IBM, etc.) 28, 69, 91
Application of a data storage
network (IP, FC, FCoE,
etc.) 28, 87, 91
Cassandra–based large data
storage architecture 35,
87, 91
Docker and Kubernetes con-
tainer management sys-
tems 26, 34, 91
CI/CD (continuous delivery
and integration) 77, 91
BCM Awareness Programs – 157,
157, 156, 160, 163, 335
British Business Continuity
Institute (BCI) 157, 157,
156, 160
International Disaster
Recovery Institute
International, DRII 60
Institute of system admin-
istrators and security
administrators – SANS
Institute 163
BCM software 121–122,
323–326, 335, 381
BCM services 121–122, 251–252,
306, 323–326, 335, 381
BC Project Management 121–
122, 170, 205, 257, 295, 306,
335, 381

C

Cyber resilience (CR) 97–98, 103, 105, 121–122, 295, 360 339, 368, 381

Cyber resilience lifecycle (NIST SP 800-160) 105, 109, 121–122

D

Disaster Recovery 257, 295, 295–296, 360, 362, 396, 406

Disaster recovery plan (DRP) 257, 295, 295–296

Defining BC strategies 309–310, 360, 362, 396, 398, 406

Real–time data centers 360, 362, 396, 406

Metro cluster 360, 362

Geo–distributed data processing center cluster 360, 362

Backup data processing centers 360, 362, 396, 406

Grid gain redundancy platform 360, 362, 396, 406

Hadoop redundancy platform 360, 362, 396, 406

Hot standby systems 360, 362, 396, 406

"Warm" reserve systems 360, 362, 396, 406

Cold reserve systems 360, 362, 396, 406

Outsourcing or mutual agreements 338

E

Enterprise continuity program (ECP) 16, 17, 18, 40, 175, 257, 256–257, 381

ECP lifecycle 165–166, 175, 250, 257, 256–257, 381

ECP Program Maturity Assessment 168, 205, 256–257, 306, 338

ECP Practice 175, 250, 256–257, 340, 381, 459–461

ECP Development Samples 175, 257, 306, 340, 459–461

Emergency recovery team 250, 257, 340, 341–342, 455–456

I

Intellectual Cyber Resilience Orchestration 256–257, 268–269, 272, 285–286, 288

ISO/TC 292 "Security and resilience" 137, 140–141, 145, 153

ISO/IEC 27001:2013 (A.17) 191, 250, 257

ISO/IEC 27031:2011 250

R

Risk management (RM) 179, 191, 306, 338

Risk assessment (RA) 168, 179, 191, 306, 338

Risk Management Practices 179, 180–181, 185–186, 191, 199, 338

ISO Family of Standards 180–181, 185–186, 191

NIST SP 800-30 191, 193

AS/NZS 5050:2010 Standard 179, 181, 191

OCTAVE Methodology 191, 197

MG– Lifecycle 191, 198

SA–CMM Maturity Model 191, 200–201

About the Author

Prof. Sergei A. Petrenko
Innopolis University, Russia

He was born in 1968 in Kaliningrad (the Baltic). In 1991, he graduated with honors from the Leningrad State University with a degree in mathematics and engineering (in 1997 – adjuncture; in 2003 – doctorate).

He is the designer of information security systems of critical information objects:

- three national *Centers for Monitoring Information Security Threats and two Situational-Crisis Centers (RCCs)* of domestic state;
- three operators of special information security services *MSSP (Managed Security Service Provider)* and *MDR (Managed Detection and Response Services)* and two virtual trusted communication operators *MVNO*;
- more than 10 state and enterprise segments of the *System for Detection, Prevention and Elimination of the Effects of Computer*

Attacks (SOPCA) and the *System for Detection and Prevention of Computer Attacks (SPOCA)*;

- five monitoring centers for information security threats and responding to information security incidents *CERT (Computer Emergency Response Team)* and *CSIRT (Computer Security Incident Response Team)* and two *industrial CERT industrial Internet IIoT/IoT.*

He is the Head of the State Scientific School *"Mathematical and Software Support of Critical Objects of the Russian Federation,"* an expert of the *Section on Information Security Problems of the Scientific Council under the Security Council of the Russian Federation,*

Scientific editor of the magazine *"Inside. Data protection,"* and Doctor of Technical Sciences, Professor.

It is part of the management of the Interregional Public Organization Association of Heads of Information Security Services (*ARSIB*), an independent non-profit organization Russian Union of IT Directors (*SODIT*).

He is the author and co-author of 12 monographs and more than 350 articles on information security issues (Proceedings of ISA RAS and SPIIRAS, journals *"Cybersecurity issues," "Information security problems," "Open systems," "Inside: Information protection," "Security systems," "Electronics," "Communication Bulletin," "Network Journal," "Connect World of Connect,"* etc.), including, monographs and practical manuals of publishing houses *"River Publishers," "Springer Nature Switzerland AG," "Peter," "Athena,"* and *"DMK-Press"*: *"Big Data Technologies for Monitoring of Computer Security: A Case Study of the Russian Federation," "Cyber Security Innovation for the Digital Economy: A Case Study of the Russian Federation," "Methods of information protection in the Internet," "Methods and technologies of information security of critical objects of the national infrastructure," "Methods and technologies of cloud security," "Audit of enterprise Internet/Internet security," "Information Risk Management," "Information Security Policies,"* and others.

He was awarded the *"Big ZUBR"* and *"Golden ZUBR"* in 2014 for the national projects of the Russian Federation in the field of information security.